AF816450

Numerical Methods and Modeling Applied for Composite Structures

Numerical Methods and Modeling Applied for Composite Structures

Guest Editors

Pawel Wysmulski
Katarzyna Falkowicz
Patryk Rozylo

MDPI

Basel • Beijing • Wuhan • Barcelona • Belgrade • Novi Sad • Cluj • Manchester

Guest Editors
Pawel Wysmulski
Machine Design and Mechatronics
Lublin University of Technology
Lublin
Poland

Katarzyna Falkowicz
Machine Design and Mechatronics
Lublin University of Technology
Lublin
Poland

Patryk Rozylo
Machine Design and Mechatronics
Lublin University of Technology
Lublin
Poland

Editorial Office
MDPI AG
Grosspeteranlage 5
4052 Basel, Switzerland

This is a reprint of the Special Issue, published open access by the journal *Materials* (ISSN 1996-1944), freely accessible at: www.mdpi.com/journal/materials/special_issues/9L636Y0U3Z.

For citation purposes, cite each article independently as indicated on the article page online and as indicated below:

Lastname, A.A.; Lastname, B.B. Article Title. *Journal Name* **Year**, *Volume Number*, Page Range.

ISBN 978-3-7258-3906-3 (Hbk)
ISBN 978-3-7258-3905-6 (PDF)
https://doi.org/10.3390/books978-3-7258-3905-6

© 2025 by the authors. Articles in this book are Open Access and distributed under the Creative Commons Attribution (CC BY) license. The book as a whole is distributed by MDPI under the terms and conditions of the Creative Commons Attribution-NonCommercial-NoDerivs (CC BY-NC-ND) license (https://creativecommons.org/licenses/by-nc-nd/4.0/).

Contents

materials

MDPI

Article

Failure Mechanism of Tensile CFRP Composite Plates with Variable Hole Diameter

Pawel Wysmulski

Department of Machine Design and Mechatronics, Faculty of Mechanical Engineering, Lublin University of Technology, Nadbystrzycka 36, 20-618 Lublin, Poland; p.wysmulski@pollub.pl

Abstract: Real thin-walled composite structures such as aircraft or automotive structures are exposed to the development of various types of damage during operation. The effect of circular hole size on the strength of a thin-walled plate made of carbon fibre-reinforced polymer (CFRP) was investigated in this study. The test object was subjected to tensile testing to investigate the strength and cracking mechanism of the composite structure with variable diameter of the central hole. The study was performed using two independent test methods: experimental and numerical. With increasing diameter of the central hole, significant weakening of the composite plate was observed. The study showed qualitative and quantitative agreement between the experimental and numerical results. The results confirmed the agreement of the proposed FEM model with the experimental test. The novelty of this study is the use of the popular XFEM technique to describe the influence of the hole size on the cracking and failure of the composite structure. In addition, the study proposes a new method for determining the experimental and numerical damage and failure loads of a composite plate under tension.

Keywords: CFRP composite; damage mechanics; crack propagation; tensile analysis; finite element method

1. Introduction

The design of modern structures with optimised strength and stiffness parameters requires the use of advanced technologies. This particularly applies to high-tech aerospace or automotive structures in which the most beneficial solutions in terms of operation and durability can be obtained by e.g., replacing previously used materials with advanced composite materials [1–3]. These primarily include polymer laminates reinforced with continuous fibres, predominantly carbon fibre-reinforced plastics (CFRPs) and glass fibre-reinforced plastics (GFRPs). Due to very favourable mechanical properties of these materials in relation to their own weight, it has become possible to use fibre composites in the production of carrying elements in thin-walled structures (e.g., for covering reinforcement profiles) [4–8]. Laminates make it possible to shape the mechanical properties of designed components in terms of their ability to carry the desired type of load. As a result, it is possible to achieve very advantageous construction designs; this, however, requires the use of modern testing methods that enable the structural performance to be analysed over the full range of loading conditions [9–12]. Previous studies on composite structures reported in the literature mostly focus on analytical and numerical considerations, with analyses conducted on structures with typical cross-sections, operating under ideal conditions and subjected to simple loading cases: compression, shear, or simple bending. Only to a limited extent are such considerations verified by experimental tests on real construction elements [13–15].

In layered composites (laminates), the state of stress is a complex issue because it depends on the fibre configuration and varies from layer to layer [16]. Therefore, the stresses induced by a hole in the laminate vary from layer to layer, and the classical Kirsch problem for isotropic materials cannot be applied in such cases [17]. The currently

Citation: Wysmulski, P. Failure Mechanism of Tensile CFRP Composite Plates with Variable Hole Diameter. *Materials* **2023**, *16*, 4714. https://doi.org/10.3390/ma16134714

Academic Editor: Karim Benzarti

Received: 2 June 2023
Revised: 26 June 2023
Accepted: 28 June 2023
Published: 29 June 2023

Copyright: © 2023 by the author. Licensee MDPI, Basel, Switzerland. This article is an open access article distributed under the terms and conditions of the Creative Commons Attribution (CC BY) license (https://creativecommons.org/licenses/by/4.0/).

popular numerical methods could help in such cases [18,19]. As is known from scientific publications, the occurrence of holes in thin-walled structures is unavoidable, if only for technological reasons [20–22]. The complexity of the above issue, resulting from the possibility of designing the material properties of laminated composites, makes this topic still valid for researchers. The literature [23–26] offers solutions to the problem of hole formation in composite materials.

The extended finite element method (XFEM) eliminates the need for a conformal finite element mesh [27]. The extended finite element method was first introduced by Belytschko and Black [28]. It is an extension of the conventional finite element method based on Melenko and Babuska's concept of partition of unity [29], which allows local enrichment functions to be easily incorporated into the finite element approximation. The presence of discontinuities is provided by special enrichment functions in combination with additional degrees of freedom. However, the finite element structure and its properties, such as sparsity and symmetry, are retained. The use of the XFEM method makes it possible to study crack initiation and propagation along any path without numerical model remeshing [27]. Moving cracks are modelled using one of two alternative approaches: the cohesive segment approach or the linear elastic fracture mechanics (LEFM) approach [27,30,31]. Using these techniques, crack initiation is defined up to the onset of cohesive degradation in the enriched component, and the degradation stage occurs when the stresses or strains meet the specified crack initiation criteria [32].

The present study analysed the effect of variable diameter of the central hole on the behaviour of a thin-walled composite plate under tension. The study was carried out on plates weakened by a central hole with diameters of 2 mm, 4 mm, and 8 mm. The tensile tests were carried out over the full range of loading, from failure initiation through crack propagation to complete failure of the composite structure [33–36]. The research was carried out using two independent methods: experimental and numerical using FEM. This approach made it possible to develop numerical models that closely reproduced real plates [37–40].

The novelty of the research problem undertaken in this study is that it describes the fracture and failure of a hole-weakened composite plate using the currently popular XFEM technique. The effect of hole size on the strength of the composite plate is investigated. The state-of-the-art Aramis measuring system is used in experiments. In addition, the study proposes a new method for determining the crack initiation and failure loads of the composite plate under tension, and the results obtained thereby are verified numerically. A review of the literature shows that many studies have used XFEM for isotropic materials [41–44]. However, there is a lack of studies describing the failure of real sandwich composites (CFRPs) using the numerical XFEM technique. A numerical model created based on experimental results can be used to analyse the failure process for such composite elements.

2. Object and Methodology

The test object was a thin-walled laminated composite plate. Holes were cut in the plate with diameters of 2 mm, 4 mm, and 8 mm. The holes were used to weaken the structure and to cause the composite to crack in a specified area during the tensile test. The test object was made from a unidirectional HeXPly prepreg strip (from Hexcel) of carbon fibre-reinforced composite in an epoxy matrix. The polymerisation process took place in an autoclave. The curing process was carried out in a package vacuum of 0.08 MPa, overpressure of 0.4 MPa, and autoclave temperature of 135 °C for 2 h. The laminate structure had a symmetrical fibre arrangement of the composite layers $[0/90/0/90_2/0/90/0]$—Figure 1b.

Table 1 shows the mechanical properties of a single layer of CFRP laminate in three orthotropic directions. The properties of the carbon–epoxy composite were determined experimentally in compliance with the ISO standard and as described in [45]. This allowed for obtaining real mechanical properties of the produced material, as they may differ from the ideal properties specified by the manufacturer. The determined properties were used to define the material model in a numerical analysis conducted by the finite element method using Abaqus.

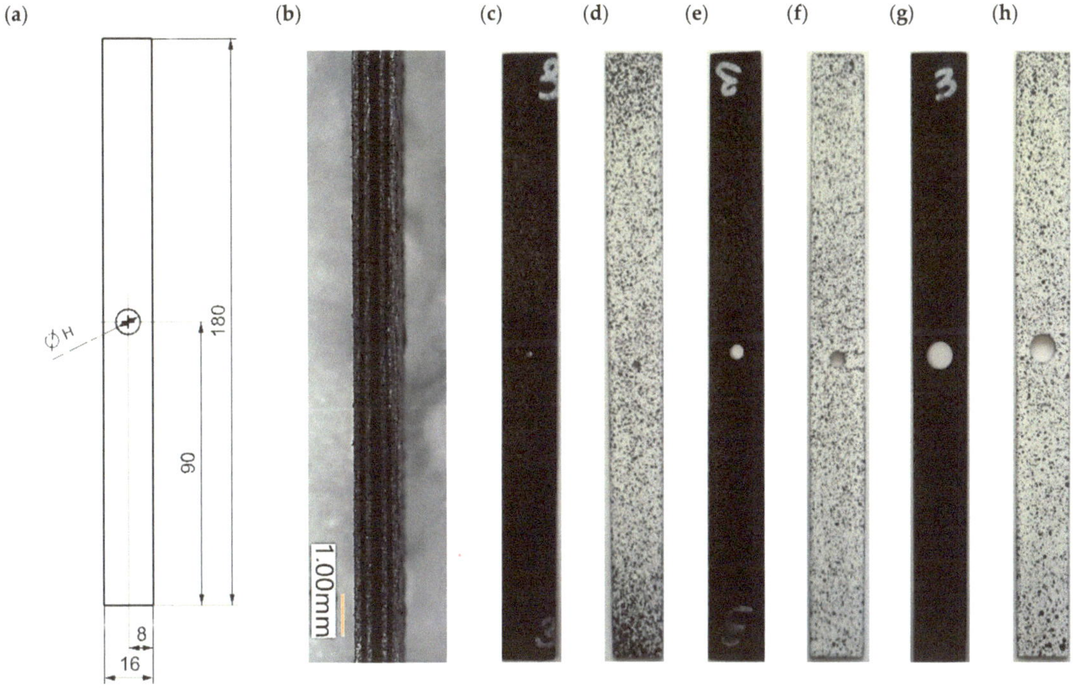

Figure 1. Real composite plate with a drilled hole: (**a**) sketch with dimensions, (**b**) section layup, (**c**) hole 2 mm, (**e**) hole 4 mm, (**g**) hole 8 mm, (**d**,**f**,**h**) specimen with contrasting pattern.

Table 1. Mechanical properties of CFRP.

Tensile Modulus		Shear Modulus	Poisson's Ratio	Tensile Strength		Shear Strength	Compressive Strength	
E_1	$E_{2,3}$	$G_{12,13,23}$	$\nu_{12,13,23}$	F_{T1}	F_{T2}	F_S	F_{C1}	F_{C2}
GPa	MPa	MPa	–	MPa	MPa	MPa	MPa	MPa
130.71	6360	4180	0.32	1867.2	25.97	100.15	1531	214

Figure 1a shows the test object, which consisted of 3 plates with dimensions of 16 mm (width) × 180 mm (length) × 1.048 mm (overall thickness). For each specimen, an oval hole with a diameter of H = 2 mm (Figure 1c,d), H = 4 mm (Figure 1e,f), and H = 8 mm (Figure 1g,h) was made in the centre of the plate. The specimens were painted in contrasting patterns (Figure 1d,f,h). The samples were measured during the experiments using a non-contact optical measuring method.

The ARAMIS optical measuring system is designed for non-contact displacement measurements in planar and spatial components under load. It consists of a set of cameras recording changes in the shape of the object under test and a suitably adapted and programmed computer storing and processing the recorded images. Depending on the configuration, i.e., the number and speed of cameras, the system can be used to analyse displacement and deformation fields of flat or spatial elements under static or dynamic loading.

The measurement principle is the same as in photogrammetry, i.e., on the basis of the images, the spatial coordinates of selected points are determined. Measurement proceeds as follows. A photo of the object in its undeformed state is taken, followed by a series of photos corresponding to the successive loading stages of the object. Each of the photographs is then compared with the output and a set of displacement values of selected points on the surface of the object is created. The selected points are points of interest on the surface of the object. They can be things such as spots, dots, or other colour changes naturally occurring on the surface. If the surface is low-contrast without visible colour changes, it is first painted with white paint with sufficient adhesion and then tinted, preferably with black spray paint to create an irregular pattern. Using this irregular pattern, the area analysis programme creates a grid of analysed points. These points are the centres of so-called "facets", i.e., the centres of small areas into which the entire analysed area has been divided. The programme records the coordinates of these points, then determines changes in their position and further determines deformations on this basis. The program records the coordinates of these points and determines the changes in their position and, on this basis, determines the deformations, logarithmic or Green's.

The manufactured plates with a central oval hole were subjected to axial tensile testing. The Cometech QC-505 M2F (Taichung City, Tajwan) universal testing machine (Figure 2 item 1) equipped with a load cell with a range up to 50 kN and an accuracy class of 0.5% (Figure 2 item 3) was used in the experiments. Specially designed wedge grips with facings for flat specimens with a thickness ranging from 0.2 to 11 mm were attached to the pivots of the measuring machine (Figure 2 item 4). They were used to constrain the test specimen (Figure 2 item 5), which was inserted axially, 30 mm each into the upper and lower grips. This made it possible to obtain a 16 × 120 mm test area of the plate. During the experimental tensile test, the load and elongation of the plate hole were measured with a constant upper crosshead speed (Figure 2 item 2) of 1 mm/min. In addition, the displacement of the composite structure in the frontal plane of the plate over time was recorded using the Aramis non-contact optical measuring system (Figure 2 item 6). This system is equipped with a 20 M resolution camera (5472 × 3648 px) and has a working area from 20 × 15 mm^2 to 5000 × 4000 mm^2, which allows for the sample to be observed with images captured at up to 17 Hz. The use of this system made it possible to examine the behaviour of the plate during the tensile test, causing it to crack and fail. The experiments were conducted in accordance with the ASTM D5766 Standard Test Method for Open-Hole Tensile Strength of Polymer Matrix Composite Laminates [46].

The numerical analysis was performed by the finite element method using the commercial version of the Abaqus system [47]. An adequate FEM model corresponding to the experimental sample was prepared. To that end, a CAD model of a 16 × 120 object with a thickness of 0.131 mm was designed, with oval holes made at the centre point of the plate with diameters of 2 mm, 4 mm, and 8 mm. The mechanical properties of the material of the numerical model were assigned in accordance with Table 1. The structure of the laminate was made by modelling the layers as separate solids. The FEM model consisted of 8 solids of 16 mm (width) × 120 mm (length) × 0.131 mm (thickness) stacked layers. For each solid, the fibre stacking orientation was assigned according to the laminate configuration [0/90/0/90/90/0/90/0]. In order to speed up the FEM computational time, the numerical model was limited to the test area of the specimen used in the experiment (16 mm × 120 mm × 1.048 mm). Therefore, boundary conditions were defined for the cross-sections of the FEM model. The lower cross-section was fully restrained by taking away all degrees of freedom as Ux = Uy = Uz = URx = URy = URz = 0, while the upper cross-section was blocked with two translational degrees of freedom as Ux = Uz = 0 and with three rotational degrees of freedom as URx = URy = URz = 0. To apply tension, the upper cross-section was assigned a displacement of 5 mm, as shown in Figure 3a. In addition, the longitudinal edges of the plate were defined as Uz = 0. In this multilayer FE model, each layer was solid and treated as a unidirectional continuous layer [48]. These orthotropic laminae were connected through Tie relations to form a lamina, thus creating perfectly connected but discontinuous interfaces between the laminae. Furthermore, this multilayer FEM model allowed the behaviour of the lamina interfaces to be made explicit in order to simulate the damage and failure process of the FRP composite laminate. When configuring the XFEM analysis, the contact interaction property was selected for each layer of the laminate in order to define the tensile crack surface behaviour. The interactions of the FEM model are represented schematically in Figure 3b. The numerical model was discretised with hexahedral solid elements of C3D8R type, having linear interpolation with 8 nodes and reduced integration. For structural meshing, partitions of the FEM model were made and the finite element mesh density was increased around the circumference of the circular hole, as shown in Figure 3c,d. The finite element size was adopted based on a preliminary numerical analysis, which proved that reducing the finite element size did not affect the quality of the numerical results. Reducing the size of the elements near the holes did not affect the results, but largely increased the CPU time.

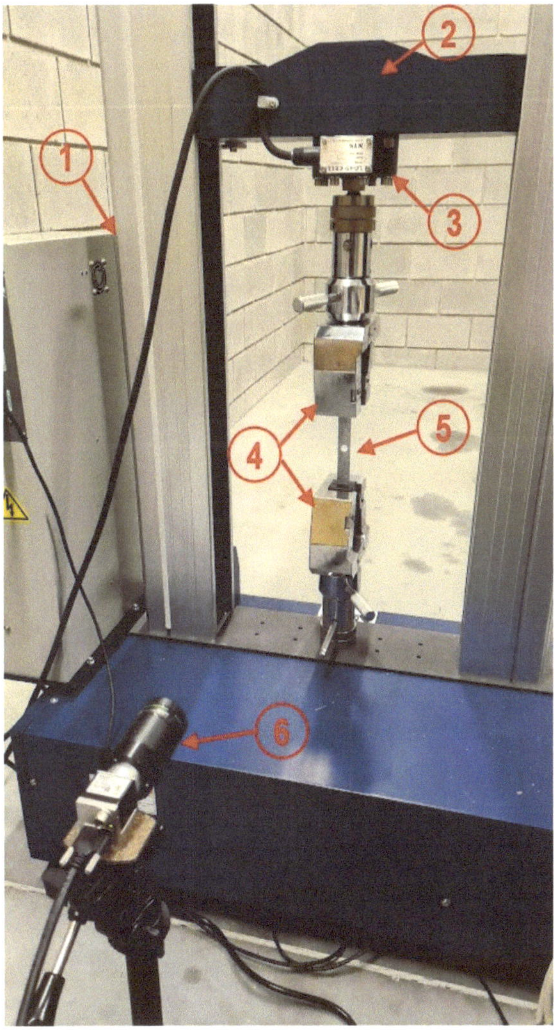

Figure 2. Experimental test stand: 1—testing machine, 2—upper crosshead, 3—measuring head, 4—wedge grips, 5—test specimen, 6—Aramis system.

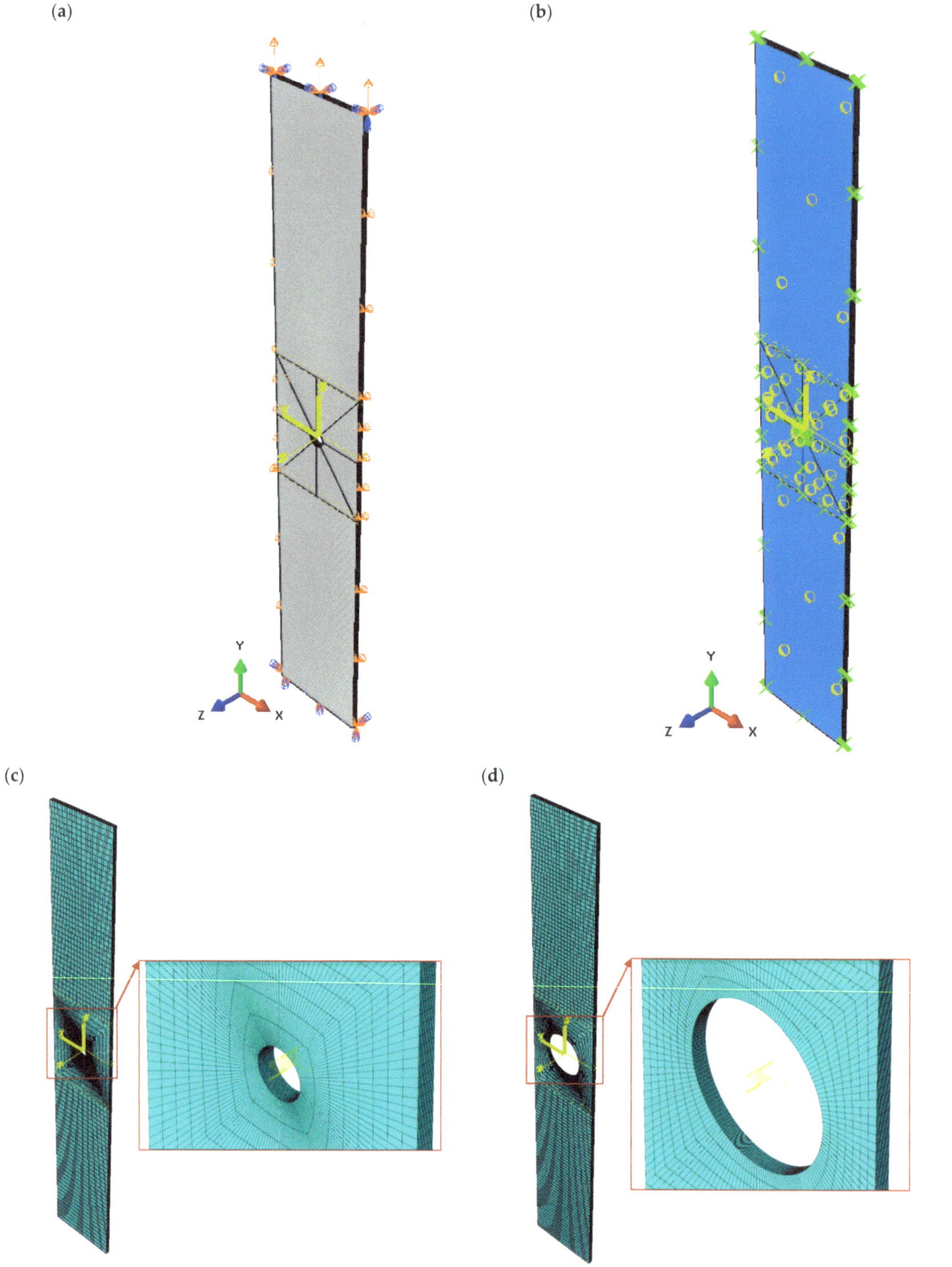

Figure 3. Numerical model: (**a**) implementation of boundary conditions, (**b**) interlayer interactions, (**c**) discretisation hole 2 mm, (**d**) discretisation hole 8 mm.

Theory of XFEM

The use of the extended finite element method (XFEM) allows the study of crack initiation and propagation without the need to re-mesh the model [27]. For crack analysis, enrichment functions typically consist of asymptotic near-tip functions that capture the singularity around the crack tip and a discontinuous function that represents the displacement spike on the crack surfaces. The nodal displacement vector enrichment function u is expressed as [49,50]

$$u = \sum_{I=1}^{N} N_I(X)\left[u_I + H(X)A_I + \sum_{\alpha=1}^{4} F_\alpha(X)B_I^\alpha\right] \tag{1}$$

where $N_I(X)$ is the nodal shape function, u_I is the nodal displacement vector of the continuous part of the finite element solution, $H(X)$ is the discontinuous jump function across the crack surface; A_I is the nodal vector of degrees of freedom; $F_\alpha(X)$ is the elastic asymptotic crack tip function; B_I^α is the nodal vector of degrees of freedom. While the first segment of the formula applies to all nodes in the model, the second segment is valid for the nodes whose shape function support is intersected by the crack interior, and the third segment is only used for the nodes whose shape function support is intersected by the crack tip. Figure 4 illustrates the tangential and normal directions (with respect to the crack) at various points along the crack interior as well as the crack apex. It also illustrates the local polar coordinate system at the crack tip.

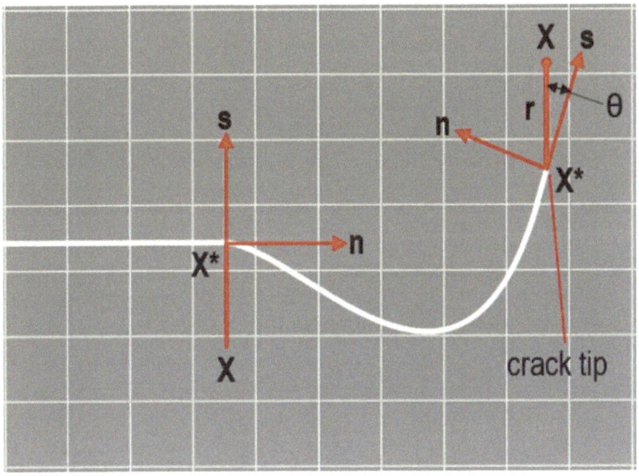

Figure 4. Normal and tangential vectors for a smooth fracture.

Asymptotic singularity functions are only taken into account when modelling stationary cracks in Abaqus/Standard. Within the moving cracks $F_\alpha(X) = 0$ and the nodal enrichment function, the displacement vector u is as follows:

$$u = \sum_{I=1}^{N} N_I(X)[u_I + H(X)A_I] \tag{2}$$

The discontinuous jump function across the fracture surface $H(X)$ could be represented as follows [50]:

$$H(X) = \begin{cases} 1\ if\ (X - X^*)\cdot n \geq 0 \\ -1\ otherwise \end{cases} \tag{3}$$

where X is the sample point (Gauss), X^* is the point on the crack closest to X, n the unit outward normal to the crack at X^*.

Moving cracks are modelled in Abaqus using one of two alternative approaches: the cohesive segment approach or the linear elastic fracture mechanics (LEFM) approach [27]. Crack initiation is defined to the onset of cohesive degradation in the enriched component. In contrast, the degradation stage occurs when the stresses or strains meet certain crack initiation criteria. One of these criteria is the maximum principal stress criterion (MAXPS), which is expressed as follows:

$$F = \frac{\langle \sigma_{MAX} \rangle}{\sigma^0_{MAX}} \tag{4}$$

where σ^0_{MAX} is the maximum permissible principal stress.

The maximum principal stress ratio $\langle \sigma_{MAX} \rangle$ shown in the Macaulay brackets assumes that the damage begins when the value equals 1:

$$\begin{cases} \langle \sigma_{MAX} \rangle = 0 \quad \text{if} \quad \sigma_{MAX} < 0 \\ \langle \sigma_{MAX} \rangle = \sigma_{MAX} \quad \text{if} \quad \sigma_{MAX} \geq 0 \end{cases} \tag{5}$$

3. Results and Discussion

An analysis of the effect of central hole diameter on the strength of the tensile plate was performed in four stages. The process of plate deformation, failure initiation, crack propagation, and failure of the CFRP composite material was described. The study proposes a new method for determining the experimental and numerical damage and failure loads of a composite plate under tension. The analysis was carried out using two independent methods simultaneously: experimental and numerical.

3.1. Plate Deformation

The experimental tensile testing of a composite plate with variable-diameter hole was conducted using the Aramis non-contact optical measuring system to measure the displacement of the specimen during the whole test. In addition, this measuring system made it possible to generate graphical displacement maps superimposed on real plates. Figure 5 shows the elongation analysis results for the plates with 2 mm, 4 mm, and 8 mm diameter holes. The proposed experimental method allowed the elongation of the specimens to be measured before complete failure. The highest elongation of 1.175 mm was obtained for the plate with a central hole diameter of 2 mm, which accounted for 1.1% of the elongation of its length (Figure 5a–c). On the other hand, the lowest elongation of 0.91 mm (0.6% elongation) was obtained for the plate with an 8 mm diameter hole (Figure 5h–j). In addition, the elongation of the hole was measured using the Aramis system, yielding an elongation of 8.8% for the plate with a 2 mm diameter hole, 6.6% for the plate with a 4 mm diameter hole, and 4% for the plate with an 8 mm diameter hole. The experimental findings showed that increasing the diameter of the hole resulted in a decrease in the total elongation of the plate. The experimental results were then compared with the results of the numerical analysis. The deformations obtained with both test methods used were found to be consistent. The maximum numerical elongation before failure occurred for the same plate and was 1.175 mm (Figure 5c), while the minimum occurred for the 8 mm diameter hole and amounted to 0.92 mm, which agreed with the experimental measurements. An analysis of the results showed qualitative and quantitative agreement between the experimental and numerical findings. The results confirmed the agreement of the proposed FEM model with the experimental test.

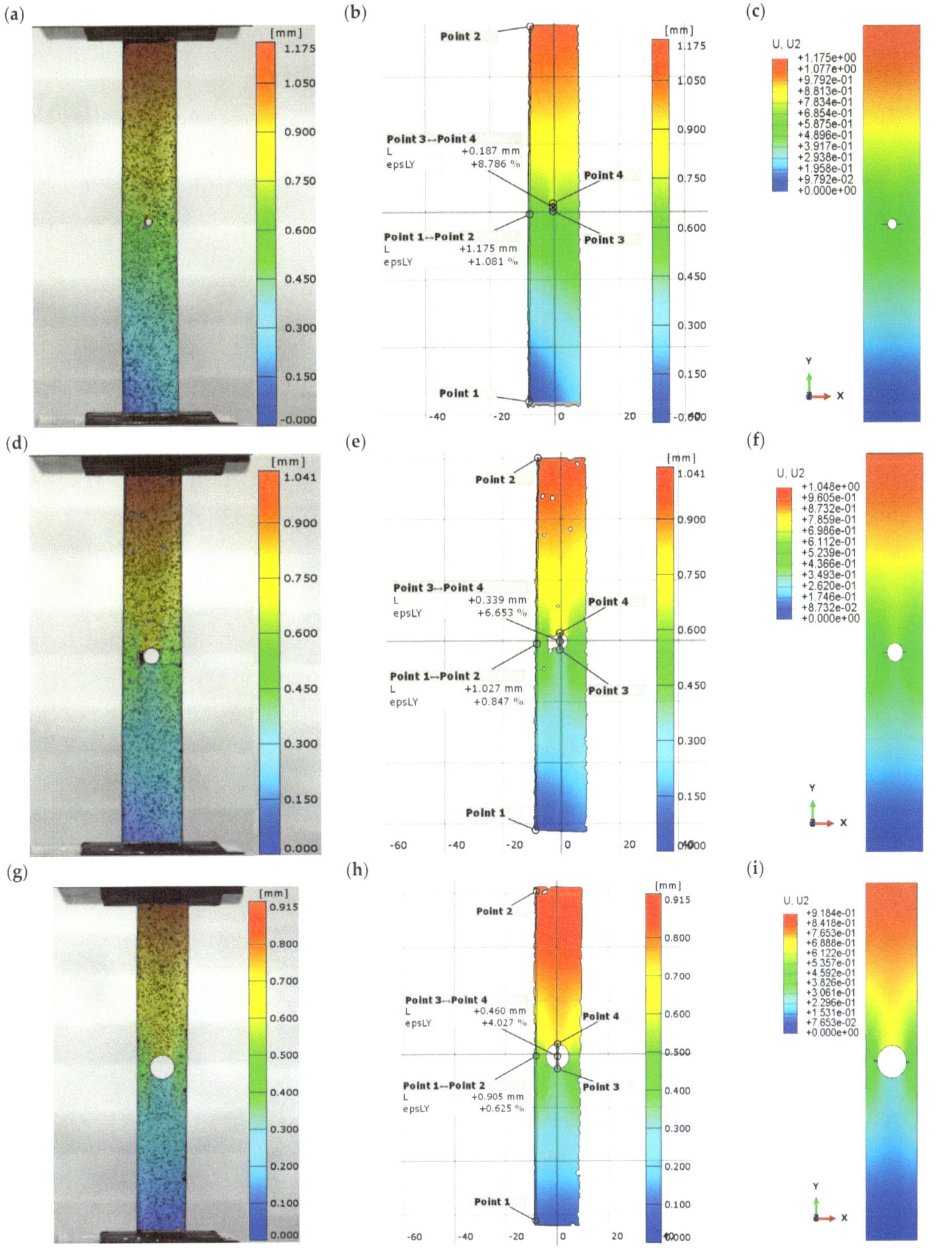

Figure 5. Hole elongation maps: (**a**,**b**) EXP_H_2 mm, (**c**) FEM_H_2 mm, (**d**,**e**) EXP_H_4 mm, (**f**) FEM_H_4 mm, (**g**,**h**) EXP_H_8 mm, (**i**) FEM_H_8 mm.

3.2. Crack Initiation

Figure 6 presents the onset of the cracking process in the laminate structure for all tested plates. The numerical cracking process was determined by XFEM. The cracking started when a value of 1 was reached according to the maximum principal stress ratio criterion. For all tested plates, the damage of the composite structure initiated with transverse cracking of the outer layer with a 0° fibre orientation in the area of the circular hole, as shown in Figure 6a–c. It should be added that the damage criterion initiated cracking of other laminate layers with a 0° fibre orientation.

(**a**)

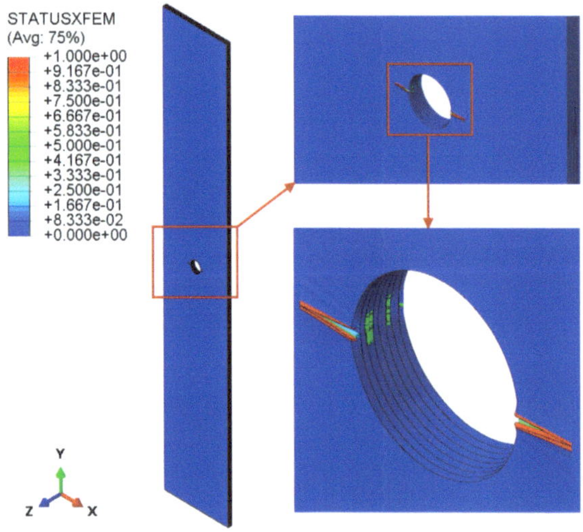

(**b**)

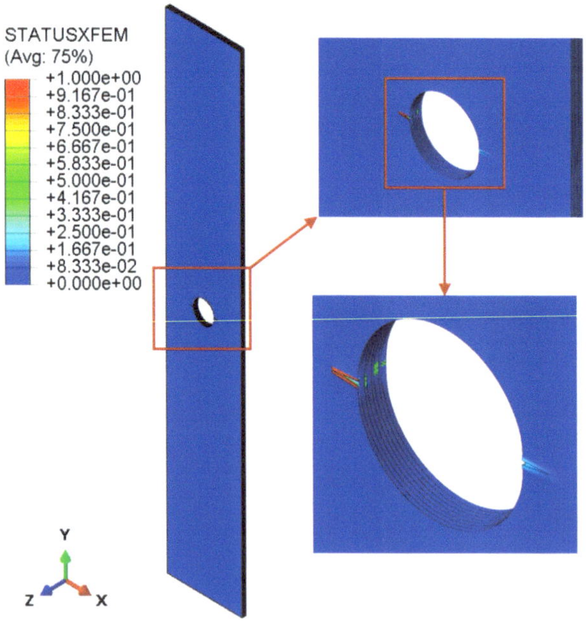

Figure 6. *Cont.*

(**c**)

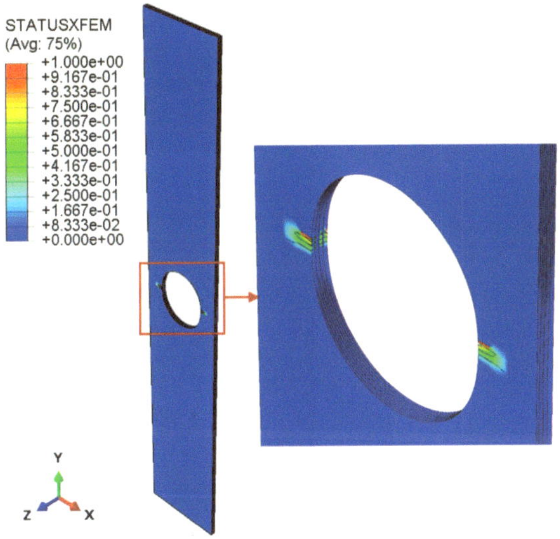

Figure 6. Damage and crack propagation in the composite structure: (**a**) FEM_H_2 mm, (**b**) FEM_H_4 mm, (**c**) FEM_H_8 mm.

3.3. Failure of the Composite

The experimental and numerical investigation was carried out over the full range of tensile loading until complete failure of the composite. Figure 7a,c,e show the experimental failure mode of the analysed plates with variable-diameter holes. The size of the hole did not affect the mode of cracking; for all cases, the real specimen cracked in the expected area where it had previously been weakened by the hole. The crack path passed across the plate halfway along its length. The parallel numerical analysis showed the same failure mode, as presented in Figure 7b,d,f. The crack propagation in the FEM model (Figure 6) initiated in the same area as that observed in the experiment (Figure 7a,c,e) and proceeded in the transverse direction. The strength of the plate with holes depended on the strength of the layers with a 0° fibre orientation, which were the most stressed in the tensile test.

(**a**)

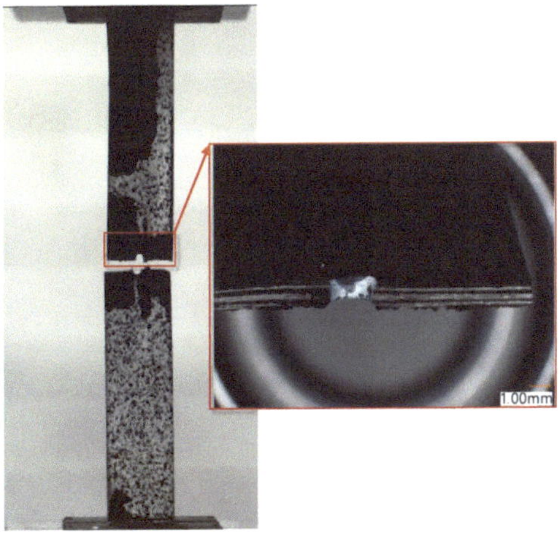

Figure 7. *Cont.*

(**b**)

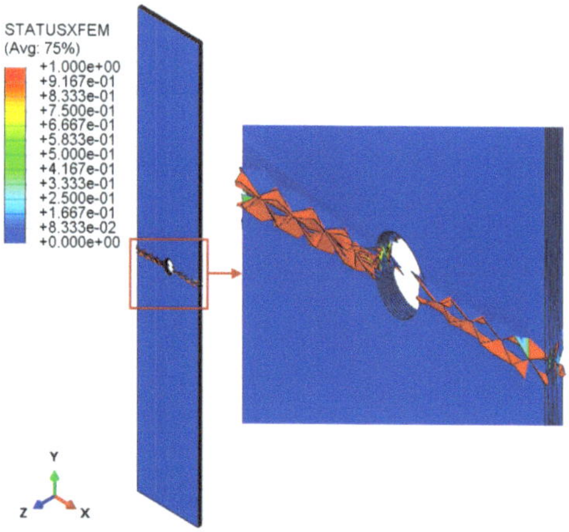

(**c**)

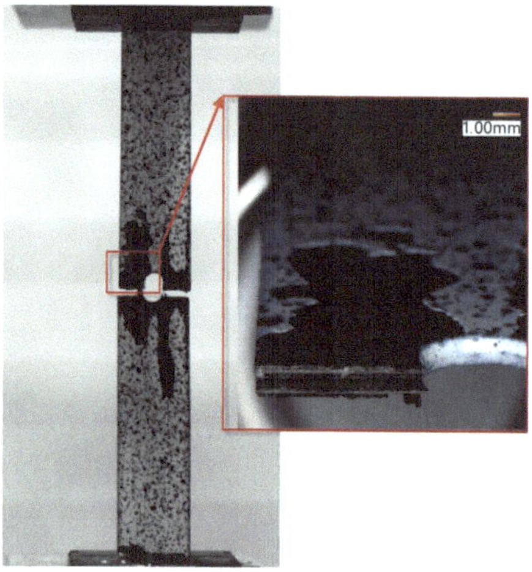

(**d**)

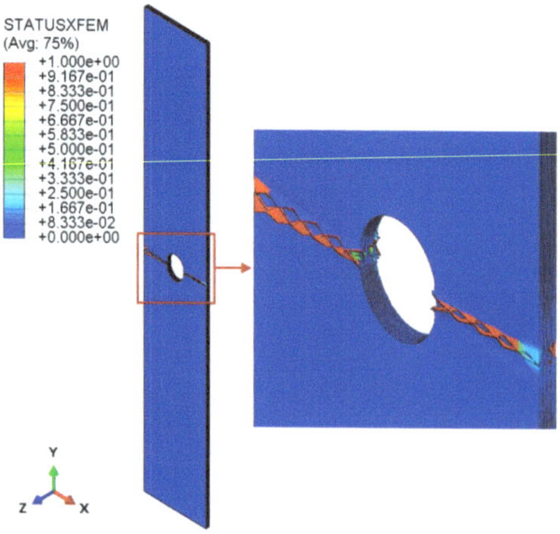

Figure 7. *Cont.*

(e)

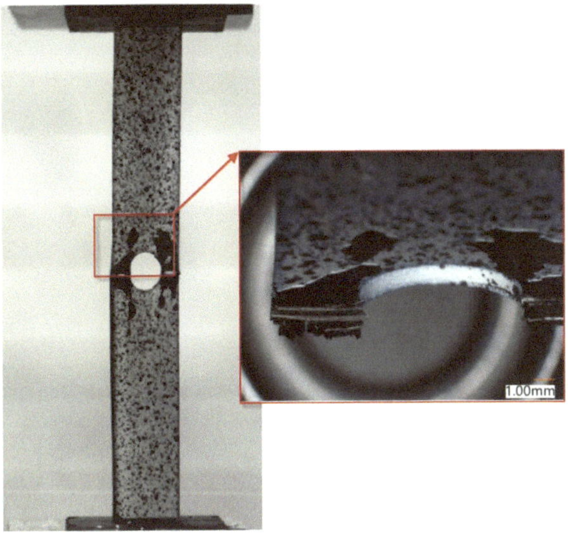

(f)

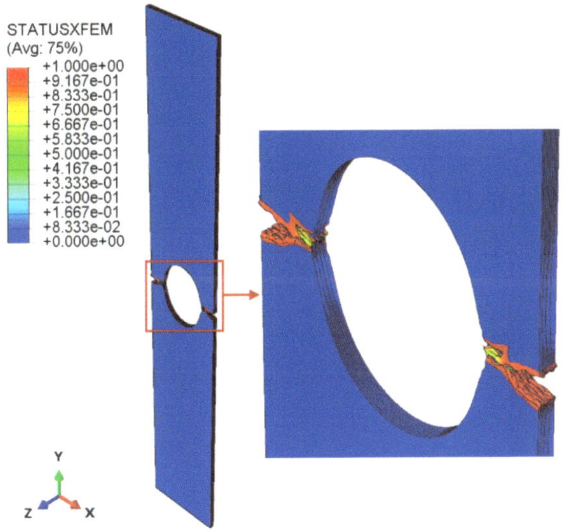

Figure 7. Complete failure of the laminate structure: (**a**) EXP_H_2 mm, (**b**) FEM_H_2 mm, (**c**) EXP_H_4 mm, (**d**) FEM_H_4 mm, (**e**) EXP_H_8 mm, (**f**) FEM_H_8 mm.

3.4. Damage and Failure Loads

Tensile load as a function of specimen elongation was measured experimentally. The experiments were extended to include a plate without a hole. This made it possible to determine the working paths of the tensile plates over the full range of loading until failure. The same characteristics were determined for the numerical model. This allowed validation of the experimental and numerical working paths, which are summarised in Figure 8a–d. For all cases, the numerical working path is stiffer than the experimental one, which is due to the fact that the numerical model was not exposed to material imperfections that may occur in the real plate. This approach allowed the damage load corresponding to the initiation of laminate cracking and the failure load causing complete failure of the composite structure to be determined graphically for all cases under study. The damage load $P_{d(EXP)}$ corresponded to the first sudden increase in elongation measured along the working path, while the failure load $P_{f(EXP)}$ was determined at the point of sudden decrease in the tensile load (Figure 8a–d). In the numerical analysis, $P_{d(FEM)}$ and $P_{f(FEM)}$ were determined in the same way as in the experiment.

(**a**)

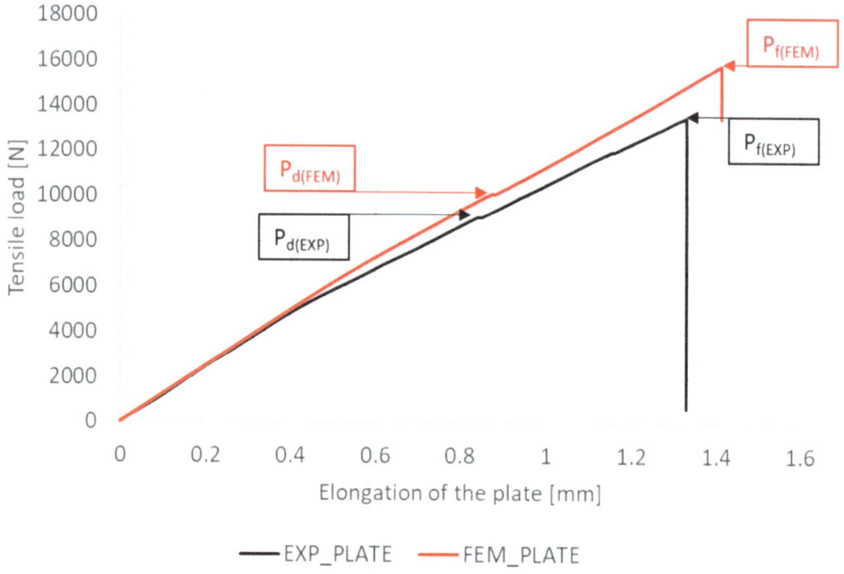

(**b**)

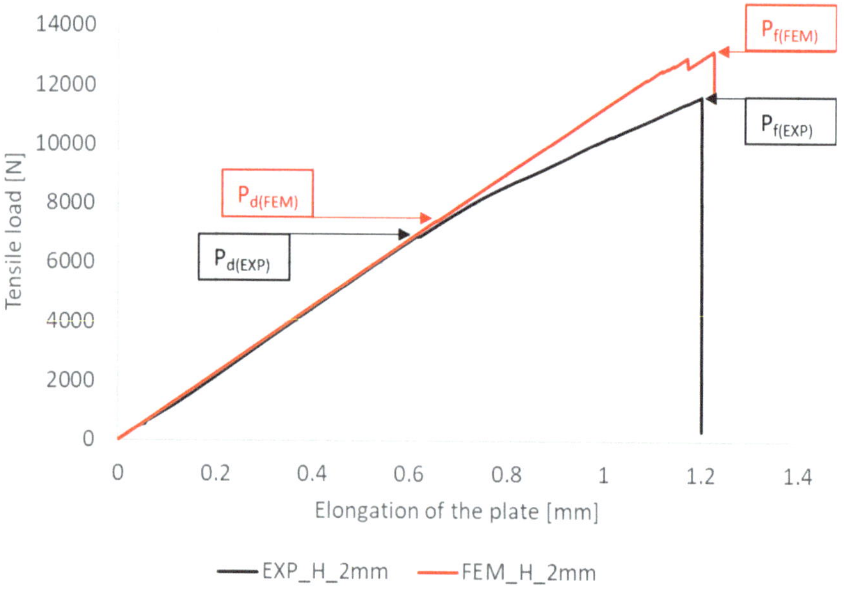

Figure 8. *Cont.*

(**c**)

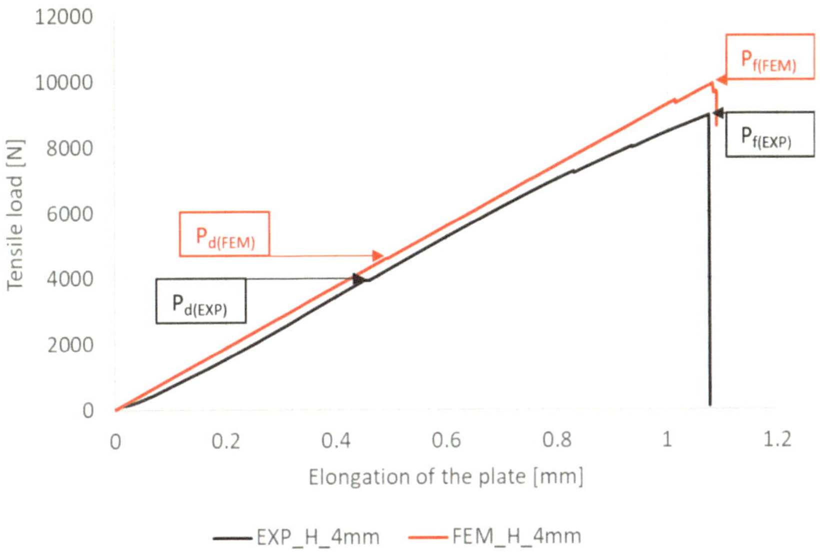

(**d**)

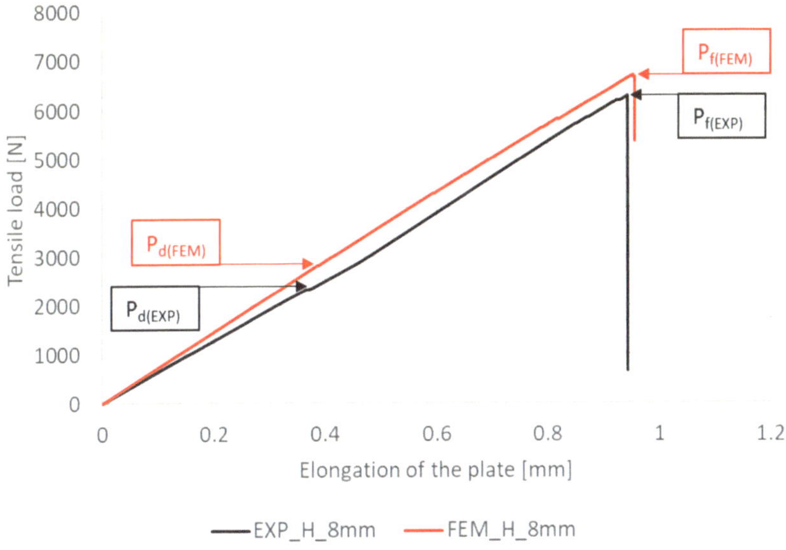

Figure 8. Work paths of the structure: (**a**) plate without a hole, (**b**) H_2 mm, (**c**) H_4 mm, (**d**) H_8 mm.

Table 2 presents the experimental and numerical damage and failure load values measured for the plate without a hole and for the plate with a hole with diameters 2 mm, 4 mm, and 8 mm. For all cases, the numerical damage initiation load $P_{d(FEM)}$ and the failure load corresponding to complete failure of the composite due to cracking $P_{f(FEM)}$ were higher than the corresponding experimental loads $P_{d(EXP)}$ and $P_{f(EXP)}$. The highest stiffness was obtained for the plate without a hole, for which $P_{d(EXP)}$ was 9031 N and $P_{f(EXP)}$ = 13,484 N. In contrast, the lowest stiffness was obtained for the plate with an 8 mm diameter hole, for which $P_{d(EXP)}$ = 2393 N and $P_{f(EXP)}$ = 6299 N. The error in prediction between the numerical and experimental values of the P_d load was in the range of <7% ÷ 15%>. The error in prediction between the experimental and numerical failure load was in the range of <6% ÷ 14%>. Based on the results, the percentage increase in the composite structure failure load $P_{f(EXP)}$ relative to the failure initiating load $P_{d(EXP)}$ was determined. It was found that after reaching the damage load value, the tensile real structure could still carry a load increased by <49% ÷ 163%>. The largest increase in the failure load relative to the damage load was obtained for the plate with an 8 mm diameter hole and the lowest for the plate without a hole.

In order to demonstrate the influence of the hole on the tensile behaviour of the composite plate, the experimental and numerical working paths for all tests are compared in Figure 9. The experimental and numerical working paths show the expected agreement between the results. The experimental paths reveal a decrease in the stiffness of the plate with increasing hole diameter. The highest decrease in stiffness was observed for the plate with an 8 mm diameter hole. For this case, the damage load $P_{d(EXP)}$ (initiating cracking) decreased by 73% and the failure load $P_{f(EXP)}$ by 53% compared to the plate without a hole. An analysis of the numerical working paths showed a similar effect of hole size on the decrease in plate stiffness—Figure 9. The maximum decrease in the numerical damage load $P_{d(FEM)}$ was 72%, and the maximum decrease in the cracking load $P_{f(FEM)}$ was 57% and was obtained for the plate with the largest hole diameter. The results showed qualitative and quantitative agreement between the experiment and the numerical analysis. The results also confirm the relevance of the developed numerical FEM/XFEM model to the experiment.

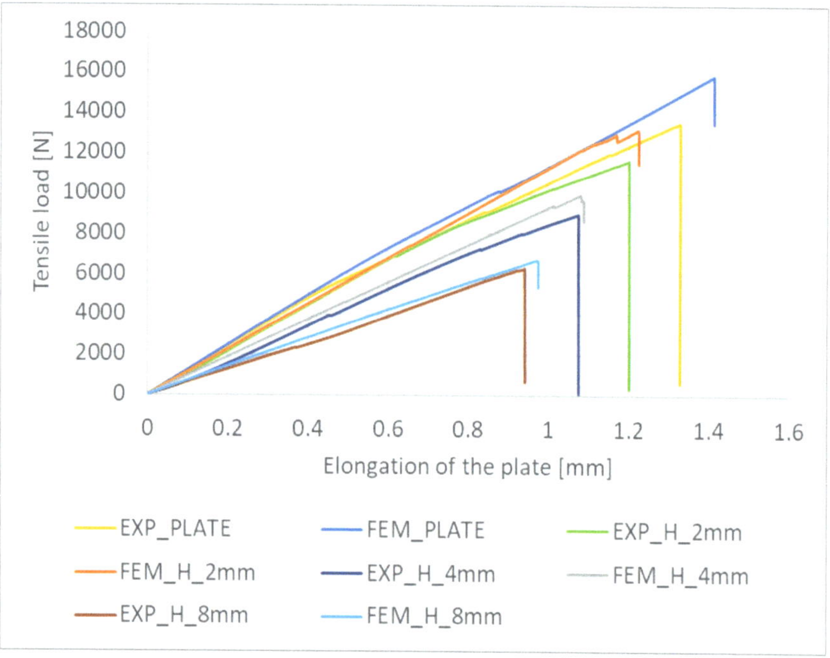

Figure 9. Comparison of experimental and numerical working paths.

Table 2. Damage load and failure load of the composite structure.

	PLATE	H_2 mm	H_4 mm	H_8 mm
$P_{d\,(EXP)}$ [N]	9031	6894	3938	2393
$P_{d\,(FEM)}$ [N]	10,153	7430	4659	2801
$P_{d\,(error\ in\ prediction)}$ [%]	11%	7%	15%	15%
$P_{f\,(EXP)}$ [N]	13,484	11,635	8954	6299
$P_{f\,(FEM)}$ [N]	15,672	13,179	9650	6683
$P_{f\,(error\ in\ prediction)}$ [%]	14%	12%	7%	6%
$P_{d\,(EXP)} \div P_{f\,(EXP)}\uparrow$ [%]	49.31%	68.77%	127.37%	163.23%
$P_{d\,(FEM)} \div P_{f\,(FEM)}\uparrow$ [%]	54.36%	77.38%	107.13%	138.59%

4. Conclusions

The proposed experimental method made it possible to measure the elongation of the specimens before they underwent complete failure. The highest elongation of 1.175 mm was obtained for the plate with a 2 mm diameter central hole, which accounted for 1.1% of its elongation. On the other hand, the lowest elongation of 0.91 mm (0.6% of the specimen elongation) was obtained for the plate with an 8 mm diameter hole. In addition, the Aramis system was used to measure hole elongation, yielding an elongation of 8.8% for the plate with a 2 mm diameter hole, 6.6% for the plate with a 4 mm diameter hole, and 4% for the plate with an 8 mm diameter hole. The experimental findings showed that increasing the diameter of the hole resulted in a decrease in the total elongation of the plate.

The size of the hole did not affect the mode of cracking; for all cases, the real specimen cracked in the expected area where it had previously been weakened by the hole. The crack path passed across the plate halfway along its length. The strength of the plate with 2 mm, 4 mm, and 8 mm diameter holes depended on the strength of the layers with a 0° fibre orientation, which were the most stressed in the tensile test.

The study also proposed a new method for determining the experimental and numerical damage and failure loads of a composite plate under tension. For all cases, the numerical value of the damage initiation load $P_{d(FEM)}$ and the failure load corresponding to the total failure of the composite due to cracking $P_{f(FEM)}$ were higher than the corresponding experimental loads $P_{d(EXP)}$ and $P_{f(EXP)}$. The highest stiffness was obtained for the plate without a hole, for which $P_{d(EXP)}$ was 9031 N and $P_{f(EXP)}$ = 13,484 N. In contrast, the lowest stiffness was obtained for the plate with an 8 mm diameter hole, for which $P_{d(EXP)}$ = 2393 N and $P_{f(EXP)}$ = 6299 N. The error in prediction of the numerical and experimental values of the P_d load was in the range of <7% ÷ 15%>. The error in prediction of the experimental and numerical failure load was in the range of <6% ÷ 14%>. The tensile real structure was still able to carry a load increased by <49% ÷ 163%> after reaching the damage load value. At the same time, the largest increase in the failure load with respect to the damage load was recorded for the plate with an 8 mm diameter hole, while the smallest was for the plate without a hole.

The largest decrease in stiffness was observed for the plate with an 8 mm diameter hole. For this case, the damage load $P_{d(EXP)}$ (initiating cracking) decreased by 73% and the failure load $P_{f(EXP)}$ by 53% compared to the plate without a hole. In contrast, the maximum decrease in the numerical damage load $P_{d(FEM)}$ was 72%, and in the failure load $P_{f(FEM)}$, it was 57% and occurred for the plate with the largest hole diameter. The results showed qualitative and quantitative agreement between the experiment and the numerical analysis. The results also confirmed the adequacy of the developed numerical FEM/XFEM model to the experiment.

Future research will investigate the influence of composite material layer configuration and the number of layers on crack propagation. As part of the research, experiments will be performed and their results will be validated numerically using the popular finite element method (FEM). The research methodology and conclusions described in this paper will be used in the future study.

Funding: This research received no external funding.

Informed Consent Statement: Not applicable.

Data Availability Statement: Not applicable.

Conflicts of Interest: The author declares no conflict of interest.

References

1. Campbell, F.C. *Manufacturing Technology for Aerospace Structural Materials*, 1st ed.; Elsevier: Amsterdam, The Netherlands, 2006.
2. Kassapoglou, C.; American Institute of Aeronautics and Astronautics. *Design and Analysis of Composite Structures: With Applications to Aerospace Structures*, 1st ed.; John Wiley & Sons: Reston, VA, USA, 2010.
3. Mlýnek, J.; Petrů, M.; Martinec, T.; Koloor, S.S.R. Fabrication of High-Quality Polymer Composite Frame by a New Method of Fiber Winding Process. *Polymers* **2020**, *12*, 1037. [CrossRef]
4. Jain, P.; Kumar, A. Postbuckling response of square laminates with a central circular/elliptical cutout. *Compos. Struct.* **2004**, *65*, 179–185. [CrossRef]
5. Parlapalli, M.R.; Soh, K.C.; Shu, D.W.; Ma, G. Experimental investigation of delamination buckling of stitched composite laminates. *Compos. Part A Appl. Sci. Manuf.* **2007**, *38*, 2024–2033. [CrossRef]
6. Turvey, G.; Zhang, Y. A computational and experimental analysis of the buckling, postbuckling and initial failure of pultruded GRP columns. *Comput. Struct.* **2006**, *84*, 1527–1537. [CrossRef]
7. Singer, J.; Arbocz, J.; Weller, T. Buckling Experiments. Experimental Methods in Buckling of Thin-walled Structure. In *Basic Concepts, Columns, Beams, and Plates*; John Wiley & Sons Inc.: New York, NY, USA, 1998.
8. Teter, A.; Kołakowski, Z. Coupled dynamic buckling of thin-walled composite columns with open cross-sections. *Compos. Struct.* **2013**, *95*, 28–34. [CrossRef]
9. Kubiak, T.; Mania, R.J. Hybrid versus FR laminate channel section columns—Buckling and postbuckling behaviour. *Compos. Struct.* **2016**, *154*, 142–149. [CrossRef]
10. Gliszczynski, A.; Kubiak, T.; Wawer, K. Barely visible impact damages of GFRP laminate profiles—An experimental study. *Compos. Part B Eng.* **2019**, *158*, 10–17. [CrossRef]
11. Gliszczynski, A.; Wiącek, N. Experimental and numerical benchmark study of mode II interlaminar fracture toughness of unidirectional GFRP laminates under shear loading using the end-notched flexure (ENF) test. *Compos. Struct.* **2021**, *258*, 113190. [CrossRef]
12. Kolakowski, Z.; Kubiak, T. Interactive dynamic buckling of orthotropic thin-walled channels subjected to in-plane pulse loading. *Compos. Struct.* **2007**, *81*, 222–232. [CrossRef]
13. Wong, P.; Wang, Y. An experimental study of pultruded glass fibre reinforced plastics channel columns at elevated temperatures. *Compos. Struct.* **2007**, *81*, 84–95. [CrossRef]
14. Orifici, A.; Thomson, R.; Degenhardt, R.; Kling, A.; Rohwer, K.; Bayandor, J. Degradation investigation in a postbuckling composite stiffened fuselage panel. *Compos. Struct.* **2008**, *82*, 217–224. [CrossRef]
15. Kim, S.-C.; Kim, J.S.; Yoon, H.-J. Experimental and numerical investigations of mode I delamination behaviors of woven fabric composites with carbon, Kevlar and their hybrid fibers. *Int. J. Precis. Eng. Manuf.* **2011**, *12*, 321–329. [CrossRef]
16. Debski, H.; Kubiak, T.; Teter, A. Experimental investigation of channel-section composite profiles' behavior with various sequences of plies subjected to static compression. *Thin-Walled Struct.* **2013**, *71*, 147–154. [CrossRef]
17. Timoshenko, S.; Gere, J. *Theory of Elastic Stability, International Student Edition*, 2nd ed.; McGraw-Hill: New York, NY, USA, 1985.
18. Cheng, B.; Zhao, J. Strengthening of perforated plates under uniaxial compression: Buckling analysis. *Thin-Walled Struct.* **2010**, *48*, 905–914. [CrossRef]
19. El-Sawy, K.M.; Martini, M.I. Elastic stability of bi-axially loaded rectangular plates with a single circular hole. *Thin-Walled Struct.* **2007**, *45*, 122–133. [CrossRef]
20. Jayashankarbabu, B.S.; Karisiddappa. Stability of Square Plate with Concentric Cutout. *Int. J. Civ. Environ. Eng.* **2014**, *8*, 259–267. [CrossRef]
21. Ghannadpour, S.; Najafi, A.; Mohammadi, B. On the buckling behavior of cross-ply laminated composite plates due to circular/elliptical cutouts. *Compos. Struct.* **2006**, *75*, 3–6. [CrossRef]
22. Lorenzini, G.; Helbig, D.; Silva, C.; Real, M.; Santos, E.; Rocha, L. Numerical evaluation of the effect of type and shape of perforations on the buckling of thin steel plates by means of the constructal design method. *Int. J. Heat Technol.* **2016**, *34*, S9–S20. [CrossRef]
23. Kaltakci, M. Stress concentrations and failure criteria in anisotropic plates with circular holes subjected to tension or compression. *Comput. Struct.* **1996**, *61*, 67–78. [CrossRef]
24. Durão, L.M.P.; Matos, J.E.; Loureiro, N.C.; Esteves, J.L.; Fernandes, S.C.F. Damage Propagation by Cyclic Loading in Drilled Carbon/Epoxy Plates. *Materials* **2023**, *16*, 2688. [CrossRef]

25. Mathavan, J.J.; Hassan, M.H.; Xu, J.; Franz, G. Hole Quality Observation in Single-Shot Drilling of CFRP/Al7075-T6 Composite Metal Stacks Using Customized Twist Drill Design. *J. Compos. Sci.* **2022**, *6*, 378. [CrossRef]
26. Wang, H.; Wu, Y.; Zhang, Y.; Zhang, X. Influence of the Temperature-Dependent Characteristics of CFRP Mechanical Properties on the Critical Axial Force of Drilling Delamination. *Polymers* **2023**, *15*, 680. [CrossRef] [PubMed]
27. Elguedj, T.; Gravouil, A.; Combescure, A. Appropriate extended functions for X-FEM simulation of plastic fracture mechanics. *Comput. Methods Appl. Mech. Eng.* **2006**, *195*, 501–515. [CrossRef]
28. Belytschko, T.; Black, T. Elastic crack growth in finite elements with minimal remeshing. *Int. J. Numer. Meth. Engng.* **1999**, *45*, 601–620. [CrossRef]
29. Melenk, J.; Babuška, I. The partition of unity finite element method: Basic theory and applications. *Comput. Methods Appl. Mech. Eng.* **1996**, *139*, 289–314. [CrossRef]
30. Liu, P.; Gu, Z.; Peng, X.; Zheng, J. Finite element analysis of the influence of cohesive law parameters on the multiple delamination behaviors of composites under compression. *Compos. Struct.* **2015**, *131*, 975–986. [CrossRef]
31. Wong, K.J.; Johar, M.; Koloor, S.S.R.; Petrů, M.; Tamin, M.N. Moisture Absorption Effects on Mode II Delamination of Carbon/Epoxy Composites. *Polymers* **2020**, *12*, 2162. [CrossRef] [PubMed]
32. Rozylo, P.; Falkowicz, K.; Wysmulski, P.; Debski, H.; Pasnik, J.; Kral, J. Experimental-Numerical Failure Analysis of Thin-Walled Composite Columns Using Advanced Damage Models. *Materials* **2021**, *14*, 1506. [CrossRef]
33. Rozylo, P.; Wysmulski, P. Failure analysis of thin-walled composite profiles subjected to axial compression using progressive failure analysis (PFA) and cohesive zone model (CZM). *Compos. Struct.* **2021**, *262*, 113597. [CrossRef]
34. Rozylo, P.; Falkowicz, K. Stability and failure analysis of compressed thin-walled composite structures with central cut-out, using three advanced independent damage models. *Compos. Struct.* **2021**, *273*, 114298. [CrossRef]
35. Rozylo, P. Comparison of Failure for Thin-Walled Composite Columns. *Materials* **2021**, *15*, 167. [CrossRef] [PubMed]
36. Rozylo, P. Failure phenomenon of compressed thin-walled composite columns with top-hat cross-section for three laminate lay-ups. *Compos. Struct.* **2023**, *304*, 116381. [CrossRef]
37. Debski, H.; Samborski, S.; Rozylo, P.; Wysmulski, P. Stability and Load-Carrying Capacity of Thin-Walled FRP Composite Z-Profiles under Eccentric Compression. *Materials* **2020**, *13*, 2956. [CrossRef] [PubMed]
38. Wysmulski, P. The effect of load eccentricity on the compressed CFRP Z-shaped columns in the weak post-critical state. *Compos. Struct.* **2022**, *301*, 116184. [CrossRef]
39. Wysmulski, P.; Falkowicz, K.; Filipek, P. Buckling state analysis of compressed composite plates with cut-out. *Compos. Struct.* **2021**, *274*, 114345. [CrossRef]
40. Wysmulski, P. Non-linear analysis of the postbuckling behaviour of eccentrically compressed composite channel-section columns. *Compos. Struct.* **2023**, *305*, 116446. [CrossRef]
41. Shao, S.; Wu, Y.; Fu, H.; Feng, S.; Zhang, J. Numerical Investigation on the Mechanical Properties of Vault Void Lining and Steel Plate Strengthening. *Materials* **2023**, *16*, 789. [CrossRef]
42. Hao, R.; Wen, Z.; Xin, H.; Lin, W. Fatigue Life Prediction of Notched Details Using SWT Model and LEFM-Based Approach. *Materials* **2023**, *16*, 1942. [CrossRef]
43. Buccino, F.; Cervellera, F.; Ghidini, M.; Marini, R.; Bagherifard, S.; Vergani, L.M. Isolating the Role of Bone Lacunar Morphology on Static and Fatigue Fracture Progression through Numerical Simulations. *Materials* **2023**, *16*, 1931. [CrossRef]
44. Wang, F.; Wei, Z.; Li, P.; Yu, L.; Huang, W. Initial Crack Propagation and the Influence Factors of Aircraft Pipe Pressure. *Materials* **2019**, *12*, 3098. [CrossRef]
45. Rozylo, P. Experimental-numerical study into the stability and failure of compressed thin-walled composite profiles using progressive failure analysis and cohesive zone model. *Compos. Struct.* **2020**, *257*, 113303. [CrossRef]
46. *ASTM D5766/D5766M-11*; Test Method for Open-Hole Tensile Strength of Polymer Matrix Composite Laminates. ASTM International: West Conshohocken, PA, USA, 2018. [CrossRef]
47. Jonak, J.; Karpiński, R.; Wójcik, A.; Siegmund, M. The Effect of Undercut Anchor Diameter on the Rock Failure Cone Area in Pullout Tests. *Adv. Sci. Technol. Res. J.* **2022**, *16*, 261–270. [CrossRef]
48. Koloor, S.; Ayatollahi, M.R.; Tamin, M. Elastic-damage deformation response of fiber-reinforced polymer composite laminates with lamina interfaces. *J. Reinf. Plast. Compos.* **2017**, *36*, 832–849. [CrossRef]
49. Remmers, J.J.; Deborst, R.; Needleman, A. The simulation of dynamic crack propagation using the cohesive segments method. *J. Mech. Phys. Solids* **2008**, *56*, 70–92. [CrossRef]
50. Ventura, G.; Benvenuti, E. Equivalent polynomials for quadrature in Heaviside function enriched elements. *Int. J. Numer. Methods Eng.* **2015**, *102*, 688–710. [CrossRef]

Disclaimer/Publisher's Note: The statements, opinions and data contained in all publications are solely those of the individual author(s) and contributor(s) and not of MDPI and/or the editor(s). MDPI and/or the editor(s) disclaim responsibility for any injury to people or property resulting from any ideas, methods, instructions or products referred to in the content.

materials

MDPI

Article

Buckling Analysis of Laminated Plates with Asymmetric Layup by Approximation Method

Katarzyna Falkowicz *, Pawel Wysmulski and Hubert Debski

Department of Machine Design and Mechatronics, Faculty of Mechanical Engineering, Lublin University of Technology, Nadbystrzycka 36, 20-618 Lublin, Poland; p.wysmulski@pollub.pl (P.W.); h.debski@pollub.pl (H.D.)
* Correspondence: k.falkowicz@pollub.pl

Abstract: This study investigated thin-walled plate elements with a central cut-out under axial compression. The plates were manufactured from epoxy/carbon laminate (CFRP) with an asymmetric layup. The study involved analyzing the buckling and post-buckling behavior of the plates using experimental and numerical methods. The experiments provided the post-buckling equilibrium paths (*P-u*), which were then used to determine the critical load using the straight-line intersection method. Along with the experiments, a numerical analysis was conducted using the Finite Element Method (FEM) and using the ABAQUS® software. A linear analysis of an eigenvalue problem was conducted, the results of which led to the determination of the critical loads for the developed numerical model. The second part of the calculations involved conducting a non-linear analysis of a plate with an initial geometric imperfection corresponding to structural buckling. The numerical results were validated by the experimental findings, which showed that the numerical model of the structure was correct.

Keywords: composites; critical state; finite element method; thin-walled structures; linear and nonlinear analysis; stability of construction; matrix couplings

Citation: Falkowicz, K.; Wysmulski, P.; Debski, H. Buckling Analysis of Laminated Plates with Asymmetric Layup by Approximation Method. *Materials* **2023**, *16*, 4948. https://doi.org/10.3390/ma16144948

Academic Editor: Aniello Riccio

Received: 8 June 2023
Revised: 30 June 2023
Accepted: 7 July 2023
Published: 11 July 2023

Copyright: © 2023 by the authors. Licensee MDPI, Basel, Switzerland. This article is an open access article distributed under the terms and conditions of the Creative Commons Attribution (CC BY) license (https://creativecommons.org/licenses/by/4.0/).

1. Introduction

Carbon fibre-reinforced polymer composites (CFRPCs) are among the most widely studied lightweight materials. They are widely used in automotive, civil engineering [1,2] and aircraft structures [3–7], among others. Due to their material properties, such as a high strength-to-weight ratio [8,9], these materials are very popular as load-bearing components [8–11]. Furthermore, their mechanical properties can be shaped by designing specified characteristics and ply configurations for these materials.

Owing to their shape and the fact that they are usually thin-walled, plate elements are particularly susceptible to stability loss [12–15]. Therefore, the determination of the critical loads for the plates and the analysis of their behavior under dynamic loads are very important parts of their strength analysis, which has been undertaken in a number of studies [12,16,17].

It is worth emphasizing here that there are many numerical and experimental studies on the stability of structures made of classical isotropic engineering materials, and their results are widely reported in the literature. In contrast, there are fewer studies investigating the problem of conjugate buckling for plate structures made of composite materials. Furthermore, there are no experimental studies investigating the buckling and post-buckling behavior of asymmetric laminate plates.

In general, the analysis of layered plates is more complicated due to their anisotropy and heterogeneity. However, with the development of computer techniques and numerical programs, such as FEM, studies attempting to describe these aspects have begun to appear. To give an example, one can mention here studies investigating the performance characteristics for structures made of thin plates, both with open and closed cross-sections, with reinforcements [18,19], initial geometric parameters [20] or cut-outs [21,22]. Special focus was placed on determining the effects of shape inaccuracy [9], the boundary conditions [23], the geometric parameters of the structure [24] or the fibre arrangement [25]. Many of these theoretical considerations were validated through experiments [22]. FEM analyses of the behavior of plates were also conducted by Batoz et al. [26] and Cui et al. [27].

The stability of thin-walled plates made of composite materials was also investigated by Kolakowski and Kowal-Michalska [28], Kolakowski and Krolak [29] and Kubiak [30,31]. These studies predominantly dealt with the problem of buckling in thin-walled composite structures under compression. The above problems were solved by Koiter's asymptotic theory. Different approaches to stability analysis of laminated composite plates using the Ritz method are presented in papers [32–34].

The problem of the stability of compressed thin-walled structures, including mixed-mode buckling, was analyzed using the finite element method by Kubiak [30], as well as by Bazant and Cedolin [35]. Examples of using FEM to solve the problems of the linear and nonlinear stability of composite structures can be found in previous studies conducted by the author [36,37], as well as in studies performed by Alfano and Crisfield [38], Kreja [39], Kopecki [40], Mania et al. [41] and Teter and Kolakowski [42]. Numerical FEM simulations of the behavior of composite structures under different loads were verified experimentally by Debski et al. [43] and Banat et al. [44,45], among others.

The literature review shows that there are numerous studies on the problems of deflection and stability. However, there is a lack of experimental studies on thin-walled asymmetric plates with a cut-out and their use as elastic elements. Furthermore, there are no experimental studies investigating the buckling and post-buckling behavior of asymmetric elements, which is the novel aspect of the current work.

Therefore, the determination of the critical load value that causes the buckling of a thin-walled structure is a very important research problem. The knowledge of this value makes it possible to prevent the structure from premature failure due to a loss of stability by its elements, a problem which was discussed, among others, in [36,46–49]. It is also worth mentioning that this study involved using the FEM method, which is widely used in many fields [50–56]. However, the numerically determined critical load value may only yield an approximate estimation of the critical force because the numerical calculations assume an ideal structure, without the geometric imperfections that occur in real structures. This means that the analyzed numerical models of thin-walled structures should be further validated experimentally. To that end, it is necessary to use approximation methods [57–59] that enable the estimation of the critical load value based on the experimental results. In this study, the straight-lines intersection method [60,61] was used to estimate the critical forces.

This study investigated the buckling and early post-buckling behavior of compressed thin-walled composite plates with a cut-out. The study involved determining the critical load of a real structure using the straight-lines intersection method, based on the results obtained using the ARAMIS system and analysis of the buckling and post-buckling behavior through the finite element method (FEM). The results showed that this approach was effective for solving the problems of linear and nonlinear stability for thin-walled composite structures.

2. Object of the Research

The study was conducted on rectangular thin-walled plates fabricated from an epoxy/carbon composite material (M12/35%/UD134/AS7) through autoclaving [62]. The material properties of the CFRP laminate used for samples were determined experimentally in compliance with the relevant ISO standards, as described in [63]. The mechanical properties of the composite material are presented in Table 1. The plates were subjected to axial compression.

Table 1. Material properties of the tested laminate.

Young's Modulus [MPa]		Poisson's Ratio ν_{12}	Kirchhoff's Modulus G_{12} [MPa]
0° (E_1)	90° (E_2)	0.36	±45°
143,530	5826		3845

The tested plates had a central rectangular cut-out with variable geometric parameters and constant overall dimensions [64]. A schematic representation of the considered model with its dimensions and ply arrangement is given in Figure 1.

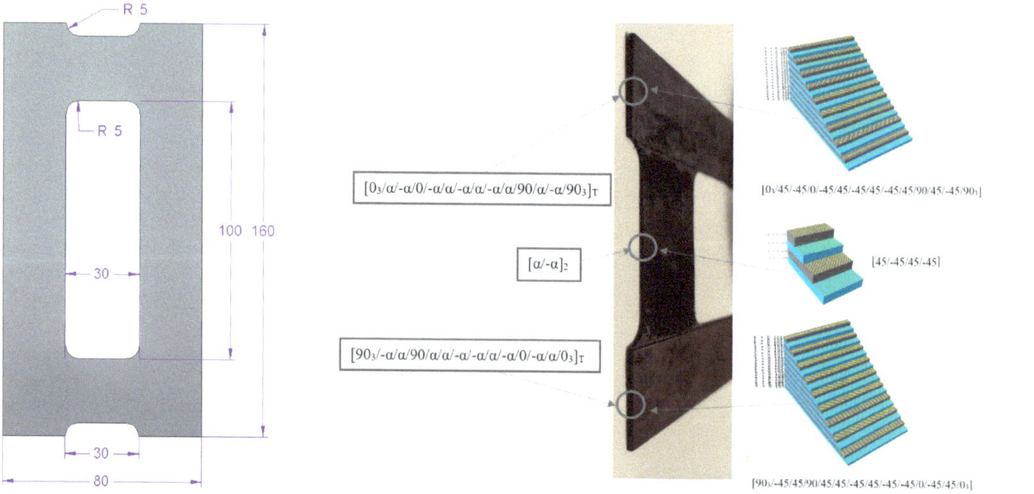

(a) (b)

Figure 1. Schematic representation of the considered model: (**a**) geometric model, (**b**) real model with ply arrangement.

Six plate models with different cut-out dimensions (20 × 100 mm, 30 × 100 mm, 40 × 100 mm, 30 × 120 mm) and three fibre arrangement angles (30°, 45°, 60°) were analyzed. The composite structure consisted of 18 plies; each 0.105 mm thick, in an asymmetric arrangement with respect to the midplane. The considered ply layup is presented in Table 2.

Table 2. Tested ply configurations.

Ply Number	Core		Strips		Final Configuration	
	Ply Orientation	Coupling	Ply Orientation	Coupling	Configuration	Coupling
18	$[\alpha/-\alpha]_2$	$A_SB_tD_S$ *	$[\alpha/-\alpha/0/-\alpha/0/\alpha/90/\alpha/-\alpha]$	$A_SB_lD_S$ *	$[0_3/\alpha/-\alpha/0/-\alpha/\alpha/-\alpha/\alpha/-\alpha/\alpha/90/\alpha/-\alpha/90_3]_T$	$A_SB_{lt}D_S$

Where: α—angle, A_S, D_S—simple laminate, B_t—extension-twisting and shearing-bending coupling, B_{lt}—extension-bending, extension–twisting and shearing–bending coupling, B_t—extension–twisting and shearing–bending coupling, D_F—twisting-bending coupling. *—it is some modification of the ply orientation tested by York [65,66], more information about the selection of this configuration was given in [67].

Each plate model consisted of a core spread over the entire plate volume and vertical strips arranged on both sides of the core along the longer edges of the plate (Figure 1b). This ply layup was selected to ensure that flexural-torsional buckling would be the lowest buckling mode of the plate, without additional forcing. The ply orientations were selected using mechanical matrix couplings based on the studies conducted by Ch. York [65,66]. This concept was comprehensively described in the authors' previous studies [68,69].

3. Methodology and Scope of the Study

The range of the conducted study included the analysis of the buckling and early post-buckling behavior of a compressed thin-walled composite plate weakened by cut-outs of various geometrical parameters and different fibre orientations. The study was performed using both experimental and numerical methods. The experiments conducted on the fabricated thin-walled laminate plates enabled the observation of the structure's behavior in the critical state and after the loss of stability. The numerical simulations conducted in parallel with the experiments were aimed at developing adequate, experimentally validated FEM models for simulating the buckling of a thin-walled laminated structure, closely reproducing the real structure's behavior.

3.1. *Experimental*

The experiments were carried out on a Zwick/Roell ZMART PRO universal testing machine with a measuring range of up to 2500 kN, coupled with the ARAMIS system [70], which enabled the collection of data in a graphical form and analysis of the deformation and displacement [71]. The experiments were performed with a constant cross-bar velocity of 2 mm/min at room temperature. The axially compressed thin-walled composite plates were loaded with approximately 150% of the numerical critical load value. During the compression process, the plate element was simply supported by specially designed and manufactured grips mounted in the testing machine. The test stand with the mounted plate sample is shown in Figure 2.

During the tests, the compressive force and the shortening of the sample in the perpendicular direction to its cross-section were measured. The shortening of the sample was measured at the top edge of the plate. The ARAMIS optical system was used to measure the plate shortening. This system uses a series of digital images to read the displacements taken during measurements at regular intervals by two cameras positioned at an appropriate distance from the tested object. The cameras are placed on a special tripod. This system is characterized by high resolution and high measurement accuracy. After preparing the sample and setting up the tripod with the cameras, the system was calibrated using a template with marked reference points. After the calibration, the first photo of the sample was taken, which was the zero load state, as well the reference state, against which all calculations were made for the subsequent images. The images were captured in the Start/Mid/Stop Trigger mode until the measurement was completed. After the measurement was completed, in order to start the analysis of the captured images, the surface area (calculation mask) on which the calculations would be carried out was determined based on the size of the facets (Figure 3). Each facet was assigned a unique structure and coordinates, thanks to which they could be recognized in the images captured during the loading process. The last step before the analysis was to manually select a starting point from which the calculation process would begin.

Figure 4 presents a sample frame from a film generated based on the data obtained with the ARAMIS system. By analyzing the frame, it is possible to determine:

- the values of the strains/main displacements in the specimen;
- changes in the specimen shape during the loading process.

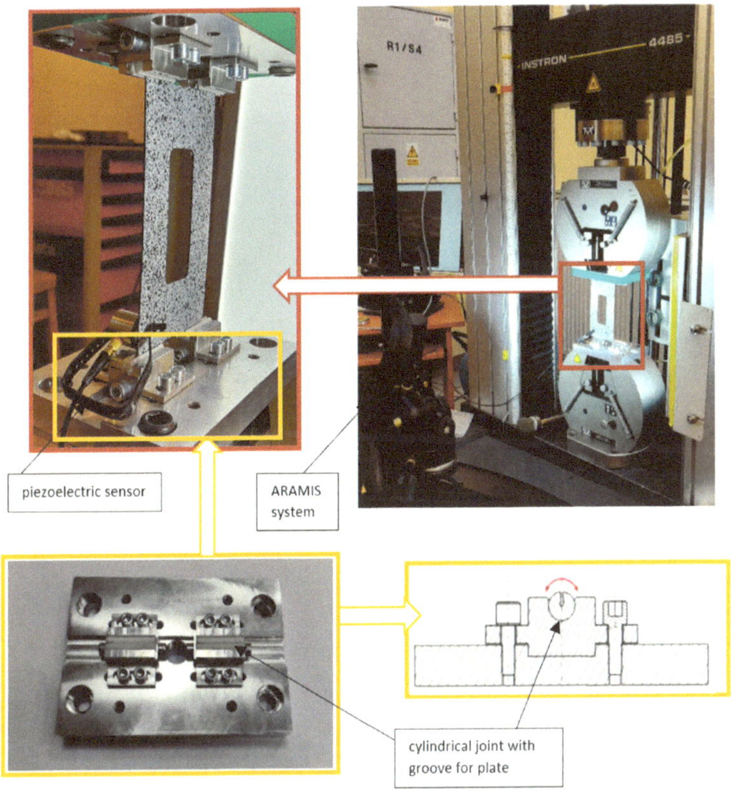

Figure 2. Test stand.

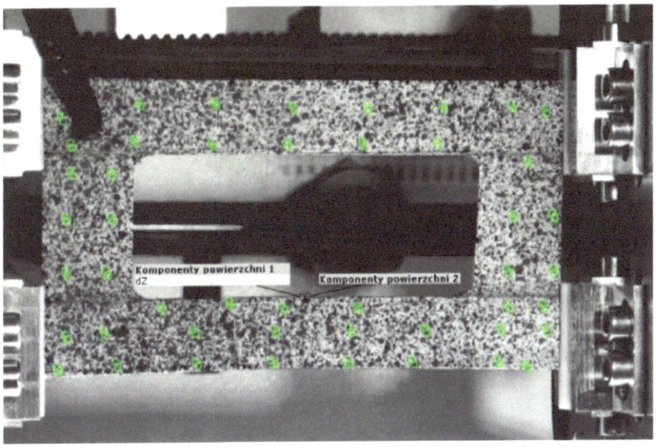

Figure 3. View of the facets in the measuring area.

The post-critical equilibrium paths obtained from the measurements, illustrating the relationship between the load and plate shortening, *P-u*, made it possible to determine the critical load value, and thus to evaluate the structure's performance in the early post-critical range.

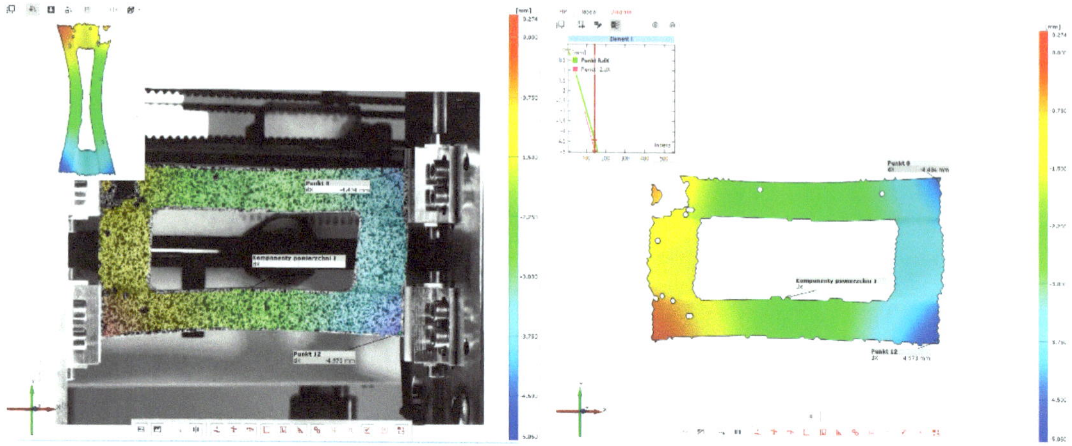

Figure 4. Examples of results generated by the ARAMIS system.

3.2. *Experimental Determination of the Buckling Load*

Inaccuracies occurring during the experiments due to different factors—such as boundary conditions and geometric imperfections of the structure, design of the test stand or application of the load—make it difficult to precisely determine the value of the buckling load. Therefore, it is necessary to use approximation methods that enable the estimation of the buckling load value based on the experimental measurements. In this study, the straight-lines intersection method was used to estimate the critical forces [61,72].

The application of the straight-lines intersection method consisted of the approximation of the post-buckling equilibrium path, which describes the relationship between the load and sample shortening measured perpendicularly to the cross-section. To estimate the approximate value of the critical force by solving two linear functions, the post-buckling equilibrium path *P-u* in the early post-critical range—which were determined experimentally—were used. The post-buckling equilibrium paths *P-u* were approximated in selected intervals by two linear functions having the form [72]:

$$\begin{cases} P_{cr} = P\frac{a_1}{a_0}u + P \\ P_{cr} = P\frac{a_2}{a_0}u + P \end{cases} \rightarrow (P_{cr};\ u) \tag{1}$$

where a_1, a_2 are unknown function parameters, P is the applied load value, P_{cr} is an unknown critical load value, u is the shortening of the plate corresponding to the critical load.

The critical (buckling) load is determined based on the intersection point of the approximation function L1 with the second linear function L2, projected by the horizontal L3 on the coordinate system vertical axis of the post-buckling characteristic of the structure *P-u* (Figure 5). The results obtained using approximation methods are not always unambiguous. The degree of the approximated curve linearity is strictly dependent on the range of data involved in the process of determining the critical loads. In addition, the result significantly depends on the number of points with specific coordinates subjected to the approximation stage.

In this study, the key determinant of the accuracy of the approximation process was the correlation coefficient R^2. This coefficient is used to determine the convergence level between the approximating function and the selected range of the approximated experimental curve. A higher value of the correlation coefficient ensured a higher accuracy of the approximation process. In this experimental approximation of the post-buckling paths, the minimum value of the correlation coefficient was assumed to be $R^2 \geq 0.85$.

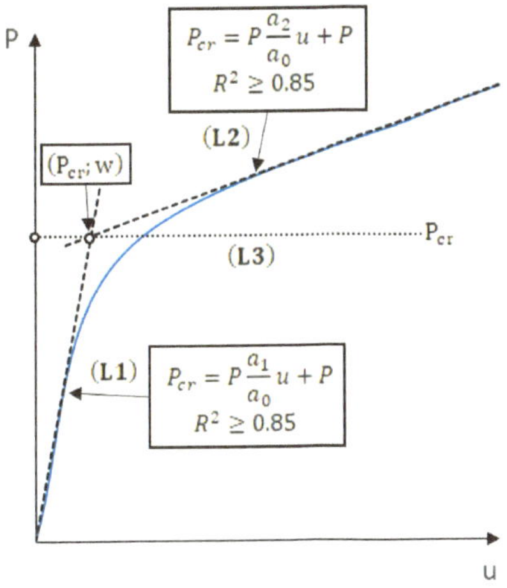

Figure 5. Straight-lines intersection method.

3.3. Numerical Model

The experimental investigation of the stability and post-buckling behavior was conducted in parallel with the numerical modelling using the finite element method. The numerical analysis was performed using the commercial ABAQUS® software. The scope of the analysis included the investigation of the buckling and early post-buckling behavior up to a value of ~150% of the lowest critical load value. The calculations for the critical state included the solution of a linear eigenvalue problem, which led to the determination of the lowest critical load value and the corresponding buckling mode. The maximum potential energy condition was used to solve the eigenproblem. It was solved using the following equation [30]:

$$([K] + \lambda_i[H])\{\psi\}_i = 0 \tag{2}$$

where $[K]$ is the structural stiffness matrix, $[H]$ is the stress stiffness matrix, λ_i is the i-th eigenvalue and ψ is the i-th eigenvector of displacement.

When the $\{\psi\}_i$ value equals zero, this means that the solution is trivial and the structure remains in the initial state of equilibrium. Equation (3) represents the eigenvalue problem, which can help find n multiplier λ buckling load values and the corresponding buckling mode.

$$|[K] + \lambda_i[H]| = 0 \tag{3}$$

In the second part of the study, a nonlinear static analysis was performed. The initial geometric imperfection was flexural-torsional buckling, and it was implemented with an amplitude of 0.1 of the plate thickness. Although nonlinear analyses are carried out with the progressive failure algorithm [67,73,74], this study only focused on the early post-buckling range. In effect, the relationship between the load and column shortening P-u in the early post-critical range could be determined. To solve the geometrically nonlinear problem, the incremental-iterative Newton-Raphson method was used.

Numerical model discretization was carried out by means of shell elements with six degrees of freedom at each node. In addition, 8-node shell elements (S8R) with a quadratic shape function and reduced integration were used. More details about the numerical analysis and the discretization process are given in [67,73]. A general view of the numerical model is presented in Figure 6.

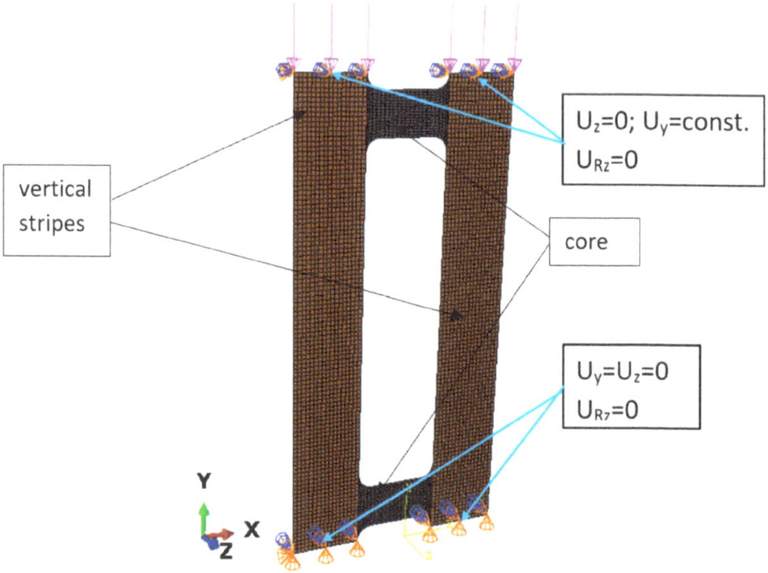

Figure 6. Discrete model of a plate and its boundary conditions.

The material properties of each layer of the CFRP were the same as those determined experimentally (Table 1). The FEM model characteristics, such as the geometry, the method of load application and the boundary conditions, were adopted as close as possible to those in the experiments (Figure 2).

The boundary conditions for the numerical model reflected a simple support of the compressed composite plate (Figure 4a). The boundary conditions were enforced by constraining the movement of the kinematic degrees of freedom of the nodes located on the top and bottom edges of the plate. The nodes located on the bottom edge had constrained movement of two translational degrees of freedom Uy = Uz = 0 and of one rotational degree of freedom URz = 0, but were allowed free rotation relative to the edge of the plate. The top edge was assigned the same boundary conditions, additionally allowing node displacement in the direction of loading, i.e., Uz = 0 and URz = 0. The vertical edges of the plate remained free during the loading process. Axial compression was applied through uniform loading of the top edge of the plate.

4. Results and Discussion

The experiments conducted on the axially compressed thin-walled plates provided information that made it possible to establish a relation between the buckling of the real structures and the external load. The experimental results allowed qualitative and quantitative analyses of the pre-buckling and buckling behavior based on the obtained test parameters. The buckling state was identified based on the obtained buckling mode and its corresponding critical load. The experimental critical loads were used to validate the numerical results.

Examples of the experimental and numerical flexural-torsional buckling modes obtained for the tested plates are shown in Figure 7.

The results show good agreement in qualitative terms and confirm the stable behavior of the tested plates in the post-buckling range. They also confirm that the selected asymmetric configuration with couplings was correct. Examples of the post-buckling flexural-torsional modes obtained for the compressed composite plate with a 40 × 100 mm cut-out and 45° fibre arrangement angle and deflection maps are given in Figure 8.

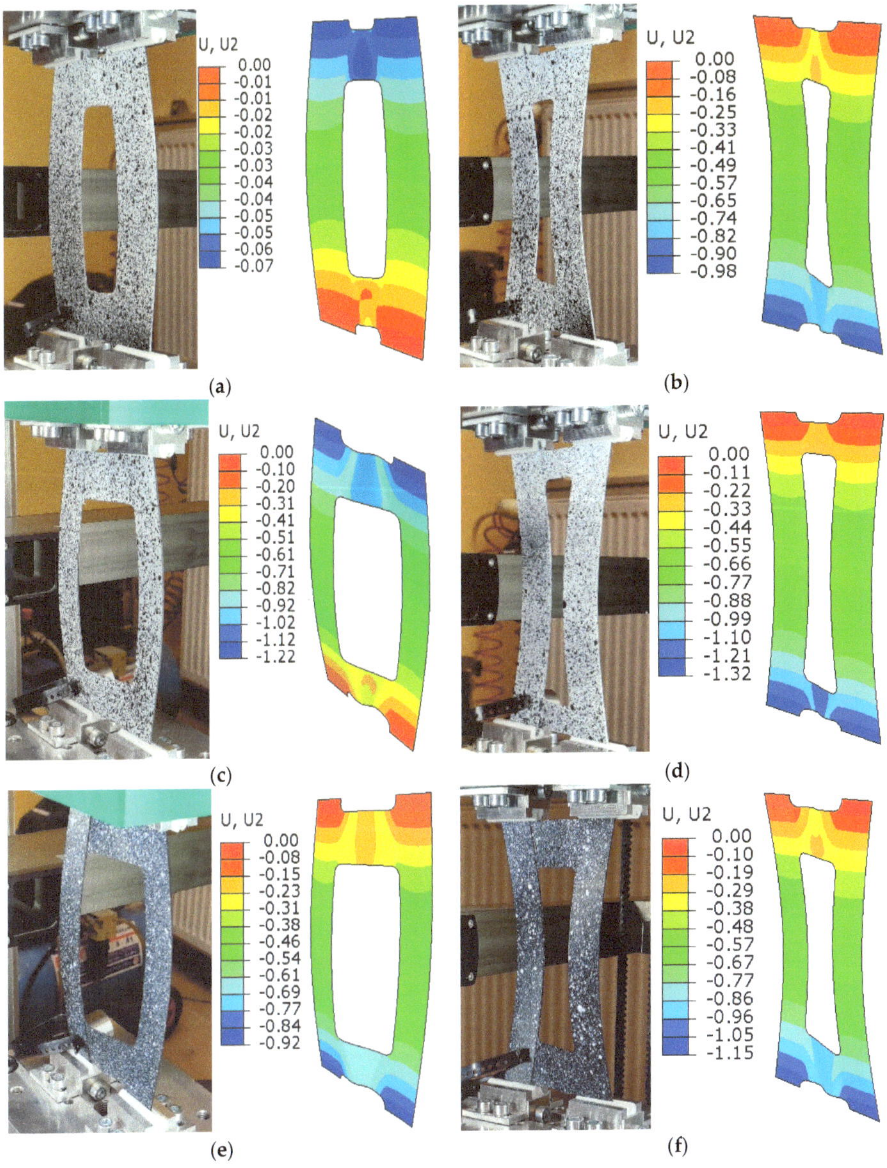

Figure 7. Flexural-torsional buckling modes obtained for the analyzed plates: (**a**) 45°_20 × 100 mm, (**b**) 45°_30 × 100 mm, (**c**) 45°_40 × 100 mm, (**d**) 45°_30 × 120 mm, (**e**) 30°_40 × 100 mm, (**f**) 60°_40 × 100 mm.

The measurements of the plate shortening made it possible to determine the post-buckling equilibrium paths, which describe the relationship between the load and shortening, *P-u* (Figure 9). As mentioned above, all samples were tested until failure, according to the progressive failure method, which is described in detail in [67,73,74], where some of the results are reported. However, in the current work, the focus is on the analysis of the buckling state and the determination of the experimental critical forces using the approximation method. Therefore, the behavior in the low post-critical range is sufficient for the analysis (Figure 10). As shown in Figure 10, the experimental results and FEM curves show good agreement in terms of both quantity and quality.

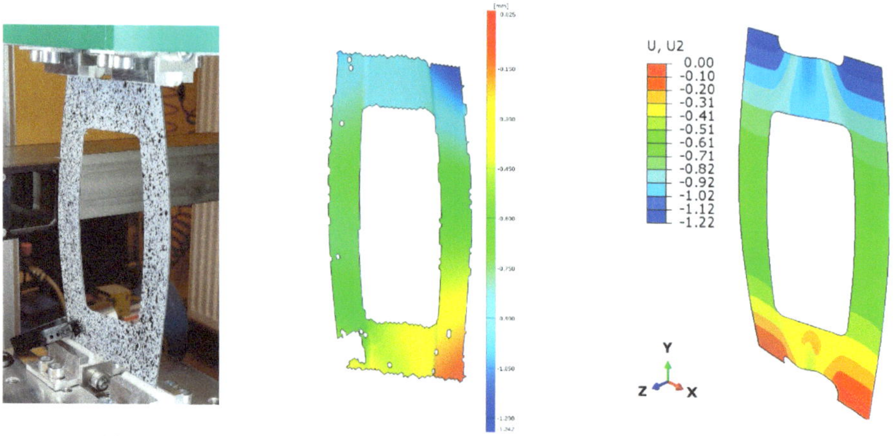

(**a**) (**b**) (**c**)

Figure 8. Post-buckling of a plate with a 40 × 100 cut-out and a 45 fibre arrangement angle, together with obtained deflection maps (**a**) experimental, (**b**) ARAMIS, (**c**) Abaqus.

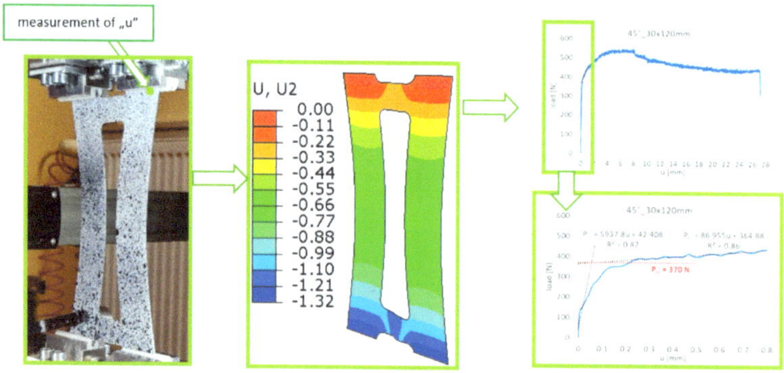

Figure 9. Determination of the critical load from an early post-critical equilibrium path.

The experimental post-critical equilibrium paths served as a basis for determining the critical load using the straight-lines intersection method. The key problem with this approach is that the measuring range must be selected correctly in order to describe the post-buckling equilibrium path, as this has a direct impact on the results. If the approximation procedures are inappropriate, the experimental critical loads will significantly differ from the numerical values. In addition, for a sufficient compliance of the approximation function with the experimental curve, the approximation range should be selected in such a way as to maintain a high value of the correlation coefficient R^2 ($R^2 \geq 0.85$).

In the straight-lines intersection method, the post-buckling equilibrium path P-u was approximated using two linear functions. The first one was used in the initial interval of the experimental path (buckling), while the other was applied after a visible change in the characteristic (post-buckling). The critical load value was determined as a horizontal line, projecting the intersection point of the approximation functions onto the vertical axis of the diagram (load axis). The critical load values obtained using the straight-lines intersection method are given in Figure 11.

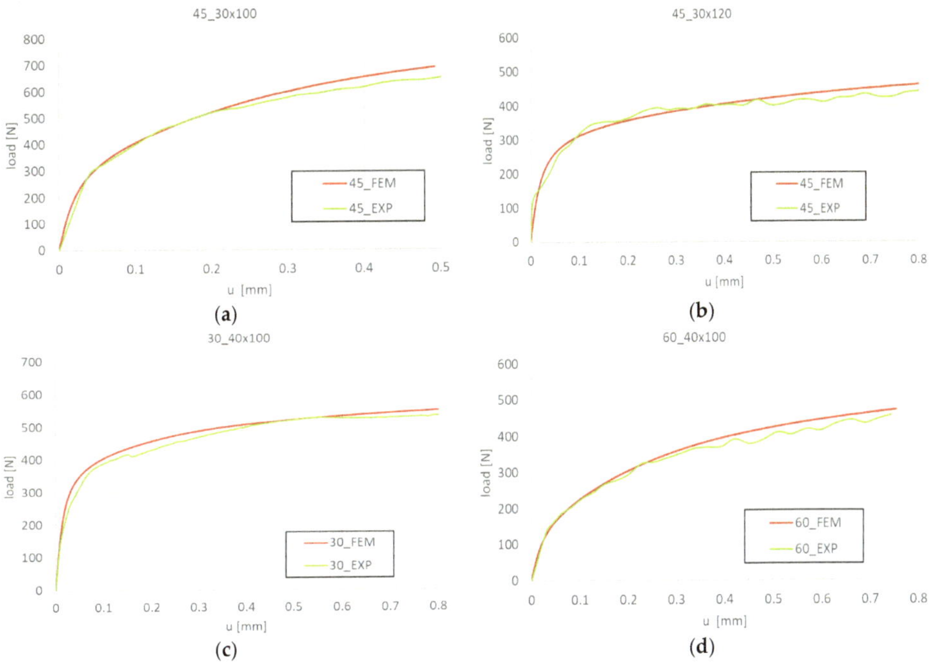

Figure 10. FEM and experimental post-buckling equilibrium paths obtained for the tested plates: (**a**) 45_30 × 100 mm, (**b**) 45_30 × 120 mm, (**c**) 30_40 × 100 mm, (**d**) 60_40 × 100 mm.

The critical loads obtained using the approximation method were compared with the eigenvalues determined through numerical analysis. The experimental and FEM critical loads are listed in Table 3.

Table 3. Comparison of FEM and experimental critical loads.

Method	45°_20 × 100	45°_30 × 100	45°_40 × 100	45°_30 × 120	30°_40 × 100	60°_40 × 100
FEM [N]	735	444	333	394	533	247
straight-lines intersection [N]	705	422	330	370	494	244
Difference [N]	20	22	3	24	39	3

Figure 12 shows a bar chart with the FEM and experimental critical loads obtained for the tested samples. The highest agreement between the experimental and numerical results was obtained for the 45°_40 × 100 plate, while the lowest for the 30°_40 × 100 plate.

It must be remembered that the numerical load values obtained through solving the eigenproblem are upper estimates of the critical load. The agreement between the approximation and numerical critical loads causing a buckling of the thin-walled flat plates weakened by a central cut-out ranges between 0.9% and 7.32% (Table 4). The greatest difference between the experimental and numerical results was obtained for the 30°_40 × 100 mm plate. The smallest difference, of 0.9%, was obtained for the 45°_40 × 100 sample. The average error of the critical loads obtained using the FEM/EXP methods is ~4%, which is an acceptable value as the experimental results represent a lower estimate of the buckling state. The numerical and experimental results show high quantitative agreement, which proves the correctness of the applied approximation method.

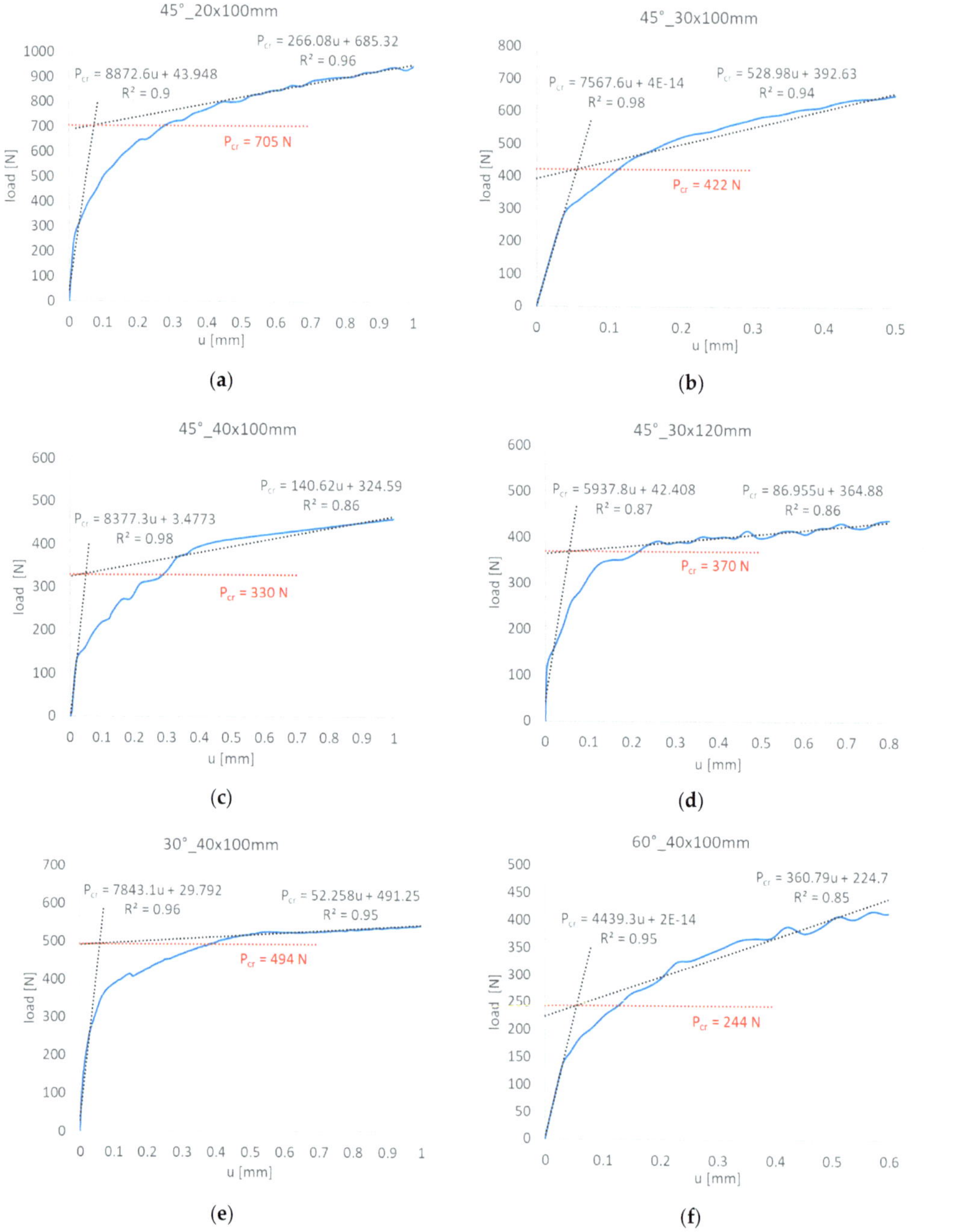

Figure 11. Buckling loads determined for the tested plates by the straight-lines intersection method: (**a**) 45°_20 × 100 mm, (**b**) 45°_30 × 100 mm, (**c**) 45°_40 × 100 mm, (**d**) 45°_30 × 120 mm, (**e**) 30°_40 × 100 mm, (**f**) 60°_40 × 100 mm.

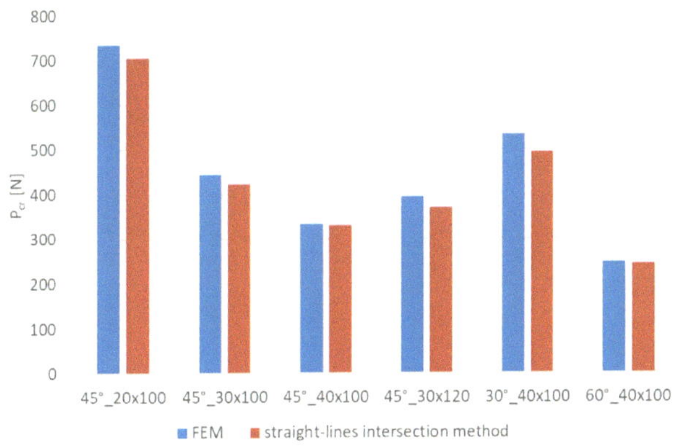

Figure 12. Bar chart showing FEM and experimental critical loads.

Table 4. Difference [in %] between numerical and experimental critical loads.

Difference [%]	45°_20 × 100	45°_30 × 100	45°_40 × 100	45°_30 × 120	30°_40 × 100	60°_40 × 100
FEM/ straight-lines intersection	4.08%	4.95%	0.9%	6.09%	7.32%	1.21%

5. Conclusions

In this study, the behavior of axially compressed thin-walled plate elements weakened by a central cut-out was investigated. An attempt was made to determine the critical load value based on the experimental post-buckling paths obtained using the straight-lines intersection method. The experimental results were then compared with the numerical critical load value determined using the finite element method. The comparison showed high agreement between the experimental and numerical critical loads. This agreement confirms that the proposed procedure can be employed to determine the critical load values for real structures. The correctly determined critical load value of a thin-walled flat plate is of vital importance for operational reasons because it helps prevent structural buckling.

The study has shown that the accuracy of the results strongly depends on the applied approximation parameters. This particularly concerns the selection of an appropriate approximation range and a high value of the correlation coefficient R^2 in order to ensure agreement between the experimental structural characteristics and approximation function.

The results provide important information about modelling thin-walled structures made of composite materials. At the same time, they confirm that the numerical models are designed correctly and are thus effective for both eigenproblem calculations and nonlinear static analysis of an early post-buckling response. The results confirm that the numerical model was designed correctly and thus made it possible to simulate the buckling and post-buckling behavior of the compressed plates with a central cut-out.

Author Contributions: Conceptualization, K.F.; methodology, K.F., P.W. and H.D.; software, K.F.; validation, K.F. and P.W.; formal analysis, K.F.; investigation, K.F. and P.W.; resources, K.F.; data curation, K.F.; writing—original draft preparation, K.F. and P.W.; writing—review and editing, K.F. and H.D.; visualization, K.F. and P.W.; supervision, H.D.; project administration, K.F.; funding acquisition, K.F. All authors have read and agreed to the published version of the manuscript.

Funding: This research was funded in part by the National Science Centre of Poland under project No. UMO-2022/47/B/ST8/00600.

Institutional Review Board Statement: Not applicable.

Informed Consent Statement: Not applicable.

Data Availability Statement: Data are contained within the article.

Acknowledgments: Lublin University of Technology, Faculty of Mechanical Engineering.

Conflicts of Interest: The authors declare no conflict of interest.

References

1. Chróścielewski, J.; Miśkiewicz, M.; Pyrzowski, Ł.; Rucka, M.; Sobczyk, B.; Wilde, K. Modal Properties Identification of a Novel Sandwich Footbridge—Comparison of Measured Dynamic Response and FEA. *Compos. Part B Eng.* **2018**, *151*, 245–255. [CrossRef]
2. Chróścielewski, J.; Miśkiewicz, M.; Pyrzowski, Ł.; Sobczyk, B.; Wilde, K. A Novel Sandwich Footbridge—Practical Application of Laminated Composites in Bridge Design and in Situ Measurements of Static Response. *Compos. Part B Eng.* **2017**, *126*, 153–161. [CrossRef]
3. Carlsson, L.A.; Adams, D.F.; Pipes, R.B. *Experimental Characterization of Advanced Composite Materials*, 4th ed.; CRC Press: Boca Raton, FL, USA, 2014; ISBN 978-0-429-10939-3.
4. Parlapalli, M.R.; Soh, K.C.; Shu, D.W.; Ma, G. Experimental Investigation of Delamination Buckling of Stitched Composite Laminates. *Compos. Part A Appl. Sci. Manuf.* **2007**, *38*, 2024–2033. [CrossRef]
5. Turvey, G.J.; Zhang, Y. A Computational and Experimental Analysis of the Buckling, Postbuckling and Initial Failure of Pultruded GRP Columns. *Comput. Struct.* **2006**, *84*, 1527–1537. [CrossRef]
6. Baker, A.A.; Dutton, S.; Kelly, D. (Eds.) *Composite Materials for Aircraft Structures*, 2nd ed.; Education Series; American Institute of Aeronautics and Astronautics: Reston, VA, USA, 2004; ISBN 978-1-56347-540-5.
7. Guo, S.; Cheng, W.; Cui, D. Aeroelastic Tailoring of Composite Wing Structures by Laminate Layup Optimization. *AIAA J.* **2006**, *44*, 3146–3150. [CrossRef]
8. Campbell, F.C. Introduction. In *Manufacturing Technology for Aerospace Structural Materials*; Elsevier: Amsterdam, The Netherlands, 2006; pp. 1–13. ISBN 978-1-85617-495-4.
9. ASM International (Ed.) *ASM Handbook*, 10th ed.; ASM International: Materials Park, OH, USA, 1990; ISBN 978-0-87170-377-4.
10. Tavares, S.M.O.; De Castro, P.M.S.T. Materials. In *Damage Tolerance of Metallic Aircraft Structures*; Springer Briefs in Applied Sciences and Technology; Springer International Publishing: Cham, Switzerland, 2019; pp. 29–42. ISBN 978-3-319-70189-9.
11. Kopecki, T.; Mazurek, P.; Lis, T.; Chodorowska, D. Post-Buckling Deformation States of Semi-Monocoque Cylindrical Structures with Large Cut-Outs under Operating Load Conditions. Numerical Analysis and Experimental Tests. *EiN* **2016**, *18*, 16–24. [CrossRef]
12. Singer, J.; Arbocz, J.; Weller, T. *Buckling Experiments: Experimental Methods in Buckling of Thin-Walled Structures*; John Wiley & Sons, Inc.: New York, NY, USA, 2000; Volume 1, ISBN 978-0-470-17298-8.
13. Doyle, J.F. *Nonlinear Analysis of Thin-Walled Structures: Statics, Dynamics, and Stability*; Mechanical Engineering Series; Springer: New York, NY, USA, 2001; ISBN 978-0-387-95216-1.
14. Khdeir, A.A.; Librescu, L. Analysis of Symmetric Cross-Ply Laminated Elastic Plates Using a Higher-Order Theory: Part II—Buckling and Free Vibration. *Compos. Struct.* **1988**, *9*, 259–277. [CrossRef]
15. Reddy, J.N.; Khdeir, A.A.; Librescu, L. Le'vy Type Solutions for Symmetrically Laminated Rectangular Plates Using First-Order Shear Deformation Theory. *J. Appl. Mech.* **1987**, *54*, 740–742. [CrossRef]
16. Bazant, Z.P.; Cedolin, L.; Hutchinson, J.W. Stability of Structures: Elastic, Inelastic, Fracture, and Damage Theories. *J. Appl. Mech.* **1993**, *60*, 567–568. [CrossRef]
17. Shanmugam, N.E. Openings in Thin-Walled Steel Structures. *Thin-Walled Struct.* **1997**, *28*, 355–372. [CrossRef]
18. Loughlan, J.; Yidris, N.; Cunningham, P.R. The Effects of Local Buckling and Material Yielding on the Axial Stiffness and Failure of Uniformly Compressed I-Section and Box-Section Struts. *Thin-Walled Struct.* **2011**, *49*, 264–279. [CrossRef]
19. Paulo, R.M.F.; Teixeira-Dias, F.; Valente, R.A.F. Numerical Simulation of Aluminium Stiffened Panels Subjected to Axial Compression: Sensitivity Analyses to Initial Geometrical Imperfections and Material Properties. *Thin-Walled Struct.* **2013**, *62*, 65–74. [CrossRef]
20. Taheri-Behrooz, F.; Omidi, M.; Shokrieh, M.M. Experimental and Numerical Investigation of Buckling Behavior of Composite Cylinders with Cutout. *Thin-Walled Struct.* **2017**, *116*, 136–144. [CrossRef]
21. Eiblmeier, J.; Loughlan, J. The Influence of Reinforcement Ring Width on the Buckling Response of Carbon Fibre Composite Panels with Circular Cut-Outs. *Compos. Struct.* **1997**, *38*, 609–622. [CrossRef]
22. Ye, J.; Rasmussen, K.J. Compression Strength of Unstiffened Elements in Cold-Reduced High Strength Steel. *J. Struct. Eng.* **2008**, *134*, 189–197. [CrossRef]
23. Zhong, H.; Gu, C. Buckling of Symmetrical Cross-Ply Composite Rectangular Plates under a Linearly Varying in-Plane Load. *Compos. Struct.* **2007**, *80*, 42–48. [CrossRef]
24. Hu, H.; Badir, A.; Abatan, A. Buckling Behavior of a Graphite/Epoxy Composite Plate under Parabolic Variation of Axial Loads. *Int. J. Mech. Sci.* **2003**, *45*, 1135–1147. [CrossRef]
25. Czapski, P.; Kubiak, T. Influence of Fibre Arrangement on the Buckling Load of Composite Plates—Analytical Solution. *Fibres Text. East. Eur.* **2015**, *23*, 92–97. [CrossRef]

26. Batoz, J.-L.; Bathe, K.-J.; Ho, L.-W. A Study of Three-Node Triangular Plate Bending Elements. *Int. J. Numer. Meth. Eng.* **1980**, *15*, 1771–1812. [CrossRef]
27. Cui, S.; Hao, H.; Cheong, H.K. Numerical Analysis of Dynamic Buckling of Rectangular Plates Subjected to Intermediate-Velocity Impact. *Int. J. Impact Eng.* **2001**, *25*, 147–167. [CrossRef]
28. Kolakowski, Z.; Kowal-Michalska, K. Interactive Buckling Regarding the Axial Extension Mode of a Thin-Walled Channel under Uniform Compression in the First Nonlinear Approximation. *Int. J. Solids Struct.* **2011**, *48*, 119–125. [CrossRef]
29. Kolakowski, Z.; Krolak, M. Modal Coupled Instabilities of Thin-Walled Composite Plate and Shell Structures. *Compos. Struct.* **2006**, *76*, 303–313. [CrossRef]
30. Kubiak, T. *Static and Dynamic Buckling of Thin-Walled Plate Structures*; Springer International Publishing: Berlin/Heidelberg, Germany, 2013; ISBN 978-3-319-00653-6.
31. Bohlooly Fotovat, M.; Kubiak, T.; Perlikowski, P. Mixed Mode Nonlinear Response of Rectangular Plates under Static and Dynamic Compression. *Thin-Walled Struct.* **2023**, *184*, 110542. [CrossRef]
32. Vaseghi, O.; Mirdamadi, H.R.; Panahandeh-Shahraki, D. Non-Linear Stability Analysis of Laminated Composite Plates on One-Sided Foundation by Hierarchical Rayleigh–Ritz and Finite Elements. *Int. J. Non-Linear Mech.* **2013**, *57*, 65–74. [CrossRef]
33. Chen, X.; Nie, G.; Wu, Z. Application of Rayleigh-Ritz Formulation to Thermomechanical Buckling of Variable Angle Tow Composite Plates with General in-Plane Boundary Constraint. *Int. J. Mech. Sci.* **2020**, *187*, 106094. [CrossRef]
34. Belardi, V.G.; Fanelli, P.; Vivio, F. Application of the Ritz Method for the Bending and Stress Analysis of Thin Rectilinear Orthotropic Composite Sector Plates. *Thin-Walled Struct.* **2023**, *183*, 110374. [CrossRef]
35. Bažant, Z.P.; Cedolin, L. *Stability of Structures Elastic, Inelastic, Fracture and Damage Theories*; World Scientific: Singapore; Hackensack, NJ, USA, 2010; ISBN 978-981-4317-04-7.
36. Falkowicz, K. Numerical Buckling Analysis of Thin-Walled Channel-Section Composite Profiles Weakened by Cut-Outs. *Adv. Sci. Technol. Res. J.* **2022**, *16*, 88–96. [CrossRef]
37. Falkowicz, K.; Valvo, P. Influence of Composite Lay-Up on the Stability of Channel-Section Profiles Weakened by Cut-Outs—A Numerical Investigation. *Adv. Sci. Technol. Res. J.* **2023**, *17*, 108–115. [CrossRef]
38. Alfano, G.; Crisfield, M.A. Finite Element Interface Models for the Delamination Analysis of Laminated Composites: Mechanical and Computational Issues. *Int. J. Numer. Meth. Eng.* **2001**, *50*, 1701–1736. [CrossRef]
39. Kreja, I. A Literature Review on Computational Models for Laminated Composite and Sandwich Panels. *Open Eng.* **2011**, *1*, 59–80. [CrossRef]
40. Kopecki, T.; Święch, Ł. Experimental and Numerical Analysis of Post-Buckling Deformation States of Integrally Stiffened Thin-Walled Components of Load-Bearing Aircraft Structures. *J. Theor. Appl. Mech.* **2014**, *52*, 905–915. [CrossRef]
41. Mania, R.J.; Kolakowski, Z.; Bienias, J.; Jakubczak, P.; Majerski, K. Comparative Study of FML Profiles Buckling and Postbuckling Behaviour under Axial Loading. *Compos. Struct.* **2015**, *134*, 216–225. [CrossRef]
42. Teter, A.; Kolakowski, Z. Coupled Dynamic Buckling of Thin-Walled Composite Columns with Open Cross-Sections. *Compos. Struct.* **2013**, *95*, 28–34. [CrossRef]
43. Rozylo, P.; Falkowicz, K.; Wysmulski, P.; Debski, H.; Pasnik, J.; Kral, J. Experimental-Numerical Failure Analysis of Thin-Walled Composite Columns Using Advanced Damage Models. *Materials* **2021**, *14*, 1506. [CrossRef]
44. Banat, D.; Mania, R. Failure Analysis of Thin-Walled GLARE Members during Buckling and Post-Buckling Response. *AIP Conf. Proc.* **2019**, *2060*, 02000.
45. Banat, D.; Mania, R.J.; Degenhardt, R. Stress State Failure Analysis of Thin-Walled GLARE Composite Members Subjected to Axial Loading in the Post-Buckling Range. *Compos. Struct.* **2022**, *289*, 115468. [CrossRef]
46. Banat, D.; Mania, R.J. Progressive Failure Analysis of Thin-Walled Fibre Metal Laminate Columns Subjected to Axial Compression. *Thin-Walled Struct.* **2018**, *122*, 52–63. [CrossRef]
47. Wysmulski, P.; Debski, H.; Falkowicz, K. Sensitivity of Compressed Composite Channel Columns to Eccentric Loading. *Materials* **2022**, *15*, 6938. [CrossRef]
48. Różyło, P. Failure Analysis of Beam Composite Elements Subjected to Three-Point Bending Using Advanced Numerical Damage Models. *Acta Mech. Autom.* **2023**, *17*, 133–144. [CrossRef]
49. Falkowicz, K.; Debski, H. The Work of a Compressed, Composite Plate in Asymmetrical Arrangement of Layers. *AIP Conf. Proc.* **2019**, *2078*, 20005.
50. Jonak, J.; Karpiński, R.; Wójcik, A. Numerical Analysis of the Effect of Embedment Depth on the Geometry of the Cone Failure. *J. Phys. Conf. Ser.* **2021**, *2130*, 012012. [CrossRef]
51. Jonak, J.; Karpiński, R.; Wójcik, A.; Siegmund, M. Numerical Investigation of the Formation of a Failure Cone during the Pullout of an Undercutting Anchor. *Materials* **2023**, *16*, 2010. [CrossRef] [PubMed]
52. Machrowska, A.; Karpiński, R.; Jonak, J.; Szabelski, J.; Krakowski, P. Numerical prediction of the component-ratio-dependent compressive strength of bone cement. *Appl. Comput. Sci.* **2020**, *16*, 88–101. [CrossRef]
53. Jonak, J.; Karpiński, R.; Siegmund, M.; Wójcik, A.; Jonak, K. Analysis of the Rock Failure Cone Size Relative to the Group Effect from a Triangular Anchorage System. *Materials* **2020**, *13*, 4657. [CrossRef]
54. Rogala, M.; Gajewski, J. Crashworthiness Analysis of Thin-Walled Square Columns with a Hole Trigger. *Materials* **2023**, *16*, 4196. [CrossRef] [PubMed]

55. Nikopour, H.; Selvadurai, A.P.S. Concentrated Loading of a Fibre-Reinforced Composite Plate: Experimental and Computational Modeling of Boundary Fixity. *Compos. Part B Eng.* **2014**, *60*, 297–305. [CrossRef]
56. Chrysanidis, T. Experimental and Numerical Research on Cracking Characteristics of Medium-Reinforced Prisms under Variable Uniaxial Degrees of Elongation. *Eng. Fail. Anal.* **2023**, *145*, 107014. [CrossRef]
57. Falkowicz, K. Composite Plate Analysis Made in an Unsymmetric Configuartion. *J. Phys. Conf. Ser.* **2021**, *2130*, 012014. [CrossRef]
58. Falkowicz, K. Experimental and Numerical Analysis of Compression Thin-Walled Composite Plates Weakened by Cut-Outs. *Arch. Civ. Eng.* **2017**, *63*, 161–172. [CrossRef]
59. Van der Heijden, A.M.A. *W. T. Koiter's Elastic Stability of Solids and Structures*; Cambridge University Press: Cambridge, UK, 2012; ISBN 978-1-107-40701-5.
60. Coan, J.M. Large-Deflection Theory for Plates With Small Initial Curvature Loaded in Edge Compression. *J. Appl. Mech.* **1951**, *18*, 143–151. [CrossRef]
61. Roorda, J.; Venkataramaiah, K.R. Analysis of Local Plate Buckling Experimental Data. In Proceedings of the 6th International Specialty Conference on Cold-Formed Steel Structures, St. Louis, MO, USA, 16–17 November 1982.
62. Bienias, J.; Gliszczynski, A.; Jakubczak, P.; Kubiak, T.; Majerski, K. Influence of Autoclaving Process Parameters on the Buckling and Postbuckling Behaviour of Thin-Walled Channel Section Beams. *Thin-Walled Struct.* **2014**, *85*, 262–270. [CrossRef]
63. Różyło, P.; Smagowski, W.; Paśnik, J. Experimental Research in the Aspect of Determining the Mechanical and Strength Properties of the Composite Material Made of Carbon-Epoxy Composite. *Adv. Sci. Technol. Res. J.* **2023**, *17*, 232–246. [CrossRef]
64. Falkowicz, K.; Szklarek, K. Analytical Method for Projecting the Buckling Form of Composite Palates with a Cut-Out. *IOP Conf. Ser. Mater. Sci. Eng.* **2019**, *710*, 012021. [CrossRef]
65. York, C.B. On Tapered Warp-Free Laminates with Single-Ply Terminations. *Compos. Part A Appl. Sci. Manuf.* **2015**, *72*, 127–138. [CrossRef]
66. York, C. Unified Approach to the Characterization of Coupled Composite Laminates: Hygro-thermally Curvature-stable Configurations. *Int. J. Struct. Integr.* **2011**, *2*, 406–436. [CrossRef]
67. Falkowicz, K. Experimental and Numerical Failure Analysis of Thin-Walled Composite Plates Using Progressive Failure Analysis. *Compos. Struct.* **2023**, *305*, 116474. [CrossRef]
68. Falkowicz, K.; Samborski, S.; Valvo, P.S. Effects of Elastic Couplings in a Compressed Plate Element with Cut-Out. *Materials* **2022**, *15*, 7752. [CrossRef]
69. Falkowicz, K.; Debski, H.; Wysmulski, P. Effect of Extension-Twisting and Extension-Bending Coupling on a Compressed Plate with a Cut-Out. *Compos. Struct.* **2020**, *238*, 111941. [CrossRef]
70. Lusiak, T.; Knec, M. Use of ARAMIS for Fatigue Process Control in the Accelerated Test for Composites. *Transp. Res. Procedia* **2018**, *35*, 250–258. [CrossRef]
71. Falkowicz, K. Validation of Extension-Bending and Extension-Twisting Coupled Laminates in Elastic Element. *Adv. Sci. Technol. Res. J.* **2023**, *17*, 309–319. [CrossRef]
72. Paszkiewicz, M.; Kubiak, T. Selected Problems Concerning Determination of the Buckling Load of Channel Section Beams and Columns. *Thin-Walled Struct.* **2015**, *93*, 112–121. [CrossRef]
73. Falkowicz, K.; Ferdynus, M.; Rozylo, P. Experimental and Numerical Analysis of Stability and Failure of Compressed Composite Plates. *Compos. Struct.* **2021**, *263*, 113657. [CrossRef]
74. Rozylo, P.; Falkowicz, K. Stability and Failure Analysis of Compressed Thin-Walled Composite Structures with Central Cut-out, Using Three Advanced Independent Damage Models. *Compos. Struct.* **2021**, *273*, 114298. [CrossRef]

Disclaimer/Publisher's Note: The statements, opinions and data contained in all publications are solely those of the individual author(s) and contributor(s) and not of MDPI and/or the editor(s). MDPI and/or the editor(s) disclaim responsibility for any injury to people or property resulting from any ideas, methods, instructions or products referred to in the content.

materials

MDPI

Article

Buckling Analysis of Thin-Walled Composite Structures with Rectangular Cross-Sections under Compressive Load

Patryk Rozylo *, Michal Rogala and Jakub Pasnik

Department of Machine Design and Mechatronics, Faculty of Mechanical Engineering, Lublin University of Technology, Nadbystrzycka 36, 20-618 Lublin, Poland; m.rogala@pollub.pl (M.R.); j.pasnik@pollub.pl (J.P.)
* Correspondence: p.rozylo@pollub.pl

Abstract: The purpose of this research was the analysis of the stability of compressed thin-walled composite columns with closed rectangular cross-sections, subjected to axial load. The test specimens (made of carbon–epoxy composite) were characterized by different lay-ups of the composite material. Experimental tests were carried out using a universal testing machine and other interdisciplinary testing techniques, such as an optical strain measurement system. Simultaneously with the experimental studies, numerical simulations were carried out using the finite element method. In the case of FEA simulations, original numerical models were derived. In the case of both experimental research and FEM simulations, an in-depth investigation of buckling states was carried out. The measurable effect of the research was to determine both the influence of the cross-sectional shape and the lay-up of the composite layers on the stability of the structure. The novelty of the present paper is the use of interdisciplinary research techniques in order to determine the critical state of compressed thin-walled composite structures with closed sections. An additional novelty is the object of study itself—that is, thin-walled composite columns with closed sections.

Keywords: buckling; closed composite profiles; experimental studies; numerical simulations; axial compression

Citation: Rozylo, P.; Rogala, M.; Pasnik, J. Buckling Analysis of Thin-Walled Composite Structures with Rectangular Cross-Sections under Compressive Load. *Materials* **2023**, *16*, 6835. https://doi.org/10.3390/ma16216835

Academic Editor: Giovanni Garcea

Received: 22 September 2023
Revised: 19 October 2023
Accepted: 20 October 2023
Published: 24 October 2023

Copyright: © 2023 by the authors. Licensee MDPI, Basel, Switzerland. This article is an open access article distributed under the terms and conditions of the Creative Commons Attribution (CC BY) license (https://creativecommons.org/licenses/by/4.0/).

1. Introduction

Thin-walled composite materials—carbon-epoxy laminates—are a special group of structures that are used in the aerospace, automotive, or construction industries. Most often, these thin composite materials are made using carbon fiber–epoxy resin (CFRP) [1,2] or glass fiber–epoxy resin (GFRP) [3,4] configurations and are characterized by both open [5,6] and closed cross sections [7–11]. The above-mentioned composite materials are characterized by a certain behavior that occurs due to compression load [12,13]. The issue is commonly known as loss of stability (buckling) [14,15] associated with the accompanying deformation of the column. It is possible to distinguish several stages of compression of thin-walled columns made of composites. Initially, the walls of the construction are only compressed (pre-buckling stage), after which buckling occurs due to further loading (buckling stage), and then, when the equilibrium path is stable, increasing loading is accompanied by an increase in deflection (post-buckling stage) [16]. The issue of loss of stability has been addressed in many scientific papers and is still relevant due to the possibility of modifying the properties of the composite material [17–19].

Analysis of the critical state shows that the values of the failure load can even be several times higher than the critical load [20–22]. The correct orientation of the fibers and the number of layers can provide the thin-walled composite materials with a different range of stiffness, which translates into the behavior characteristics of the construction [23–25]. Accurate analysis of the critical state allows us to determine the form of buckling and the corresponding value of the critical load. In experimental studies, the value of critical load is determined based on approximation methods presented in many scientific papers [26,27].

These methods involve estimating the value of the critical load on the basis of experimental equilibrium post-buckling paths. A detailed description of the methods for determining the approximate value of the critical load is presented in many scientific papers, where a group of results and the methods of analysis are presented [28]. For numerical simulations, the critical load and the form of buckling are determined from the linear eigenproblem solution [27].

The aim of analyzing the process of axial compression of composite columns with closed cross-sections requires the use of several independent test methods. The evaluation of the behavior of the structure in the case of experimental testing was based on a universal testing machine, an acoustic emission testing system, and an optical system for measuring the deformation of thin-walled composite materials [29,30]. Coupled tests based on several independent methods make it possible to determine the limit states of the construction, the deformations obtained, and the values of critical forces [31,32]. The current paper contains a comparative stability analysis of two types of columns with closed cross-sections.

The novelty of the present research mainly includes:

- The use of interdisciplinary testing methods for structural stability assessment (testing machine, optical deformation measurement system, numerical FEA simulations);
- Manufacture of a new object of research in the form of thin-walled carbon-epoxy composite materials with closed sections;
- Study of the influence of the lay-up of the composite layers and the shape of the cross-section of the composite materials on the critical state.

The manufactured thin-walled composite materials with closed sections made of CFRP composite were developed through a project from the National Science Centre (Poland)—project number 2021/41/B/ST8/00148.

2. Subject of the Study

The study focused on thin-walled composite profiles made of carbon fiber-reinforced polymer (CFRP). Each profile consisted of eight layers of CFRP [33]. This paper describes two different types of profiles, denoted as B and C, with the following dimensions: 30 mm × 50 mm and 20 mm × 60 mm, respectively, with a wall thickness of 1.2 mm. The profiles had a maximum height of 200 mm. The following stacking sequences were utilized: B1/C1—$[0°/45°/-45°/90°]_s$, B2/C2—$[0°/90°/0°/90°]_s$, B3/C3—$[45°/-45°/90°/0°]_s$, B4/C4—$[90°/-45°/45°/0°]_s$. The sequences of layer configurations were derived from preliminary numerical simulations (which made it possible to predetermine critical loads and the form of buckling in order to preserve variety in the study of construction stability). For each of the layup configurations, three specimens were made. Note that every layout was symmetrical with respect to the center surface, as indicated by the subscripts next to the layout of the layer sequence. The columns were manufactured with autoclave technology using prepreg tapes with the trade name: CYCOM 985-42%-HS-135-305 (Solvay, Tempe, AZ, USA). For the production of the prepreg, epoxy resin type 985 was used, while the reinforcement was high-strength (HS) carbon fibers with a density of 135 g/m^2. The volume fraction of the resin in the prefabricated material was 42%. Profiles were made by winding a 305 mm wide prepreg tape at the desired angle, corresponding to the sequence of layers in the final product, on a properly prepared inner core. The parameters of the autoclaving curing process were set at a temperature of 177 °C and a pressure of 0.6 MPa and monitored throughout the course of the process. The production of the profiles was carried out by an external company specializing in making composite parts using an autoclave technique. The expertise of the contractor resulted in top-quality profiles with high repeatability of mechanical properties and dimensions. The quality of the profile fabrication was checked by using several techniques, including the use of the Keyence VHX 970F digital microscope (Keyence, Mechelen, Belgium) [34]. This microscope, equipped with a dedicated mobile head, allowed thorough observation of the structure and digital image capture. Figure 1 shows examples of ready-made profiles for experimental studies.

(a) (b)

Figure 1. Experimental specimens: (**a**) B—column, (**b**) C—column.

To obtain the material properties of the CFRP, test specimens for the determination of material data were made in accordance with the ISO standards [35]. Static tensile tests were carried out under the requirements and restrictions outlined in PN-EN ISO 527-5 (of 2010) [36] of which ASTM D 3039 [37] was the equivalent. Subsequent tests were performed as static shear tests based on PN-EN ISO 14129 (of 2000) [38]—the equivalent of ASTM D 3518 [39]. Finally, static compression tests were performed in accordance with PN-EN ISO 14126 (of 2002) [40]; the American Standard equivalent was ASTM D 3410 [41]. The process of manufacturing the specimens, their preparation for testing, and the tests themselves are described in detail in the paper [42]. The above-mentioned paper presents the methodology for determining the required material parameters of CFRP extensively. The data derived from these tests are shown in Table 1 [42,43].

Table 1. Material properties of the carbon–epoxy composite—average values (with standard deviation).

Mechanical Parameters		**Strength Parameters**	
Young's modulus E_1 [MPa]	103,014.11 (2145.73)	Tensile Strength F_{TU} (0°) [MPa]	1277.41 (56.23)
Young's modulus E_2 [MPa]	7361.45 (307.97)	Compressive Strength F_{CU} (0°) [MPa]	572.44 (46.20)
Poisson's ratio v_{12} [-]	0.37 (0.17)	Tensile Strength F_{TU} (90°) [MPa]	31.46 (9.64)
Kirchhoff modulus G_{12} [MPa]	4040.53 (167.35)	Compressive Strength F_{CU} (90°) [MPa]	104.04 (7.34)
-	-	Shear Strength F_{SU} (45°) [MPa]	134.48 (2.71)

3. Experimental Study

Interdisciplinary research methods were used to perform the experimental tests. Experimental studies were conducted in order to determine the stability of composite materials [42]. All the above-mentioned tests were conducted on a Zwick Z100 universal testing machine (ZwickRoell GmbH & Co. KG, Ulm, Germany) [22,29]. The next stage of the research was to run axial compression tests on thin-walled composite structures at room temperature. The crosshead of the testing machine was moving at a rate of 1 mm/min. The effect of the tests was to obtain the critical state by observing the formation of the buckling of the profile and the subsequent determination of the critical load using approximate methods [16,28]. To determine the critical force, one of the approximation methods was chosen—the method of intersection of straight lines [26]. To determine the approximate value of the critical load using this method, a load-displacement or, in other words, a load-shortening curve for the chosen structure was required. The chosen method involves approximating with a linear function two appropriately selected areas of the experimental curve, one before the point of change in "stiffness" within the force-displacement curve and the other after the change in "stiffness". The selected areas cannot be arbitrary; the requirement for the correct determination of the critical force by the method of intersection of straight lines is the selection of the areas of the force-displacement characteristics that are most nearly aligned with the straight line. Making the convergence between the two lines as high as possible means keeping the correlation coefficient R^2 as close as possible to the value of 1. In practice, the value of the coefficient R^2 cannot decrease below 0.95. The closer to the value of 1 one is, the better the obtained results will be. Ideally, this coefficient is 1. In order to correctly determine the critical force, the matrix method (determinant method) was used.

As basic geometric relationships indicate, two lines that are not parallel to each other intersect at a certain point. The point of intersection is located on both lines at the same time, so the coordinates must concurrently satisfy the equations of both lines. These coordinates can be obtained by solving a simple system of two linear equations:

$$\begin{cases} A_1x + B_1y + C_1 = 0 \\ A_2x + B_2y + C_2 = 0 \end{cases} \tag{1}$$

where A_1 and A_2 are the values of the directional coordinates of the lines at x, B_1 and B_2 are the values of the coefficients at y, while C_1 and C_2 are the numerical values that determine the so-called free expression of the function.

For determining the intersection point, Equation (1) must be rearranged to the form depicted in Equation (2):

$$\begin{cases} A_1x + B_1y = -C_1 \\ A_2x + B_2y = -C_2 \end{cases} \tag{2}$$

The system of first-degree equations in the form shown in Equation (2) with two unknowns may be solved employing the method of determinants of matrices as follows:

$$W = \begin{vmatrix} A_1 & B_1 \\ A_2 & B_2 \end{vmatrix} = A_1 \cdot B_2 - A_2 \cdot B_1 \tag{3}$$

$$W_x = \begin{vmatrix} -C_1 & B_1 \\ -C_2 & B_2 \end{vmatrix} = (-C_1) \cdot B_2 - (-C_2) \cdot B_1 \tag{4}$$

$$W_y = \begin{vmatrix} A_1 & -C_1 \\ A_2 & -C_2 \end{vmatrix} = A_1 \cdot (-C_2) - A_2 \cdot (-C_1) \tag{5}$$

Under the initial assumptions that the above-mentioned lines are nonparallel, and for $W \neq 0$, the system of equations is marked and has exactly a single solution:

$$\begin{cases} x = \frac{W_x}{W} \\ y = \frac{W_y}{W} \end{cases} \quad (6)$$

where x and y are the coordinates of the intersection point of two straight lines.

Consequently, the approximation method made it possible to determine the approximate value of the critical load within the experimental load-shortening curve.

Moreover, experimental studies also allow one to determine the path of post-buckling equilibrium. Such studies are carried out until the complete failure of the specimen and provide an opportunity to capture the ultimate failure force, i.e., the maximum load that the profile can carry. These tests were conducted on a universal testing machine, as mentioned elsewhere. The total number of specimens tested was 24 (12 specimens of type B and 12 specimens of type C). Axial compression tests were performed using special heads with flat working surfaces that were parallel to each other. These heads were rigidly attached to the bottom of the testing machine and to the top crosshead. Figure 2a illustrates the test stand with the heads installed on the machine. In addition, a vision-based system for measuring the deformation of the profile at the very moment of critical load application—the ARAMIS 2D digital image correlation system [44,45]—was used. The use of the referred device enables, in particular, the observation and measuring of deformations at the moment of the loss of stability of the structure (buckling). Figure 2b presents the test stand with the vision system employed. In order to obtain valid deformation values using the ARAMIS 2D system, dedicated non-reflective, red-colored mats were used as a background for the tests. When too much illumination is applied to the specimen during the test, unwanted overexposed areas appear within the composite profile, which have an adverse effect on the deformation registration of the structure. The use of a non-reflective background eliminated the problem with overexposed areas due to the fact that the mats absorb excess illumination and neutralize this unwanted effect. In order to obtain accurately captured images of profiles in the axial compression test, proper lighting is required, which was achieved using LED lamps.

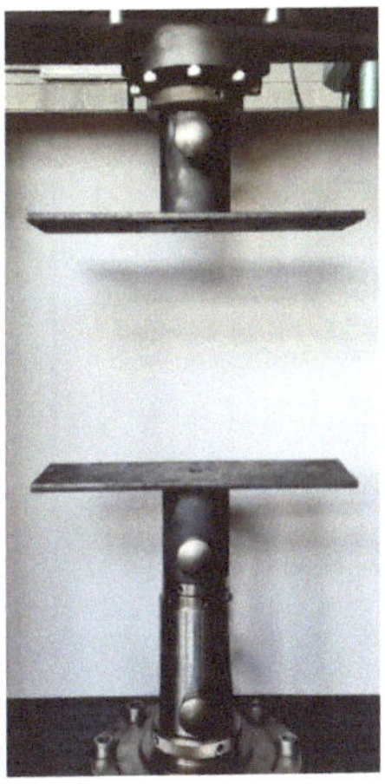

(a)

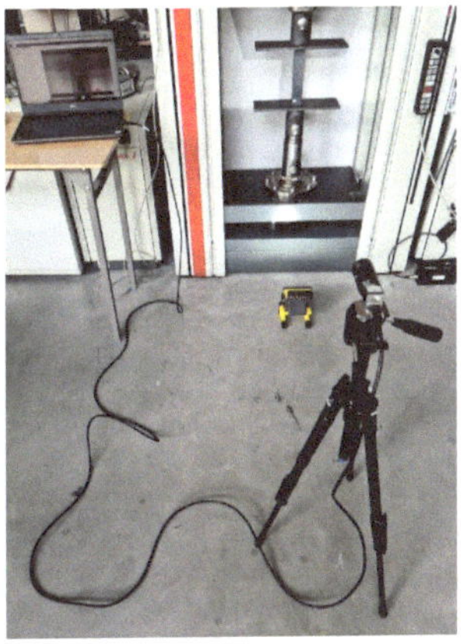

(b)

Figure 2. Experimental test stand: (**a**) experimental test heads, (**b**) general view of the test stand—Zwick Z100 testing machine with Aramis 2D system.

In addition, the AMSY-5 acoustic emission measurement system was also used in the experimental studies. By recording signals such as number of counts, number of hits, amplitude and energy, the state of the structure and its damage could be assessed.

Experimental studies made it possible to determine both the values of critical loads and the structure's buckling forms. The former was determined by means of the method of the intersection of straight lines while the latter was established through the structure's deformations obtained using a digital image correlation system during the tests.

4. Numerical Simulations

Numerical studies were based on the finite element method and were conducted using Abaqus software (Abaqus 2023, Dassault Systemes Simulia Corporation, Velizy Villacoublay, France). The numerical studies used a Lamin-type material model, the data of which was described in more detail during the presentation of the research subject. All numerical studies were carried out in two steps. The first stage was determining the linear stability of the structure (buckling) within the framework of which the linear eigenproblem was solved, based on the criterion of minimum potential energy. In view of the above, the buckling form of the thin-walled composite column was determined, along with the determination of the value of the critical load, corresponding to the obtained buckling form. The value of the critical load was determined by defining the unit load of the structure, which made it possible to determine the critical state [29]. The following is the relationship that allows the calculation of the critical load (7), it comes directly from the documentation of the FEM software (Abaqus 2023):

$$\left(K_0^{NM} + \lambda_i K_\Delta^{NM}\right) v_i^M = 0 \tag{7}$$

where K_0^{NM} is structural stiffness matrix relating to the baseline (includes preload effects P^N), K_Δ^{NM} refers to the differential matrix of initial stress and load stiffness caused by the incremental loading pattern (Q^N), λ_i illustrates the eigenvalues, v^M is the buckling mode (known as the eigenvectors), ^{M} and ^{N} refer to degrees of freedom M and N of the whole model, and i refers to the $_I$ th buckling mode. Furthermore, the critical buckling loads represent then $P^N + \lambda_i Q^N$. Additionally, v^M is normalized vectors (do not reflect the actual quantities of strain at critical load). They are normalized so that the maximum component of displacement is 1.0. When all components of displacement are zero, the maximum component of rotation is normalized to 1.0. Once damage is initiated, further loading of the composite structure will degrade the stiffness parameters of the material.

The numerical model consisted of a composite structure and non-deformable plate elements, which allowed correct modelling of the boundary conditions. The composite column with rectangular cross-section consisted of eight layers of composite material (CFRP) of equal thickness for both B- and C-type specimens. The numerical model included four different arrangements of fiber composite orientation shown in Figure 3. The composite structure had the same geometric parameters regardless of the arrangement of the composite material layers used. Both experimental studies and numerical simulations considered the following cases of arrangement of composite material layers: *B*1 and C1—[0°/45°/−45°/90°]s, *B*2 and C2—[0°/90°/0°/90°]s, *B*3 and C3—[45°/−45°/90°/0°]s, *B*4 and C4—[90°/−45°/45°/0°]s, as shown in Figure 3.

The discrete model was formulated using Continuum Shell elements (with a physical representation of the thickness of the composite material, which included eight layers of composite material), whereas the plate elements serving as supports were modelled using Shell elements. The composite structure consisted of SC8R-type finite elements (8-node quadrilateral continuous general-purpose shell in-plane, reduced integration with hourglass control, finite membrane deformations, having three translational degrees of freedom per computational node). In contrast, the supports were defined by non-deformable finite elements of type R3D4 (4-node three-dimensional rigid quadrilateral, having six degrees of freedom (three translational and three rotational) per computational node). A mesh density of 2 mm was used for the composite structure, while 2.5 mm was used for the non-deformable plates. The discrete model consisted of 10,320 finite elements (9200 linear hexahedral elements of type SC8R and 1120 linear tetrahedral elements of type R3D4).

Contact properties representing the interaction of the contacting surfaces were reflected by using normal and tangential contact (friction coefficient 0.2). To represent the correct behaviors of the structure, boundary conditions were applied by assigning the load to reference points assigned to the lower and upper non-deformable plate, respectively. The upper plate, acting as the loading element, had all degrees of freedom locked, with the exception of the displacement relative to the Z axis, on which the load was applied. The bottom plate serving as the base had all rotational as well as translational degrees of freedom locked. The load was realized with a displacement relative to the Z axis. A discrete model of the structure with defined boundary conditions is shown in Figure 4. The numerical model presented below was used to perform a simulation using the finite element method of stability (buckling) of thin-walled structures.

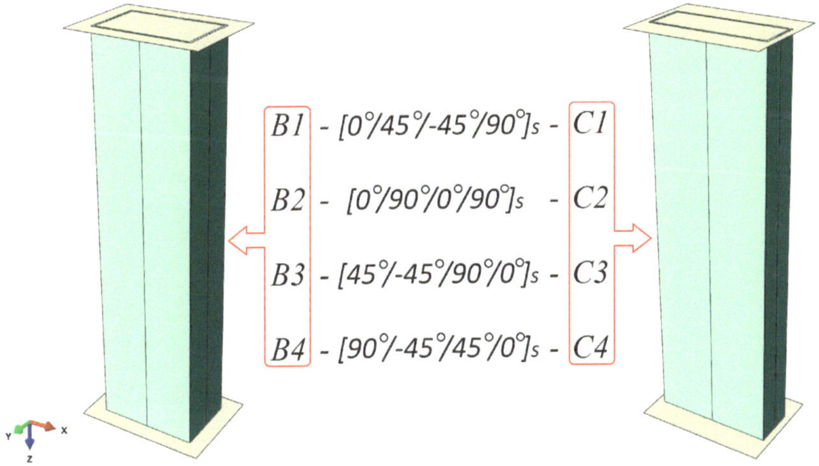

Figure 3. Numerical model with listed configurations of composite material layers for two types of columns.

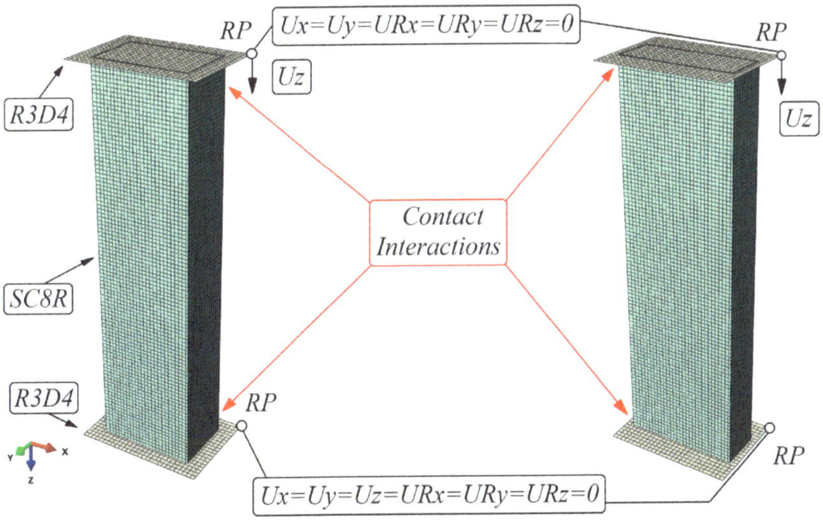

Figure 4. Discrete model with defined boundary conditions.

5. Research Results

In the course of the experimental research and numerical simulations using the finite element method, it was possible to assess the stability of thin-walled composite structures, which is important for the evaluation of composite structures for the use of such components in the aerospace or automotive industries. Experimental research used interdisciplinary testing techniques to assess the structural stability, while in the case of numerical simulations, it was possible to determine critical (buckling) states using an advanced model of the composite material.

The main purpose of the research conducted was to analyze the critical state. The research included both an experiment on physical specimens and a numerical study using the finite element method. The analysis of the critical state for physical specimens was carried out using a universal testing machine, where the occurring form of buckling was observed in axial compression of the structure using an optical strain measurement system, while the critical load values were determined based on the approximation method of intersecting straight lines. The method of determining the critical load for the described method is presented in Equations (1)–(6) in Section 3. The procedure for estimating the critical load values for all experimentally tested specimens was the same. To determine the critical load value, we relied on load-displacement curves obtained from bench tests. The effective approximation ranges for the experimental curves (the range before and after the change in the "stiffness" of the experimental curve) were approximated by using linear functions while maintaining the correct correlation coefficient between the approximation functions and the selected approximation ranges at the highest possible level of $R^2 \geq 0.95$. All tested cases obtained a coefficient value that was significantly higher, oscillating above $R^2 \geq 0.99$, which indicates the high accuracy of the realized tests. Therefore, linear approximating functions were determined, which allowed further calculation of approximate values of critical forces. The value was determined by solving a system of equations, that is, determining the point of intersection of the approximating functions. As an example of the first sample B1_1, the methodology for determining the critical load approximation is presented, in which two approximation functions are initially compared using a system of equations:

$$\begin{cases} A_1x + B_1y + C_1 = 24{,}008.79x - 1y - 1622.46 = 0 \\ A_2x + B_2y + C_2 = 16{,}798.83x - 1y + 4850.59 = 0 \end{cases} \tag{8}$$

To determine the point of intersection, the notation resulting from Equation (8) must be transformed to another form, consistent with the following notation (9):

$$\begin{cases} A_1x + B_1y = -C_1 \leftrightarrow 24{,}008.79x - 1y = 1622.46 \\ A_2x + B_2y = -C_2 \leftrightarrow 16{,}798.83x - 1y = -4850.59 \end{cases} \tag{9}$$

The obtained system of first-degree equations with two unknowns is solvable by the matrix determinant method (10)–(12):

$$W = \begin{bmatrix} A_1 & B_1 \\ A_2 & B_2 \end{bmatrix} \leftrightarrow \begin{bmatrix} 24{,}008.79 & -1 \\ 16{,}798.83 & -1 \end{bmatrix} = -7209.96 \tag{10}$$

$$W_x = \begin{bmatrix} -C_1 & B_1 \\ -C_2 & B_2 \end{bmatrix} \leftrightarrow \begin{bmatrix} 1622.46 & -1 \\ -4850.59 & -1 \end{bmatrix} = -6473.05 \tag{11}$$

$$W_y = \begin{bmatrix} A_1 & -C_1 \\ A_2 & -C_2 \end{bmatrix} \leftrightarrow \begin{bmatrix} 24{,}008.79 & 1622.46 \\ 16{,}798.83 & -4850.59 \end{bmatrix} = -143{,}712{,}226.41 \tag{12}$$

With the initial assumption that the aforementioned lines are not parallel, with $W \neq 0$, the system of equations is determined and has exactly one solution (13):

$$\begin{cases} x = \frac{W_x}{W} = 0.90 \\ y = \frac{W_y}{W} = 19{,}932.46 \end{cases} \quad (13)$$

With the method described above, the approximate critical load value was determined for the first specimen of type B, designated B1_1. Thus, it was determined that the critical load value, causing loss of stability of the thin-walled composite structure, is approximately P_{cr} = 19,932 N and occurs when the structure is shortened by u = 0.90 mm (vertical displacement of the crosshead of the testing machine). The above-described method was used to derive the critical load values for all specimens in the experimental tests. Figures 5 and 6 show graphically how the critical load was determined for the six selected specimens, i.e., B_1 and C_1 (three specimens of each column type).

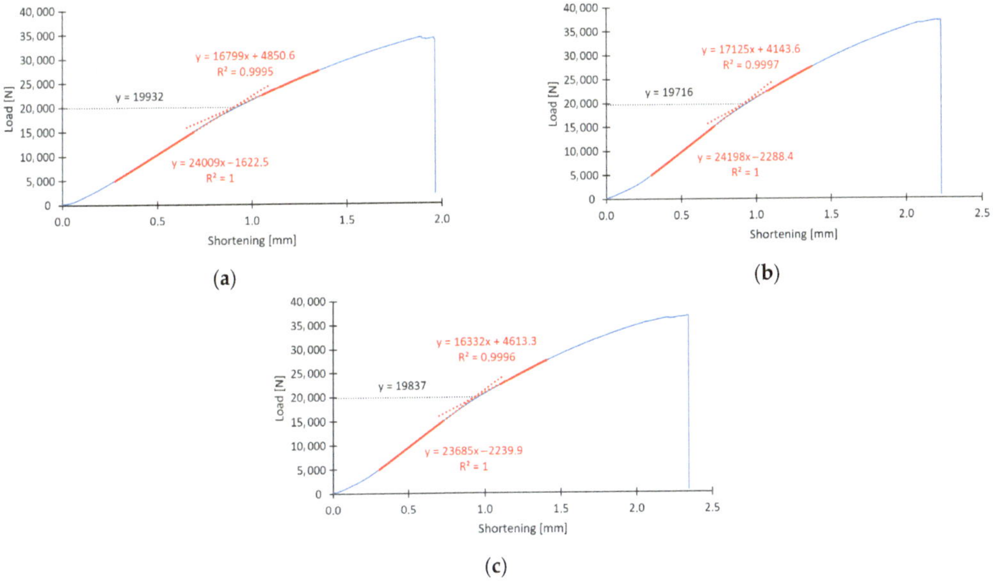

Figure 5. Experimentally determined critical load: (**a**) specimen B1_1, (**b**) specimen B1_2, (**c**) specimen B1_3.

In Figures 5 and 6, the depicted lines indicate successively: red dashed line—approximation function, blue solid line—experimental curve, red solid line—effective range of approximation, black dashed line—line representing critical load. The determined values of critical forces made it possible to compare the tested specimens in terms of the influence of the arrangement of the fiber composite layers on the stability of the structure. In order to better present the obtained experimental results, the values were presented in Tables 2 and 3 for specimen types B and C, respectively.

Table 2. Critical state results for column type B—experimental studies.

Specimen Type \ Specimen No.	1	2	3	Average Value
B1	19,932 N	19,716 N	19,837 N	19,829 N
B2	18,544 N	18,892 N	18,771 N	18,736 N
B3	21,654 N	22,054 N	22,133 N	21,947 N
B4	16,992 N	17,665 N	16,666 N	17,108 N

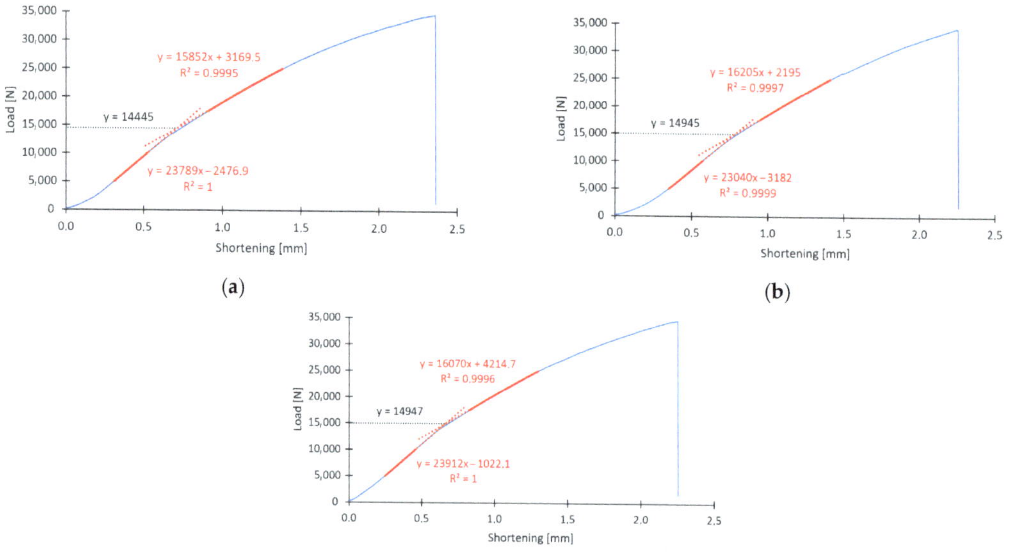

(a) (b) (c)

Figure 6. Experimentally determined critical load: (**a**) specimen C1_1, (**b**) specimen C1_2, (**c**) specimen C1_3.

Table 3. Critical state results for column type C—experimental studies.

Specimen Type \ Specimen No.	1	2	3	Average Value
C1	14,445 N	14,945 N	14,947 N	14,779 N
C2	13,818 N	13,352 N	13,284 N	13,485 N
C3	16,487 N	18,041 N	17,075 N	17,201 N
C4	13,864 N	13,656 N	13,091 N	13,537 N

It was determined that the highest critical load values were obtained by the B3 and C3 type profiles—characterized by composite material layer arrangements [45°/−45°/90°/0°]s, where the average critical load value was P_{cr} = 21,947 N for the B3 model and P_{cr} = 17,201 N for the C3 model. The composite columns with the lowest critical load were characterized by B4 [90°/−45°/45°/0°]s and C2 [0°/90°/0°/90°]s, where the average load values were P_{cr} = 17,108 N and P_{cr} = 13,485 N, respectively. In describing the type C column, it is worth noting that models C2 and C4 had very similar values of critical loads. In the case of specimens C2_3 and C4_3, it was the C4 column that obtained a lower value of critical load, according to Table 3. Based on the results of the average values of critical load, it was determined that specimens of type B3 showed about 1.28 times higher load than specimens of type B4, in the case of model C it was 1.28 for specimens of type C3 and C2, respectively. Analyzing the extreme results, i.e., the highest value of critical load (sample B3_3) and the lowest value of critical load (sample B4_3), it was determined that the ratio of maximum to minimum load was 1.33. A similar comparison of extreme values for column type C showed a ratio of load values of 1.38 between samples C3_2 (P_{cr} = 18,041 N) and C4_3 (P_{cr} = 13,091 N).

It was also noted that buckling of the structure occurs at different deflection values, i.e., in the case of type B3 profiles, it occurs when the structure is shortened by u = 0.95 mm, while in the case of type B4 profiles, it occurs when the structure is shortened by u = 0.81, which is about a 0.14 mm difference between the above-mentioned structure types. In the case of the type C column, the extremes of deflection at which the loss of stability occurred were u = 0.91 mm (C3) and u = 0.55 (C2) on average. Thus, it was concluded that the arrangement of fiber composite layers has a major impact on the stability of thin-walled composite structures with a closed square section. In addition, it is noticeable that there are significant differences in the values of critical loads and deflections at which stability is lost for the two types of columns analyzed (B and C). The thin-walled column with a cross-section of 20 × 60 mm (type C) was characterized by a lower critical load. The described effect is observed when comparing all layer arrangements (1–4) shown for columns B and C of Tables 2 and 3.

In addition, a qualitative evaluation of the critical condition was carried out in the experimental study. The study consisted of recording buckling forms obtained by capturing images of each type of composite profile during loss of stability (buckling), as well as recording buckling forms using an optical strain measurement system—Aramis 2D. In the case of the Aramis 2D optical system, it was necessary to use special filters applied directly in the software, highlighting the buckling form (registration of deformations in the longitudinal direction of the structure with a median filter). The registered experimental buckling forms are shown below (Figures 7 and 8).

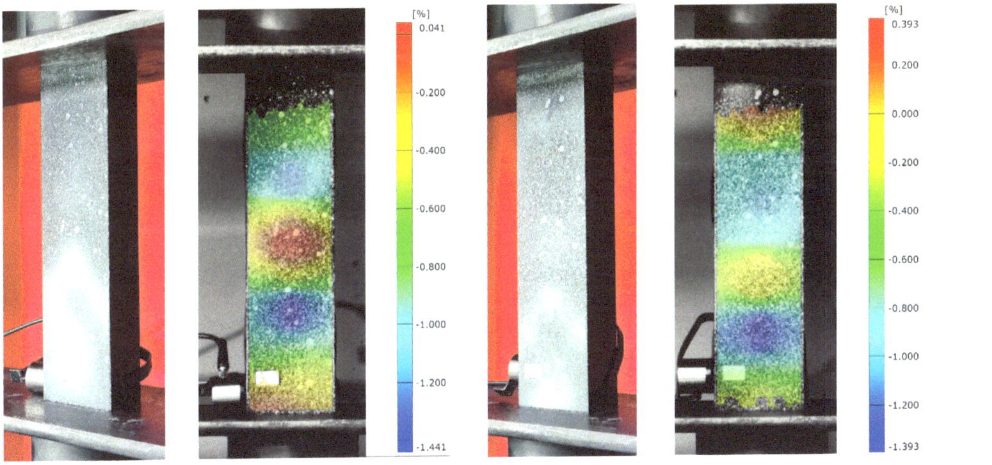

(a) (b)

Figure 7. *Cont.*

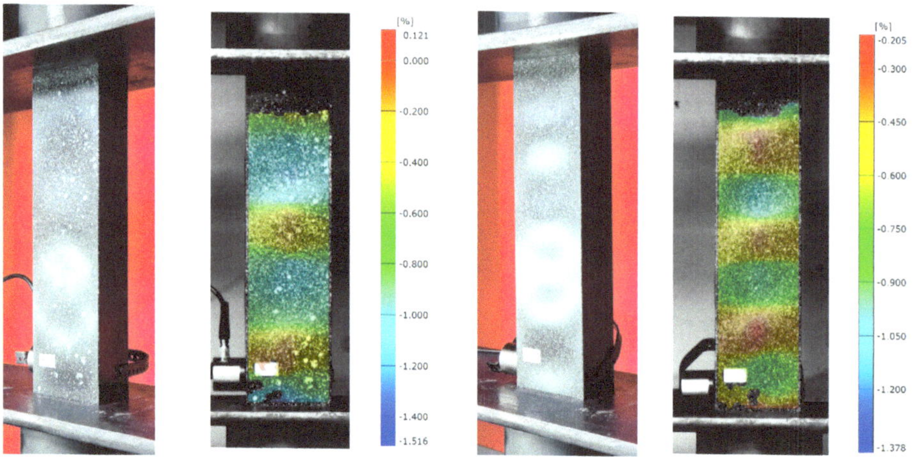

(**c**) (**d**)

Figure 7. Loss of structural stability—experimental studies: (**a**) specimen type B1, (**b**) specimen type B2, (**c**) specimen type B3, (**d**) specimen type B4.

During the execution of the experimental tests, it was observed that for the tested profiles there were specific numbers of half-waves in the longitudinal direction of the column: B1—three half-waves, B2—four half-waves, B3—five half-waves, and B4—seven half-waves. In the case of the C-type model, a different number of half-waves was observed for specific layer arrangements, whereas the values obtained reflected the results obtained with numerical simulations using FEM.

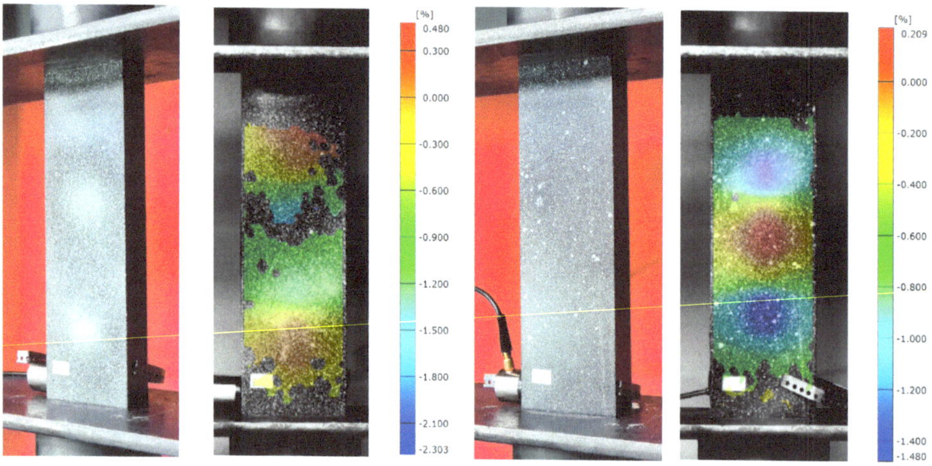

(**a**) (**b**)

Figure 8. *Cont.*

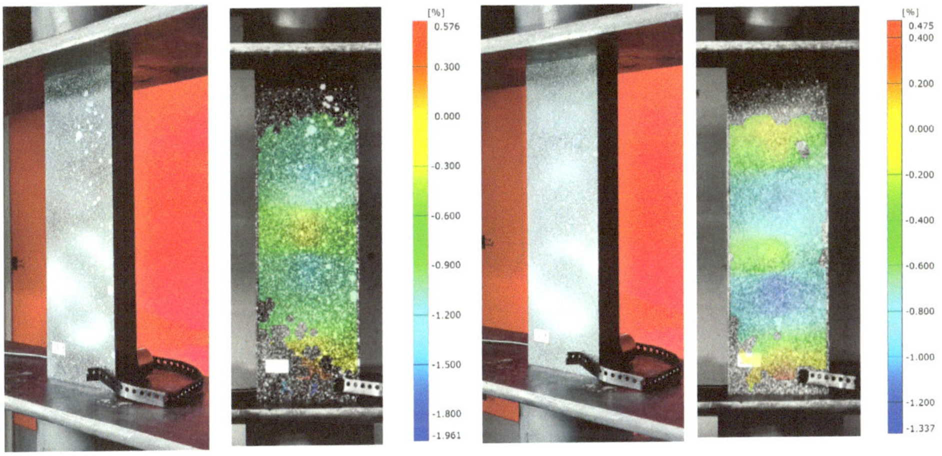

(c) (d)

Figure 8. Loss of structural stability—experimental studies: (**a**) specimen type C1, (**b**) specimen type C2, (**c**) specimen type C3, (**d**) specimen type C4.

For numerical simulations using FEM, the critical state analysis was carried out based on the solution of a linear eigenproblem. During the preparation of numerical models, the effect of mesh density on the value of critical load was made (Figure 9). The study was carried out on a sample specimen B1 that made it possible to estimate the value of critical load—the most consistent with experimental results (a mesh density of 2 mm was adopted).

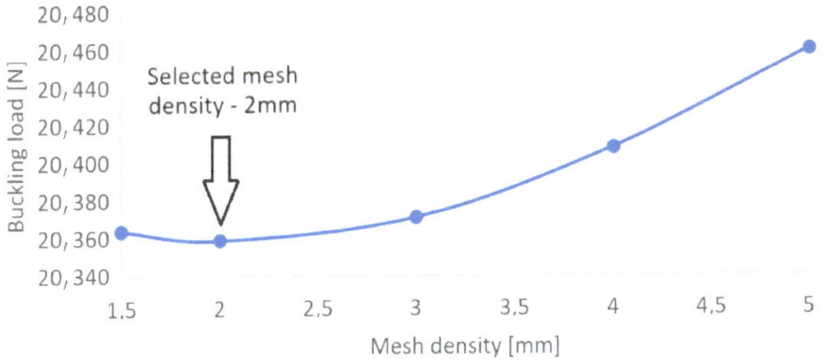

Figure 9. The influence of mesh density on the buckling load value (on the specimen B1).

The study of the critical state for numerical calculations made it possible to determine the geometric form of buckling and the corresponding critical load values for each stacking sequence of the composite material, as shown below (Figures 10 and 11).

The study of the critical state of thin-walled B- and C-type columns showed high qualitative and quantitative convergence of the findings. The results of the numerical analyses made it possible to determine the forms of buckling and the corresponding critical load values. Therefore, the following results were determined for specimens with different fiber arrangements: specimen B1—three half-waves with critical load value P_{cr} = 20,359 N, specimen B2—four half-waves with critical load value P_{cr} = 19,556 N, specimen B3—five half-waves with critical load value P_{cr} = 22,336 N, and specimen B4—seven half-waves with critical load value P_{cr} = 17,753 N.

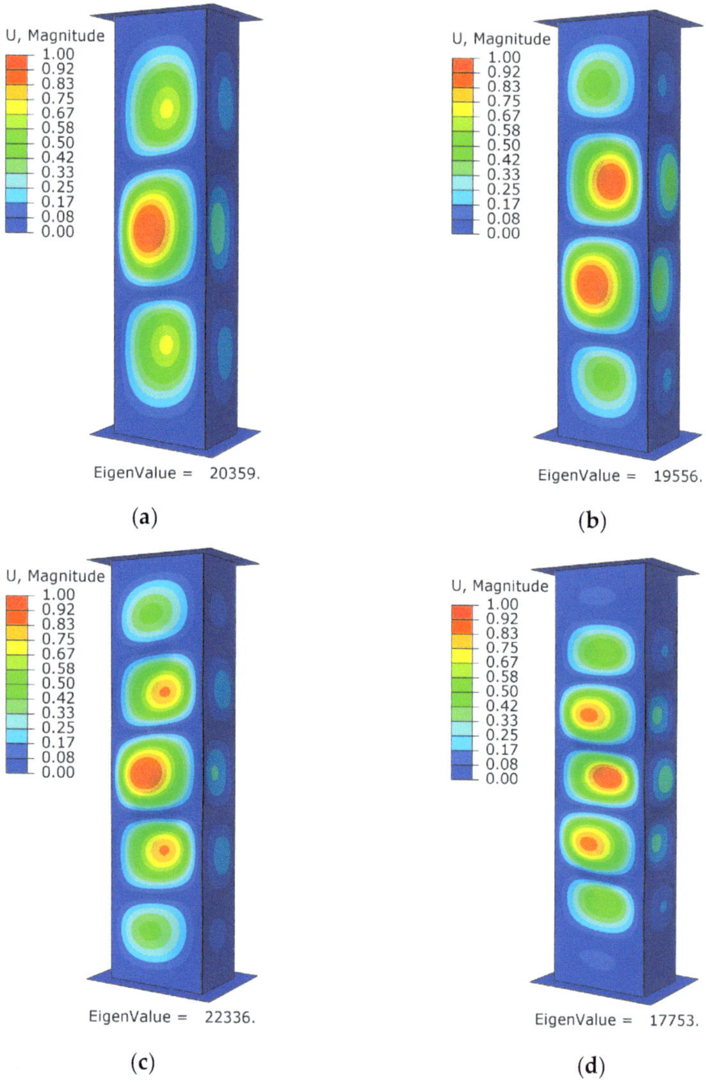

Figure 10. Loss of structural stability—numerical studies: (**a**) specimen type B1, (**b**) specimen type B2, (**c**) specimen type B3, (**d**) specimen type B4.

Similar results were obtained for C-type columns. The values of the critical forces achieved and the number of half-waves are as follows for subsequent arrangements of composite layers: specimen C1—three half-waves with critical load value P_{cr} = 15,170 N, specimen C2—three half-waves with critical load value P_{cr} = 14,037 N, specimen C3—five half-waves with critical load value P_{cr} = 18,221 N, and specimen C4—six half-waves with critical load value P_{cr} = 13,937 N. It is worth noting that the number of half-waves obtained for layer arrangement 1 and 3 was the same; however, the loss of stability for type C columns occurred at a critical load 4–5 kN lower than for type B columns.

Qualitatively, the experimental tests and numerical simulations showed a high level of agreement. The high qualitative agreement between the results of numerical simulations and bench tests is shown in Table 4.

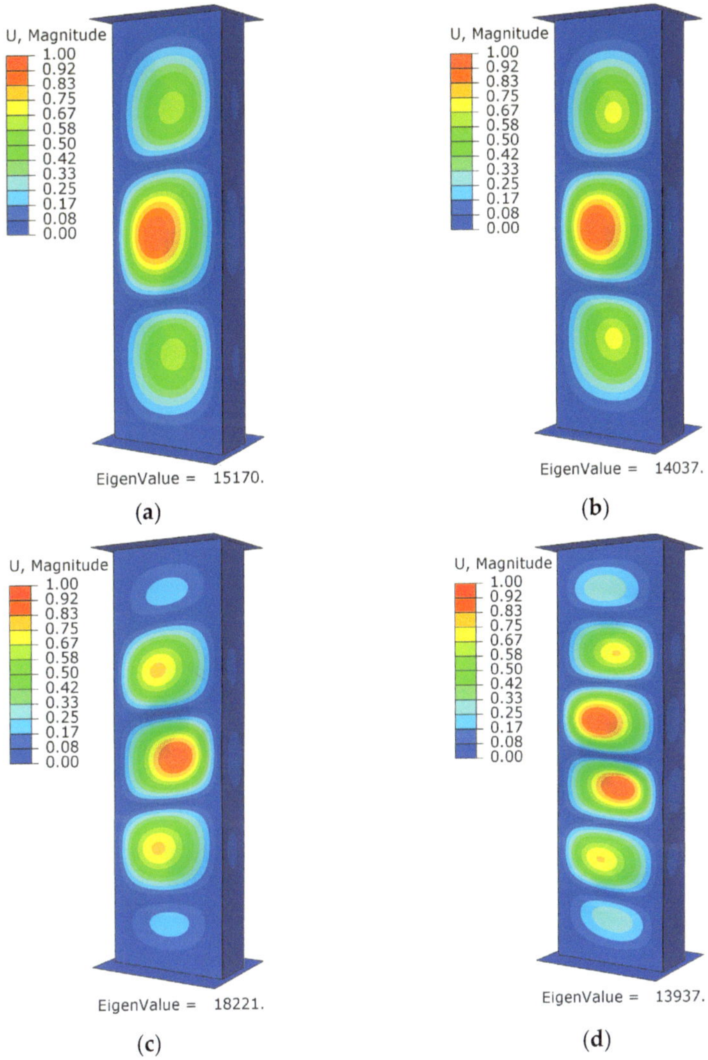

Figure 11. Loss of structural stability—numerical studies: (**a**) specimen type C1, (**b**) specimen type C2, (**c**) specimen type C3, (**d**) specimen type C4.

Based on the tests conducted, it was observed that the results of the numerical simulations slightly exceeded the value of the obtained forces in experimental tests. Higher values of critical loads in the case of simulations were due to the fact that in numerical simulations perfectly reflected physical models were considered but without geometric imperfections due to manufacturing technology. These models were characterized then by a slightly higher stiffness, which translated into the values of the obtained forces. A direct comparison of the results of the two types of analysis showed a discrepancy in the range of 2–6%. In the case of type C3 specimens, the critical load obtained in FEM simulations was 1.06 times higher than that obtained from the average result (of three specimens) from experimental tests. The remaining results had a much smaller error, indicating a high convergence of the obtained quantitative results. The highest value of critical load was observed for sample type B3: P_{cr} = 22,336 N—FEM, P_{cr} = 21,947 N—mean value EXP.

Table 4. Critical state results—comparison of experimental studies and numerical simulations.

Specimen Type	Average Value P_{cr} (EXP) [N]	P_{cr} (FEM) [N]	FEM/EXP
B1	19,829	20,359	1.03
B2	18,736	19,556	1.04
B3	21,947	22,336	1.02
B4	17,108	17,753	1.04
C1	14,779	15,170	1.03
C2	13,485	14,037	1.04
C3	17,201	18,221	1.06
C4	13,537	13,937	1.03

The findings presented in this paper were the result of research work carried out within the framework of a project financed with resources from the National Science Centre with registration number 2021/41/B/ST8/00148.

6. Conclusions

The research presented in this article constitutes a buckling analysis of thin-walled composite columns with rectangular cross-sections. The study of two types of columns (B and C) investigated four different layer arrangements (lay-ups). The analyses carried out involved physically manufactured structures as well as numerical simulations using the finite element method. The research was carried out using interdisciplinary testing techniques using a universal testing machine, an optical deformation measurement system, and numerical simulations using FEM. Evaluation of the achieved results was conducted qualitatively (percentage discrepancies) and quantitatively (several samples of profiles with the same layer stacking). The study showed that the highest stability is characterized by columns with an arrangement of layers defined by the number 3 [45°/−45°/90°/0°]s for both type B and C columns. It is worth noting that thin-walled structures with a shape closer to a square (type B) show higher values of the critical load at which buckling of the column occurs. Thin-walled structures of type B showed an average of 4–5 kN higher critical load value than type C columns. The specimens characterized by the lowest critical load values had a lay-up of [0°/90°/0°/90°]s for the type B column and [90°/−45°/45°/0°]s for the type C column. Noteworthy is that the type C column with a cross-section of 20 × 60 mm had similar critical load values for the C2 and C4 systems. All the results obtained through the numerical analyses as well as the bench tests are characterized by high quantitative and qualitative agreement. The presented results describe the critical condition of thin-walled composite columns, and this is the first stage of the work. The next stage of the work in the next article will realize the study of the load capacity of the structure using numerical simulations, taking into account the failure of composite materials such as CZM, XFEM, PFA, or LaRC05, among others [46–49].

Author Contributions: Conceptualization, P.R.; Methodology, P.R. and J.P.; Software, P.R.; Validation, P.R. and M.R.; Formal analysis, P.R.; Investigation, P.R.; Resources, P.R., M.R. and J.P.; Data curation, P.R., M.R. and J.P.; Writing—original draft, P.R., M.R. and J.P.; Writing—review & editing, P.R.; Visualization, P.R., M.R. and J.P.; Supervision, P.R.; Project administration, P.R.; Funding acquisition, P.R. All authors have read and agreed to the published version of the manuscript.

Funding: This research was funded in whole or in part by National Science Centre, Poland [2021/41/B/ST8/00148]. For the purpose of Open Access, the author has applied a CC-BY public copyright license to any Author Accepted Manuscript (AAM) version arising from this submission.

Institutional Review Board Statement: Not applicable.

Informed Consent Statement: Not applicable.

Data Availability Statement: The data that support the findings of this study are available from the corresponding author, upon reasonable request.

Conflicts of Interest: The authors declare that they have no known competing financial interests or personal relationships that could have appeared to influence the work reported in this paper.

References

1. Falkowicz, K. Validation of Extension-Bending and Extension-Twisting Coupled Laminates in Elastic Element. *Adv. Sci. Technol. Res. J.* **2023**, *17*, 309–319. [CrossRef]
2. Rozylo, P.; Falkowicz, K.; Wysmulski, P.; Debski, H.; Pasnik, J.; Kral, J. Experimental-Numerical Failure Analysis of Thin-Walled Composite Columns Using Advanced Damage Models. *Materials* **2021**, *14*, 1506. [CrossRef]
3. Berardi, V.P.; Perrella, M.; Feo, L.; Cricrì, G. Creep behavior of GFRP laminates and their phases: Experimental investigation and analytical modeling. *Compos. Part B Eng.* **2017**, *122*, 136–144. [CrossRef]
4. Fascetti, A.; Feo, L.; Nisticò, N.; Penna, R. Web-flange behavior of pultruded GFRP I-beams: A lattice model for the interpretation of experimental results. *Compos. Part B Eng.* **2016**, *100*, 257–269. [CrossRef]
5. Gliszczynski, A.; Kubiak, T. Progressive failure analysis of thin-walled composite columns subjected to uniaxial compression. *Compos. Struct.* **2017**, *169*, 52–61. [CrossRef]
6. Falkowicz, K.; Valvo, P.S. Influence of Composite Lay-Up on the Stability of Channel-Section Profiles Weakened by Cut-Outs—A Numerical Investigation. *Adv. Sci. Technol. Res. J.* **2023**, *17*, 108–115. [CrossRef]
7. Urbaniak, M.; Świniarski, J.; Czapski, P.; Kubiak, T. Experimental investigations of thin-walled GFRP beams subjected to pure bending. *Thin-Walled Struct.* **2016**, *107*, 397–404. [CrossRef]
8. Droździel, M.; Podolak, P.; Czapski, P.; Zgórniak, P.; Jakubczak, P. Failure analysis of GFRP columns subjected to axial compression manufactured under various curing-process conditions. *Compos. Struct.* **2021**, *262*, 113342. [CrossRef]
9. Czapski, P.; Kubiak, T. Numerical and experimental investigations of the post-buckling behaviour of square cross-section composite tubes. *Compos. Struct.* **2015**, *132*, 1160–1167. [CrossRef]
10. Czajka, B.; Różyło, P. The Influence of Composite Lay-Up and the Shape of the Closed Section on the Stability of the Structure. *Adv. Sci. Technol. Res. J.* **2022**, *16*, 216–224. [CrossRef]
11. Kim, N.-I.; Ku Shin, D.; Park, Y.-S. Coupled stability analysis of thin-walled composite beams with closed cross-section. *Thin-Walled Struct.* **2010**, *48*, 581–596. [CrossRef]
12. Banat, D.; Mania, R.J. Failure assessment of thin-walled FML profiles during buckling and postbuckling response. *Compos. Part B Eng.* **2017**, *112*, 278–289. [CrossRef]
13. Falkowicz, K.; Wysmulski, P.; Debski, H. Buckling Analysis of Laminated Plates with Asymmetric Layup by Approximation Method. *Materials* **2023**, *16*, 4948. [CrossRef] [PubMed]
14. Bažant, Z.P.; Cedolin, L. *Stability of Structures: Elastic, Inelastic, Fracture and Damage Theories*; World Scientific Pub. Co. Pte Ltd.: Singapore, 2010. [CrossRef]
15. Singer, J.; Arbocz, J.; Weller, T. *Buckling Experiments: Experimental Methods in Buckling of Thin-Walled Structures*; Basic Concepts, Columns, Beams and Plates; Wiley: New York, NY, USA, 1998; Volume 1.
16. Rozylo, P.; Teter, A.; Debski, H.; Wysmulski, P.; Falkowicz, K. Experimental and Numerical Study of the Buckling of Composite Profiles with Open Cross Section under Axial Compression. *Appl. Compos. Mater.* **2017**, *24*, 1251–1264. [CrossRef]
17. Li, Z.-M.; Qiao, P. Buckling and postbuckling behavior of shear deformable anisotropic laminated beams with initial geometric imperfections subjected to axial compression. *Eng. Struct.* **2015**, *85*, 277–292. [CrossRef]
18. Ragheb, W.F. Local buckling analysis of pultruded FRP structural shapes subjected to eccentric compression. *Thin-Walled Struct.* **2010**, *48*, 709–717. [CrossRef]
19. Turvey, G.J.; Zhang, Y. A computational and experimental analysis of the buckling, postbuckling and initial failure of pultruded GRP columns. *Comput. Struct.* **2006**, *84*, 1527–1537. [CrossRef]
20. Zaczynska, M.; Mania, R.J. Investigation of dynamic buckling of fiber metal laminate thin-walled columns under axial compression. *Compos. Struct.* **2021**, *271*, 114155. [CrossRef]
21. Li, W.; Cai, H.; Li, C.; Wang, K.; Fang, L. Progressive failure of laminated composites with a hole under compressive loading based on micro-mechanics. *Adv. Compos. Mater.* **2014**, *23*, 477–490. [CrossRef]
22. Banat, D.; Mania, R.J.; Degenhardt, R. Stress state failure analysis of thin-walled GLARE composite members subjected to axial loading in the post-buckling range. *Compos. Struct.* **2022**, *289*, 115468. [CrossRef]
23. Zhang, H.; Yang, D.; Ding, H.; Wang, H.; Xu, Q.; Ma, Y.; Bi, Y. Effect of Z-pin insertion angles on low-velocity impact mechanical response and damage mechanism of CFRP laminates with different layups. *Compos. Part A Appl. Sci. Manuf.* **2021**, *150*, 106593. [CrossRef]
24. Qiu, C.; Han, Y.; Shanmugam, L.; Zhao, Y.; Dong, S.; Du, S.; Yang, J. A deep learning-based composite design strategy for efficient selection of material and layup sequences from a given database. *Compos. Sci. Technol.* **2022**, *230*, 109154. [CrossRef]
25. Hu, Y.; Zhang, Y.; Fu, X.; Hao, G.; Jiang, W. Mechanical properties of Ti/CF/PMR polyimide fiber metal laminates with various layup configurations. *Compos. Struct.* **2019**, *229*, 111408. [CrossRef]

26. Bin Kamarudin, M.N.; Mohamed Ali, J.S.; Aabid, A.; Ibrahim, Y.E. Buckling Analysis of a Thin-Walled Structure Using Finite Element Method and Design of Experiments. *Aerospace* **2022**, *9*, 541. [CrossRef]
27. Kubiak, T. *Static and Dynamic Buckling of Thin-Walled Plate Structures*; Springer Science and Business Media LLC: Dordrecht, The Netherlands, 2013. [CrossRef]
28. Paszkiewicz, M.; Kubiak, T. Selected problems concerning determination of the buckling load of channel section beams and columns. *Thin-Walled Struct.* **2015**, *93*, 112–121. [CrossRef]
29. Rozylo, P. Stability and failure of compressed thin-walled composite columns using experimental tests and advanced numerical damage models. *Int. J. Numer. Methods Eng.* **2021**, *122*, 5076–5099. [CrossRef]
30. Kubiak, T.; Samborski, S.; Teter, A. Experimental investigation of failure process in compressed channel-section GFRP laminate columns assisted with the acoustic emission method. *Compos. Struct.* **2015**, *133*, 921–929. [CrossRef]
31. Wysmulski, P.; Falkowicz, K.; Filipek, P. Buckling state analysis of compressed composite plates with cut-out. *Compos. Struct.* **2021**, *274*, 114345. [CrossRef]
32. Wysmulski, P.; Debski, H.; Falkowicz, K. Sensitivity of Compressed Composite Channel Columns to Eccentric Loading. *Materials* **2022**, *15*, 6938. [CrossRef] [PubMed]
33. Wysmulski, P. Numerical and Experimental Study of Crack Propagation in the Tensile Composite Plate with the Open Hole. *Adv. Sci. Technol. Res. J.* **2023**, *17*, 249–261. [CrossRef] [PubMed]
34. Rozylo, P. Failure analysis of beam composite elements subjected to three-point bending using advanced numerical damage models. *Acta Mech. Autom.* **2023**, *17*, 133–144.
35. Ramkumar, R.; Rajaram, K.; Saravanan, P.; Venkatesh, R.; Saranya, K.; Jenaris, D.S. Determination of mechanical properties of CFRP composite reinforced with Abaca and Kenaf fibres. *Mater. Today Proc.* **2022**, *62*, 5311–5316.
36. *PN-EN ISO 527–5:2010*; Tworzywa Sztuczne—Oznaczanie Właściwości Mechanicznych Przy Statycznym Rozciąganiu—Część 5: Warunki Badań Kompozytów Tworzywowych Wzmocnionych Włóknami Jednokierunkowo. The Polish Committee for Standardization: Warsaw, Poland, 2010.
37. *ASTM D3039*; Standard Test Method for Tensile Properties of Polymer Matrix Composite Materials. ASTM International: West Conshohocken, PA, USA.
38. *PN-EN ISO 14129:2000*; Kompozyty Tworzywowe Wzmocnione Włóknem—Oznaczanie Naprężenia Ścinającego i Odpowiadającego Odkształcenia, Modułu Ścinania i Wytrzymałości Podczas Rozciągania pod Kątem +- 45 Stopni. The Polish Committee for Standardization: Warsaw, Poland, 2010.
39. *ASTM D3518*; Standard Test Method for In-Plane Shear Response of Polymer Matrix Composite Materials by Tensile Test of a ±45° Laminate. ASTM International: West Conshohocken, PA, USA.
40. *PN-EN ISO 14126:2002*; Kompozyty Tworzywowe Wzmocnione Włóknem—Oznaczanie Właściwości Podczas Ściskania Równolegle do Płaszczyzny Laminowania. The Polish Committee for Standardization: Warsaw, Poland, 2010.
41. *ASTM D 3410*; Standard Test Method for Compressive Properties of Polymer Matrix Composite Materials with Unsupported Gage Section by Shear Loading. ASTM International: West Conshohocken, PA, USA.
42. Różyło, P.; Smagowski, W.; Paśnik, J. Experimental Research in the Aspect of Determining the Mechanical and Strength Properties of the Composite Material Made of Carbon-Epoxy Composite. *Adv. Sci. Technol. Res. J.* **2023**, *17*, 232–246. [CrossRef]
43. Rozylo, P. Determined Material Properties within the Framework of the NCN Project (OPUS) No. 2021/41/B/ST8/00148. Dataset. Source: Zenodo, 2023. Available online: https://zenodo.org/records/7606942 (accessed on 19 October 2023).
44. Holmes, J.; Sommacal, S.; Das, R.; Stachurski, Z.; Compston, P. Digital image and volume correlation for deformation and damage characterisation of fibre-reinforced composites: A review. *Compos. Struct.* **2023**, *315*, 116994. [CrossRef]
45. Nowak, M.; Maj, M. Determination of coupled mechanical and thermal fields using 2D digital image correlation and infrared thermography: Numerical procedures and results. *Arch. Civ. Mech. Eng.* **2018**, *18*, 630–644. [CrossRef]
46. Li, Z.; Cen, S.; Wu, C.J.; Shang, Y.; Li, C.F. High-performance geometric nonlinear analysis with the unsymmetric 4-node, 8-DOF plane element US-ATFQ4. Internat. *J. Numer. Methods Engrg.* **2018**, *114*, 931–954. [CrossRef]
47. Camanho, P.P.; Davila, C.G.; de Moura, M.F. Numerical simulation of mixed-mode progressive delamination in the composite materials. *J. Compos. Mater.* **2003**, *37*, 1415–1438. [CrossRef]
48. Lapczyk, I.; Hurtado, J.A. Progressive damage modeling in fiber-reinforced materials. *Compos. Part A Appl. Sci. Manuf.* **2007**, *38*, 2333–2341. [CrossRef]
49. Hu, H.; Niu, F.; Dou, T.; Zhang, H. Rehabilitation effect evaluation of cfrp-lined prestressed concrete cylinder pipe under combined loads using numerical simulation. *Math. Probl. Eng.* **2018**, *2018*, 3268962. [CrossRef]

Disclaimer/Publisher's Note: The statements, opinions and data contained in all publications are solely those of the individual author(s) and contributor(s) and not of MDPI and/or the editor(s). MDPI and/or the editor(s) disclaim responsibility for any injury to people or property resulting from any ideas, methods, instructions or products referred to in the content.

materials

MDPI

Article

Early Detection and Analysis of Cavity Defects in Concrete Columns Based on Infrared Thermography and Finite Element Analysis

Fan Yang [1], Xianwang Zeng [2], Qilong Xia [1,*], Ligui Yang [2], Haonan Cai [2] and Chongsheng Cheng [2,3,*]

1 No.3 Construction Co., Ltd., Chongqing Construction Engineering Group, Chongqing 481102, China; cqwushengjiu@163.com
2 School of Civil Engineering, Chongqing Jiaotong University, Chongqing 400074, China; 622230970107@mails.cqjtu.edu.cn (X.Z.); yangligui@cqjtu.edu.cn (L.Y.); 611210080005@mails.cqjtu.edu.cn (H.C.)
3 State Key Laboratory of Mountain Bridge and Tunnel Engineering, Chongqing Jiaotong University, Chongqing 400074, China
* Correspondence: 15002368332@163.com (Q.X.); ccheng@cqjtu.edu.cn (C.C.)

Abstract: Concrete, known for its high strength, durability, and flexibility, is a core material in construction. However, defects such as voids and honeycombing often occur due to improper pouring or vibration, weakening the concrete's strength and affecting its long-term performance. These defects typically require costly repairs. Therefore, timely identification and repair of such early defects is crucial for improving construction quality. This paper proposes a method for non-destructive detection of honeycomb defects in concrete using infrared thermography (IR) during the hydration stage. By analyzing the temperature differences between defect and non-defect areas based on the temperature distribution generated during hydration, defects can be detected. Furthermore, the study uses the COMSOL finite element model to explore the relationship between defect size, ambient temperature, formwork thickness, and thermal contrast. The results show that IR technology can effectively and reliably detect honeycomb defects, especially during the hydration phase. As a convenient and feasible non-destructive testing method, IR technology has significant potential for application and development in concrete defect detection.

Keywords: infrared thermography; hydration heat; concrete structures; thermal contrast

Academic Editors: Pawel Wysmulski, Katarzyna Falkowicz and Patryk Rozylo

Received: 17 February 2025
Revised: 27 March 2025
Accepted: 3 April 2025
Published: 7 April 2025

Citation: Yang, F.; Zeng, X.; Xia, Q.; Yang, L.; Cai, H.; Cheng, C. Early Detection and Analysis of Cavity Defects in Concrete Columns Based on Infrared Thermography and Finite Element Analysis. *Materials* **2025**, *18*, 1686. https://doi.org/10.3390/ma18071686

Copyright: © 2025 by the authors. Licensee MDPI, Basel, Switzerland. This article is an open access article distributed under the terms and conditions of the Creative Commons Attribution (CC BY) license (https://creativecommons.org/licenses/by/4.0/).

1. Introduction

Concrete is a versatile material used in constructing bridges, roads, buildings, and other structures, owing to its ability to be molded into complex shapes and meet diverse construction requirements. Due to its high compressive strength, durability, and superior workability, concrete has emerged as a critical component in contemporary civil engineering practices [1,2]. However, during concrete pouring, poor fluidity, insufficient vibration, and adverse environmental conditions often result in the aggregates failing to compact fully, leading to the formation of localized defects such as honeycombs and voids. These early-stage defects compromise concrete compactness, accelerating structural degradation and performance decline. More critically, they persistently undermine integrity throughout the service lifespan, posing significant safety risks [3,4]. During the concrete pouring stage, if honeycomb defects in the concrete can be promptly identified, timely remedies (such as vibration or tamping) can be applied using non-destructive methods. However, once the concrete has hardened, repairing these defects often requires significant human and material resources, thereby substantially increasing construction costs. Therefore, quickly and accurately detecting honeycomb and void defects in concrete during the pouring stage holds significant value.

In recent years, with the advancement of non-destructive testing (NDT) technologies, numerous NDT methods for detecting defects in concrete structures have emerged [5–9]. Zhao et al. [10] combined Digital Image Correlation (DIC) technology with wavelet analysis to identify and warn of concrete structures of micro-damage. Through four-point bending experiments and the analysis of acceleration response signals, the experimental results demonstrated that this method can effectively provide early warnings for micro-damage and accurately locate and assess the extent of the damage. Janku et al. [11] measured concrete bridges and defect test specimens using various NDT techniques, providing a comprehensive comparison in terms of accuracy, operability, and cost. Jiao et al. [12] developed an automated defect detection algorithm for concrete based on Ground Penetrating Radar (GPR). The algorithm utilizes signal polarity and morphological features to identify delamination, voids, and water infiltration defects. Experimental results showed the algorithm's strong detection capabilities in both simulated and real-world data. McCabe et al. [13] utilized GPR as a non-destructive technique to detect early characteristics of honeycomb defects in concrete pavements. Experiments validated the sensitivity of GPR in identifying the size, shape, and depth of honeycombs, providing a reference for its application in pavement defect detection. Chow et al. [14] proposed an automated concrete defect detection framework based on a 360° camera and LiDAR. The framework utilizes deep learning algorithms to identify defects and integrates the results into a Building Information Model (BIM), effectively addressing the inefficiency of traditional manual inspection methods. Christoph et al. [15] proposed a concrete honeycomb defect detection method based on multi-sensor data fusion. By integrating impact echo (IE), Ultrasonic (US), and GPR technologies, the method significantly improved detection reliability and accuracy through a feature-level data fusion algorithm. Xu et al. [16] combined deep learning technology with the impact echo method, utilizing wavelet transform to extract signal features and establishing a deep learning network detection system. The results demonstrated that this system can efficiently and accurately identify defects in concrete structures, achieving high detection accuracy. However, these techniques still exhibit limitations in large-area coverage, adaptability to complex environments, and real-time monitoring capabilities. In particular, there is an urgent need for more efficient and reliable solutions for the emergent detection of concrete defects and evaluating their dynamic evolution during the early stage of concrete hardening.

Infrared thermography (IRT), as a non-destructive testing method based on temperature distribution differences, demonstrates unique advantages in concrete defect detection due to its portability, efficiency, non-contact nature, and large-area coverage capabilities [17]. IRT can rapidly capture temperature anomalies in infrared radiation, accurately locating defect areas, making it particularly suitable for real-time monitoring in complex environments and during hydration processes. Compared to other NDT techniques, IRT not only offers higher detection efficiency but also enables the rapid acquisition of inspection results without direct contact with the concrete structure. Cheng et al. [18] utilized IRT and elastic wave technology to scan concrete specimens with prefabricated defects and analyzed the size of the defects based on thermal images and elastic wave signals. Vemuri et al. [19] employed IRT to detect defects in concrete structures and concluded that IRT offers advantages such as short detection time and precise localization. Cheng et al. [20] cast concrete specimens with defects and utilized IRT to detect these defects during concrete hydration. Hong et al. [21] utilized IRT and Digital Image Correlation (DIC) technologies to detect early defects in concrete. The results indicated that IRT could accurately identify surface and internal voids, and the geometric dimensions of the voids significantly affected the temperature difference, showing a power function relationship with the product of thickness and depth. Cheng et al. [22] investigated the effect of different time windows on the detection performance of concrete delamination defects using infrared thermography (IRT). They found that the appropriate selection of time windows is critical for improving detection sensitivity and proposed the optimal time window for delamination detection under natural conditions.

Although IRT has unique advantages in concrete defect detection, its application is typically limited to the post-hydration stage. In this hardened state, remediation of identified defects incurs high complexity and cost. In contrast, during the hydration stage, the concrete is still malleable, and timely detection and intervention can effectively prevent defects from solidifying, significantly reducing repair costs and difficulty. Therefore, using IRT for defect detection during the hydration stage is of great significance. This study is the first to explore the use of IRT for detecting honeycomb defects during the early stages of concrete hydration. Unlike previous studies, this research focuses on the temperature changes caused by the release of hydration heat, using the temperature difference between defective and non-defective areas to detect honeycomb defects [23]. This method enables early intervention when defects occur, preventing high-cost and complex repairs later, and improving the overall quality of concrete construction. Garg et al. [24] experimentally studied the feasibility of using hydration heat as an internal heat source to detect voids in prestressed tendon ducts. The results demonstrated that this method is effective for void detection. Li et al. [25] proposed a method for detecting grouting defects in external prestressed tendon ducts using IRT during the hydration stage and discussed the impact of defect severity on the infrared detection capability. Wan et al. [26] experimentally studied the application of IRT in defect detection during the concrete hydration process, defining the detection time window and identifying the optimal timing for IRT application. Cai et al. [27] proposed enhancing the detection performance of IRT for interface debonding defects in CFST arch bridges by utilizing concrete hydration heat and water cooling. Experiments demonstrated that thermal contrast improved by 2–3 times after cooling excitation, and numerical simulations confirmed a linear relationship between cooling intensity and thermal contrast. This method shows potential for early debonding detection. Cheng et al. [28] experimentally investigated the effects of hydration heat rising rate, void size, and environmental factors on the detection of debonding in CFST. The results showed that the absolute temperature difference had the most significant impact on detection performance, while the hydration rate had the least effect. The interaction between hydration rate and void size had a secondary impact on detection performance, and this interaction weakened as both the hydration rate and void size decreased.

This study aims to conduct numerical and experimental investigations on IRT detection of honeycomb defects during the hydration stage of concrete under formwork conditions. First, an experimental model is developed using COMSOL (version 6.2) finite element simulations to investigate the feasibility and efficiency of infrared thermography (IRT) in detecting defects during the concrete hydration stage. Next, physical experiments are conducted to validate the simulation results, confirming the model's accuracy and reliability in replicating the hydration process. Finally, the validated model is applied to analyze the effects of ambient temperature, honeycomb defect size, and formwork thickness on the thermal contrast evolution between defective and intact regions. This study proposes a concrete honeycomb defect detection method based on infrared thermography technology, combined with finite element simulation, to thoroughly investigate the impact of various factors on detection effectiveness. It provides an innovative technical approach and methodology for detecting honeycomb defects during the concrete hydration stage. Additionally, it offers significant theoretical foundations and practical references for optimizing and developing defect detection technologies under complex working conditions.

Particularly, the integration of numerical simulation and physical experiments to verify and optimize the detection method ensures its reliability and applicability, providing strong technical support for practical engineering applications.

2. Materials and Methods

During the concrete pouring, construction issues such as insufficient vibration may result in honeycomb defects in localized areas. These defects are typically air-filled, leading to a significantly lower thermal conductivity than the surrounding dense concrete. This difference causes temperature variations between the defective and non-defective regions By analyzing the temperature differences on the surface of the concrete formwork, early defects can be effectively identified and located. Therefore, conducting an in-depth analysis of the temperature difference (i.e., thermal contrast) between defective and non-defective regions is of significant importance for the early detection of defects.

During the concrete hydration process, the temperature difference on the surface of the formwork is influenced by multiple factors, including ambient temperature, formwork thickness, the particle size of the honeycomb defect, and the initial pouring temperature of the concrete. As shown in Figure 1, this study employed the finite element method (COMSOL Multiphysics) to simulate the heat conduction process in concrete structures containing honeycomb defects under hydration heat conditions. The research procedure was as follows: (1) a three-dimensional concrete model with honeycomb defect was established; (2) boundary conditions were defined for the model; (3) the model was meshed according to the shape and size of the material; (4) the concrete hydration process was simulated; (5) real concrete specimens were cast to validate the 3D model and assessed its accuracy; and (6) the simulation results were analyzed to reveal the variation patterns of thermal contrast. Subsequently, a parametric study was conducted to quantify the impact of ambient temperature, formwork thickness, and defect size on the detection sensitivity of honeycomb defects during hydration.

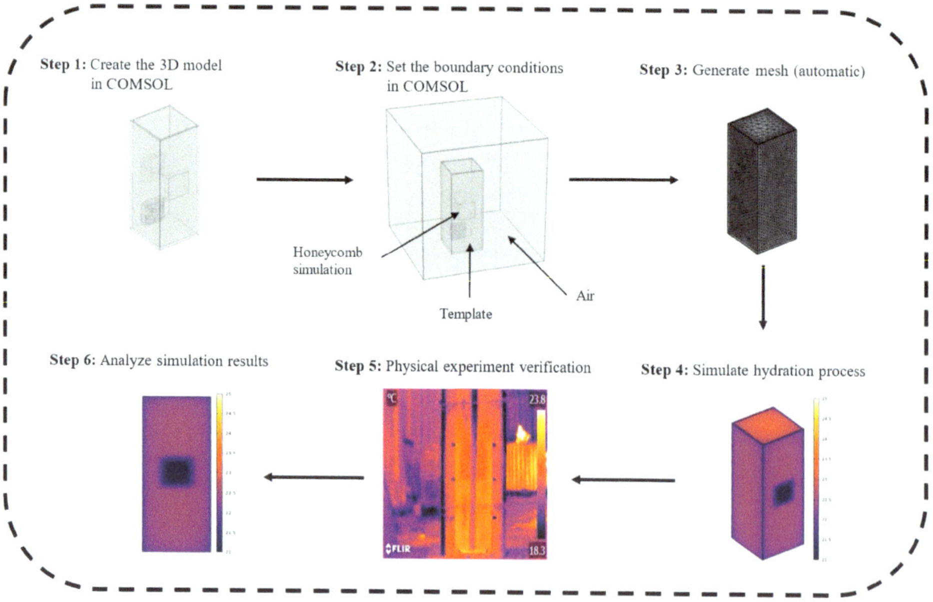

Figure 1. Research framework.

2.1. Concrete Hydration Heat Calculation

Hydration heat, a fundamental phenomenon occurring during concrete hydration, induces internal temperature variations due to exothermic chemical reactions. This thermal behavior can be accurately modeled through numerical simulations, offering a robust approach to investigating spatiotemporal temperature field evolution in hydrating concrete. In this study, the hydration heat process was simulated using COMSOL Multiphysics software (version 6.2), where heat transfer was modeled as a transient solid-phase conduction process. The results quantitatively delineated the temperature field dynamics driven by hydration heat, facilitating in-depth analysis of formwork surface temperature gradients and their implications for defect detection.

Zhu [29] explicitly pointed out that cement hydration heat is closely related to the age of the concrete. In commonly used models for calculating cement hydration heat, there are typically three main types: the exponential calculation model, the composite exponential calculation model, and the hyperbolic calculation model. Liu [30] compared the three hydration heat calculation models by comparing their calculated results with the measured data. The results showed that the calculated results of the composite exponential hydration heat model were closer to the measured values. Through a series of experimental and computational studies, Lin et al. [31,32] conducted an in-depth investigation of the hydration heat release model. They concluded that, compared to other models, the composite exponential model exhibits higher accuracy in describing the hydration heat release process. In this study, the composite exponential model was adopted to calculate the cement hydration heat, as expressed in Equation (1) [33]:

$$Q(t) = Q_0\left(1 - e^{-at^b}\right) \tag{1}$$

where $Q(t)$ is the cumulative heat of hydration of concrete at time t, and its unit is kJ/kg. Q_0 is the cumulative hydration heat as $t \to \infty$, with its value selected based on reference [33], and its unit is kJ/kg. t is the time, and its unit is days (d). a, b are constants related to cement varieties.

Rewriting Equation (1) to the unit of hours and the derivative of time yields the formula for the heat generation rate of concrete, after unit conversion: 1 w = 1 J/s = 3.6 kJ/h, as (2):

$$HENG = W\frac{dQ(t)}{dt} = \frac{ab}{24}WQ_0t^{b-1}e^{-a\left(\frac{t}{24}\right)^b} \tag{2}$$

where $Q(t)$ is the cumulative heat of hydration of concrete at time t, and its unit is kJ/kg. Q_0 is the cumulative hydration heat as $t \to \infty$, with its value selected based on reference [33], and its unit is kJ/kg. t is the time, and its unit is days (d). a, b are constants related to cement varieties. W is the cement content, and its unit is kg/m^3. $HENG$ is the heat generation rate, and its unit is $kJ/(h \cdot m^3)$.

After unit conversion, 1 W = 1 J/s = 3.6 kJ/h, the heat generation rate $HENG$ can be obtained as follows:

$$HGEN = W\frac{dQ(t)}{dt} = \frac{\frac{ab}{24}WQ_0t^{b-1}e^{-a\left(\frac{t}{24}\right)^b}}{3.6} = \frac{abWQ_0t^{b-1}e^{-a\left(\frac{t}{24}\right)^b}}{24 \times 3.6} \tag{3}$$

2.2. Numerical Modeling

This section focuses on the establishment of the finite element model. The accurate simulation of honeycomb defects forms the basis for temperature field analysis during the concrete hydration process, as the presence of honeycomb defects significantly affects the temperature field distribution on the surface of concrete formwork. To precisely simulate the detection of honeycomb defects, a concrete column model with dimensions of 400 mm × 400 mm × 1000 mm was designed. A honeycomb defect region measuring 200 mm × 200 mm was placed at the center of the inner surface of the formwork, with these defects typically filled with air. Additionally, to further investigate the sensitivity of infrared thermography to honeycomb defects with varying particle sizes, three types of honeycomb defects were designed: defects with particle sizes of 10–20 mm, 40–50 mm, and a single defect block with an overall size of 200 mm × 200 mm. The specific layout of the defects is shown in Figure 2.

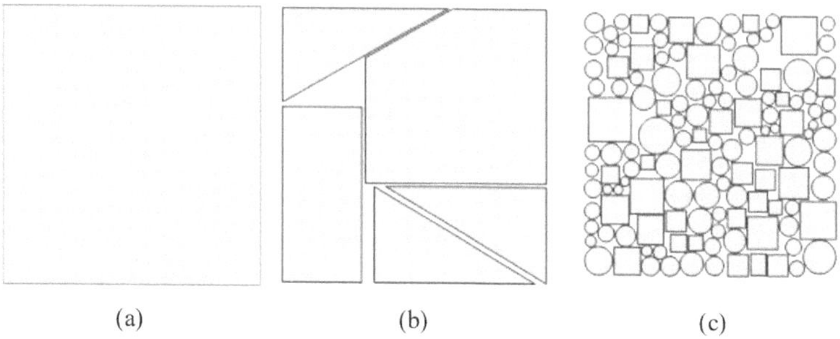

Figure 2. Honeycomb defect model design: (**a**) 200 mm; (**b**) 40–50 mm; (**c**) 10–20 mm.

To study the detection of honeycomb defects, a three-dimensional finite element model (as shown in Figure 3) was developed using the heat transfer module in COMSOL Multiphysics software to simulate the concrete hydration process. The material parameters used in the model are listed in Table 1. In this model, the formwork, concrete column, and honeycomb defect regions were finely meshed to improve computational accuracy (as shown in Figure 4). The meshing used a "physics-controlled mesh" setting, with the element size set to "finer", resulting in a total of 135,201 domain elements, 32,656 boundary elements, and 4405 edge elements.

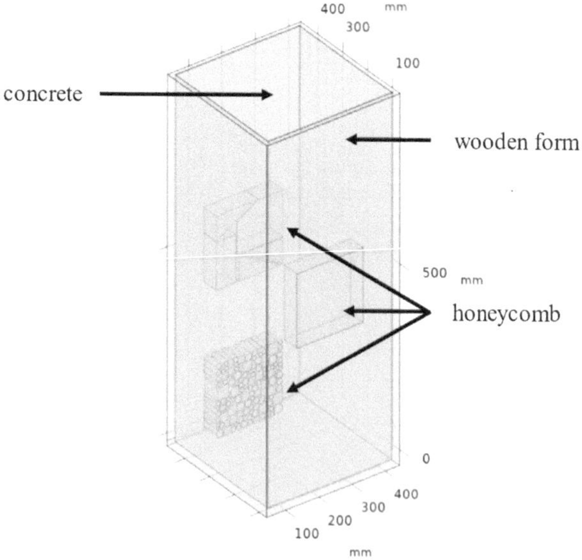

Figure 3. Geometric shape of the concrete column model.

Table 1. Material parameters of the model.

Material	Specific Heat Capacity (J/(kg·K))	Density (kg/m^3)	Thermal Conductivity (W/(m·K))
Wooden Board	2700	532	k(T)
Air	k(T)	k(T)	k(T)
Concrete	980	2450	0.17

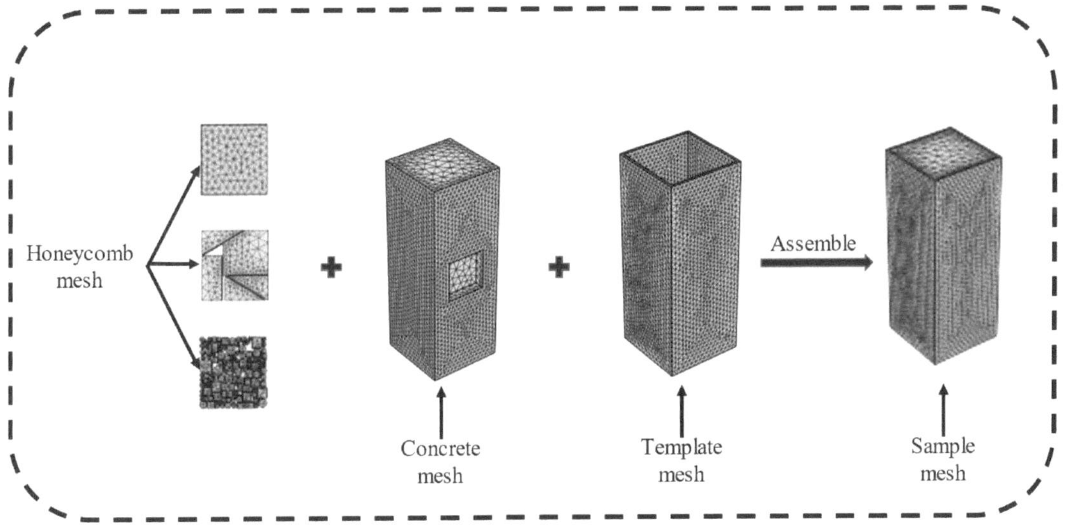

Figure 4. Mesh division of the concrete column model.

In the numerical simulation, natural convection boundary conditions were applied at the formwork-air interface to represent the heat exchange process. Fresh concrete has high fluidity and self-weight characteristics. During the pouring stage, the concrete-formwork system formed a gapless contact interface. As a result, the impact of interfacial contact thermal resistance on the heat transfer process was neglected in the model. To capture the transient nature of the heat released during hydration, an unsteady-state heat transfer model with a time step of 1 min was established. This setting ensured the accurate representation of the spatial and temporal evolution of the temperature field on the formwork surface while balancing computational accuracy and efficiency. By adopting physically justified boundary conditions and dynamic solution strategies, the study successfully simulated the temperature difference formation mechanism on the formwork surface during hydration. Furthermore, the mapping relationship between these temperature differences and defect features was quantitatively characterized.

2.3. Case Design

This study specifically examined the impact of honeycomb size, ambient temperature, and formwork dimensions on detecting honeycomb defects in concrete structures. The study designed a 200 mm × 200 mm area as the honeycomb simulation region. Apart from differences in particle size, the honeycomb particles occupied 50% of the total simulation area. Additionally, a single 200 mm × 200 mm solid honeycomb block was designed to simulate a large void in the concrete structure. The relevant working condition information is shown in Table 2. Specifically, the defect sizes were determined based on common types of concrete defects in actual engineering projects: 200×200 mm corresponds to void defects, and 10–30 mm and 30–60 mm correspond to small and large honeycomb defects, respectively, while 4–10 mm corresponds to surface pitting defects. For formwork thickness, considering that the most commonly used formwork thickness in engineering practice falls within the 10–20 mm range [34], the study selected several representative thickness values within this range for comparative analysis to investigate the effect of formwork thickness on detection performance.

Table 2. Working condition design.

Parameters	Group 1	Group 2	Group 3
L × D × H (mm)	400 × 400 × 1000	400 × 400 × 1000	400 × 400 × 1000
Ambient temperature Ta (°C)	10, 16, 22, 28, 34	22	16, 22
Honeycomb size (mm)	30–60	4–10, 10–30, 30–60, 200 × 200	30–60
Formwork thickness (mm)	14	14	11, 14, 17, 20

2.4. Evaluation Indicators

When honeycomb defects exist in a concrete structure, air entrapped in the defect area results in a significant thermal conductivity contrast between the concrete and air, creating a measurable temperature gradient between the two regions. This study analyzed the temperature difference between the honeycomb defect area and the non-defect area as an indicator of detection performance. As shown in Figure 5, the temperature difference (also referred to as absolute thermal contrast) can be defined using the following equation:

$$\Delta T = Tnd - Td \tag{4}$$

in the equation, Tnd, Td, and ΔT represent the average surface temperature of the non-defect area, the defect area, and the temperature difference between the non-defect and defect areas, respectively. When $\Delta T > 0$, it indicates that $Tnd > Td$, meaning the temperature of the non-defect area is higher than that of the defect area. Conversely, when $\Delta T < 0$, it indicates that the temperature of the non-defect area is lower than that of the defect area. To enhance the clarity and coherence of the paper, several parameters were defined for use in subsequent analysis and discussion. The details are provided in Table 3.

Table 3. Defined parameters for analysis.

Parameter Name	Parameter Explanation
Tc	The initial temperature of concrete
Ta	Ambient temperature
Ts	Formwork surface temperature
σ	Standard deviation

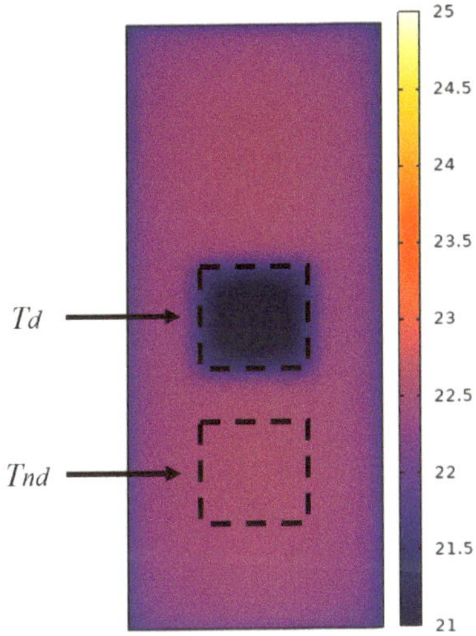

Figure 5. Schematic diagram of infrared thermal contrast calculation.

3. Experimental Validation

3.1. Experimental Design

In the physical experiment, three honeycomb defects identical to those in the simulation were set up, and their spatial configuration is shown in Figure 6. After assembling the formwork containing pre-installed defects, a formwork system enclosing the concrete was assembled. The detailed structure and arrangement of the formwork frame are shown in Figure 7.

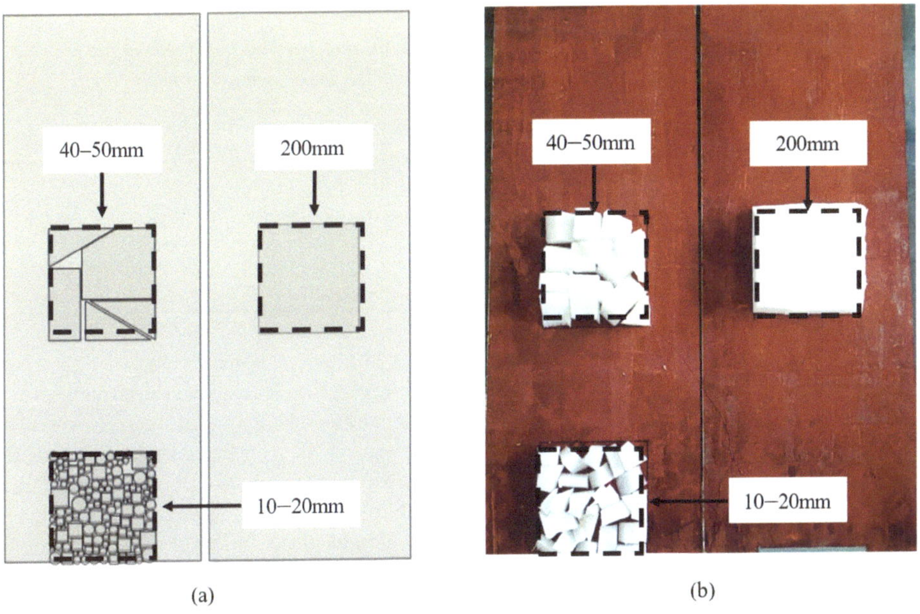

Figure 6. Honeycomb defect design diagram: (**a**) finite element simulation defect layout; (**b**) physical experiment defect layout.

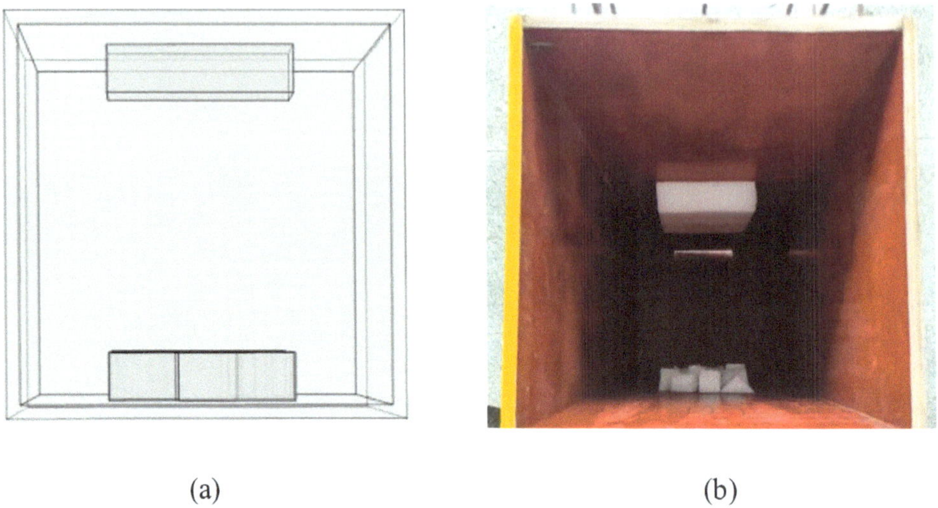

Figure 7. Concrete formwork frame design: (**a**) finite element simulation formwork frame; (**b**) physical experiment formwork frame.

This study used the FLIR A300 and MAG-F6 infrared thermal imaging cameras to perform infrared detection on the experimental specimens. The parameters of the thermal imaging cameras are listed in Table 4. The FLIR A300 camera has a resolution of 320 × 240 pixels, a spectral range of 7.5–13 μm, and an accuracy of ±2 °C or ±2%. The MAG-F6 camera has a resolution of 640 × 480 pixels, a spectral range of 7.5–13 μm, and an accuracy of ±0.7 °C or ±0.7%. Both cameras were positioned on the formwork side adjacent to the honeycomb defect, approximately 3 m from the outer surface of the specimen, with their height aligned with the center height of the specimen. Data acquisition was configured at a sampling rate of 1 frame/min, and the thermal images were stored on a connected PC. The detailed layout of the experimental site is shown in Figure 8.

Table 4. Specifications of infrared cameras.

Camera Name	**FLIR A300**	**MAG-F6**
Detector type	Uncooled microbolometer	Uncooled microbolometer
Accuracy	±2 °C or ±2%	±0.7 °C or ±0.7%
Resolution	320 × 240 pixels	640 × 480 pixels
Spectral range	7.5–13 μm	7.5–13 μm

As shown in Figure 8, the formwork was reinforced to ensure the smooth progress of the experiment and prevent the formwork from being damaged due to excessive force. Before the experiment officially began, the infrared data acquisition equipment (computer) was used to preset the acquisition frequency of the infrared thermal camera. Parameters affecting the accuracy of temperature measurements, such as measurement distance, humidity, airflow velocity, and emissivity, were also configured to ensure the precision of infrared temperature measurement. The height and focus of the cameras were precisely adjusted to ensure that the infrared image acquisition system obtains clear and accurate images. Subsequently, the temperature data acquisition system was activated simultaneously with the concrete pouring process to obtain real-time temperature distribution images of the column specimen's surface.

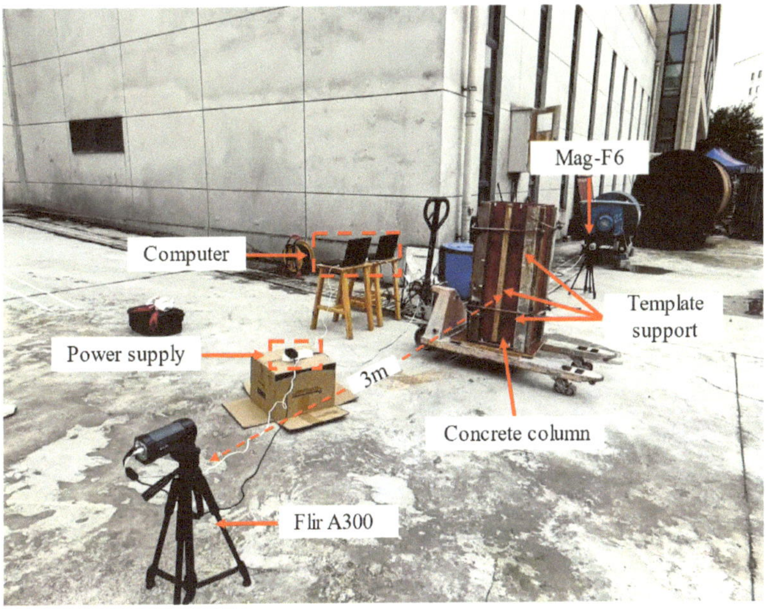

Figure 8. Experimental site layout.

The concrete pouring process is shown in Figure 9. First, the uniformly mixed concrete was loaded into a hopper, which was then moved above the concrete column structure using a forklift. Next, the hopper valve was opened to pour concrete until the formwork was fully filled. The entire process was carried out strictly by the experimental design requirements to ensure the accuracy of data collection.

Figure 9. On-site record of concrete pouring.

3.2. Experimental Materials

This study designed physical specimens with dimensions consistent with the simulation model and used experimental data to validate the numerical model. The materials required for producing concrete include cement, sand, water, and aggregates. The concrete mix design followed the "*Technical Specification for High-Strength Concrete Structures*" (CECS104:99). The material mix proportion for the concrete specimen is shown in Table 5. The sand (fine aggregates) had a maximum particle size of 2.36 mm, while the crushed stone (coarse aggregates) ranged from 4.75 mm to 9.5 mm.

Table 5. Concrete mix proportion (1 m^3).

Cement (kg)	Fly Ash (kg)	Silica Ash (kg)	Sweller (kg)	Sand (kg)	Aggregate (kg)	Water (kg)	Polycarboxylate Superplasticizer (kg)
411	89	35	59	968	731	143	7.08

In actual concrete pouring processes, honeycomb defects (air voids) typically form due to insufficient vibration, and these honeycomb defects are filled with air. Under standard atmospheric pressure at 15 °C, the thermal conductivity values are as follows: air (0.023 W/(m·K)), concrete (1.7 W/(m·K)), and polystyrene foam (0.02–0.05 W/(m·K)). The thermal conductivity of polyethylene foam is close to that of air; therefore, this experiment used polystyrene foam as the prefabricated material to simulate a honeycomb defect. The formwork for casting the concrete columns was constructed from standard wooden formwork, with thickness dimensions matching those defined in the simulation model.

3.3. Validation Results

As shown in Figure 10, five minutes after pouring, the defects began to appear. Additionally, a vertical shadow region appeared at the column center in the infrared image. This shadow resulted from the formwork frame's support structure, which physically blocked the infrared detection of the honeycomb defect region. However, excluding this elongated shadow region, significant temperature differences were still observed between the defect and non-defect areas in Figure 10a,b. At this stage, the thermal contrast (ΔT) between the non-defect area and the defect area was 0.2 °C. Thus, a thermal contrast threshold of 0.2 °C was established in this study. The results demonstrate that infrared thermography enables real-time detection of honeycomb defects in early-stage concrete hydration.

To validate the accuracy of the finite element (FE) simulations for concrete structures, experimental infrared thermograms were compared with FE-simulated temperature distributions. Figure 10c,d shows the infrared temperature images obtained by simulating the concrete hydration process using COMSOL Multiphysics software. To ensure the consistency of the validation results, the temperature data from both the experimental and simulated images were extracted using the same method. Specifically, the honeycomb defect area's average temperature was selected as the honeycomb defect's surface temperature, while the surrounding area's average temperature was selected as the surface temperature of the non-defective area. Through comparative analysis, the consistency between the simulation results and the experimental data were validated, further demonstrating the reliability of the simulation model.

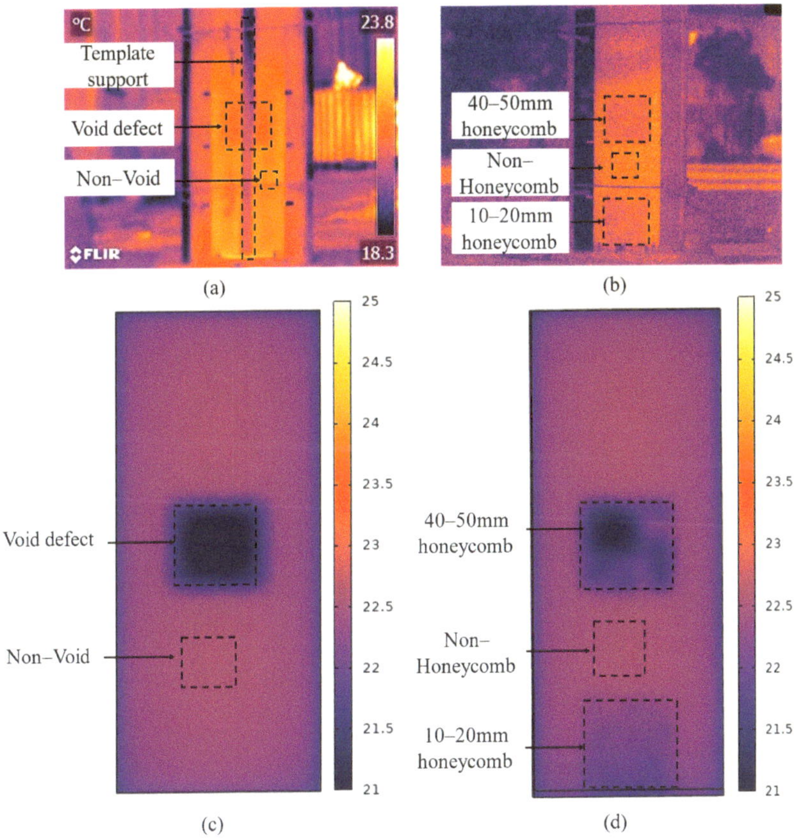

Figure 10. Comparison of infrared images of honeycomb defect: (**a**,**b**) results from physical experiments; (**c**,**d**) results from finite element simulations.

Figure 11 presents a detailed comparison of thermal contrast between the FE simulations and concrete experiments. After concrete pouring, the thermal contrast gradually increased over time. After 30 min, differences in thermal contrast began to emerge for different sizes. The figure shows that the larger the size of the honeycomb defect, the greater the thermal contrast ΔT. Experimental and simulated temperature profiles showed congruent trends, though minor deviations were observed. The mean errors between experimental and simulated curves were 0.06 °C, 0.04 °C, and 0.05 °C, respectively. These discrepancies are mainly due to three influencing factors: environmental variables, measurement system errors, and differences in simulation settings. External factors such as ambient temperature, humidity, and airflow can affect the temperature measurement accuracy of the infrared thermal camera. Furthermore, the materials used in the specimen and the idealized conditions set in the simulation model, including material properties and boundary conditions, can also introduce errors. Although there are some minor differences, these discrepancies are within an acceptable range for this study. Therefore, the simulation method used in this research demonstrates accuracy and reliability in predicting the temperature field of honeycomb defects in concrete.

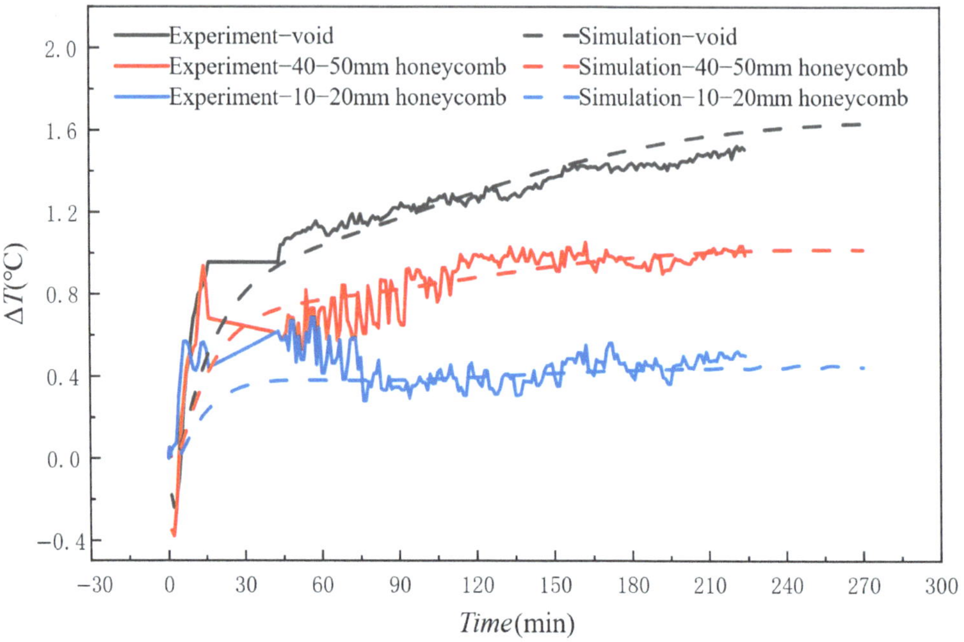

Figure 11. Comparative analysis of experimental and simulated temperature data.

4. Results and Analysis

4.1. Influence of Ambient Temperature on Thermal Contrast

This section discusses the influence of honeycomb defects under conditions where the initial concrete pouring temperature is 22.5 °C, the formwork thickness is 14 mm, the honeycomb defect size ranges from 30 mm to 60 mm, and the defect thickness is 60 mm. The analysis was conducted for ambient temperatures of 10 °C, 16 °C, 22 °C, 28 °C, and 34 °C.

Figure 12a illustrates the trend of thermal contrast (ΔT) in the area of the honeycomb defect over time at different ambient temperatures (*Ta*). From the figure, it can be observed that after concrete pouring, when Ta is 10 °C and 16 °C (below the initial concrete temperature, *Tc*), ΔT increases sharply and reaches its peak at around 50 min (Phase I), then starts to decrease (Phase II), gradually slowing down after 200 min (Phase III). When *Ta* is 28 °C and 34 °C (above the initial concrete temperature, *Tc*), ΔT increases immediately in the negative direction, reaches its peak in the negative direction at around 50 min, then starts to increase again, gradually rising after 200 min. When *Ta* is 22 °C, ΔT increases slowly, reaching its peak at $t = 200$ min, and then slowly decreases and stabilizes. Based on the above analysis, it can be concluded that the greater the temperature difference between the initial concrete temperature (*Tc*) and the ambient temperature (*Ta*), the higher the peak value of ΔT. When the ambient temperature is closer to the initial concrete temperature, the peak value of ΔT is smaller. Therefore, it can be concluded that when the temperature difference ($Tc - Ta$) is large, ΔT is mainly influenced by the ambient temperature, while when $Tc - Ta$ is small, ΔT is mainly influenced by the heat generated by the concrete hydration process.

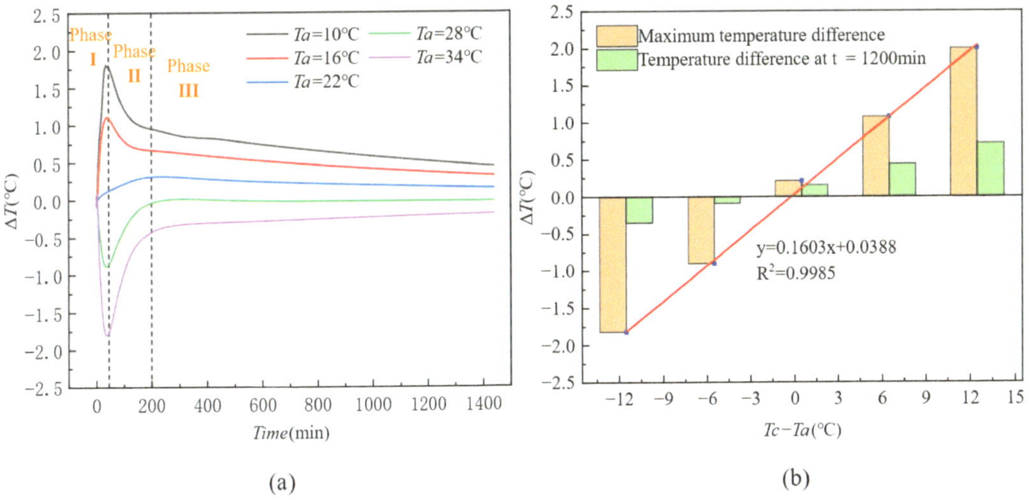

Figure 12. Influence of ambient temperature on thermal contrast: (**a**) variation in thermal contrast during the hydration process; (**b**) relationship between thermal contrast and the temperature difference between the initial concrete temperature and ambient temperature.

Figure 12b presents bar charts of the thermal contrast at the maximum value and t = 1200 min for different values of $Tc - Ta$. It can be observed from the figure that thermal contrast is positively correlated with $Tc - Ta$. By fitting the relationship between the maximum thermal contrast and $Tc - Ta$, the following equation is obtained:

$$y = 0.1603x + 0.0388$$
$$R^2 = 0.9985$$

Figure 13 illustrates the evolution patterns of infrared images of honeycomb defects under different ambient temperatures (Ta). Firstly, due to the wide range of ambient temperatures, the color scale on the right is adjusted accordingly: Ta = 10 °C and Ta = 16 °C use the first color scale on the right, Ta = 22 °C uses the second color scale on the right, and Ta = 28 °C and Ta = 34 °C use the third color scale on the right. Secondly, at ambient temperatures of 10 °C and 16 °C, the overall trend in the infrared images shows an initial temperature rise followed by a decline. The temperature in the defect area is lower than that in the non-defect area, with the thermal contrast being greater than zero. The defect features are more pronounced in the early stages of the hydration process, but as the hydration reaction progresses, the thermal contrast gradually weakens. Additionally, the lower the ambient temperature, the greater the color contrast in the infrared images of the defect area, resulting in clearer defect features. When the ambient temperature is 22 °C, the overall changes in the infrared images are relatively minor, and the temperature distribution tends to stabilize. As the hydration reaction progresses, the color contrast between the defect and non-defect areas slightly increases. However, the overall thermal contrast remains low, making the defect features relatively indistinct. At ambient temperatures of 28 °C and 34 °C, the infrared images show an initial temperature decrease followed by an increase. The temperature in the defect area is higher than that in the non-defect area, with the thermal contrast being less than zero. In the early stages of the hydration process, the defect features are more pronounced, but as the thermal contrast weakens in the later stages, the defect features gradually become indistinct. Additionally, the higher the ambient temperature, the greater the color contrast in the infrared images of the defect area, resulting in clearer defect features. Finally, at the same hydration time, the greater the difference between the initial concrete temperature and the ambient temperature, the more pronounced the thermal contrast between the defect area and the non-defect area, resulting in clearer defect features in the infrared images.

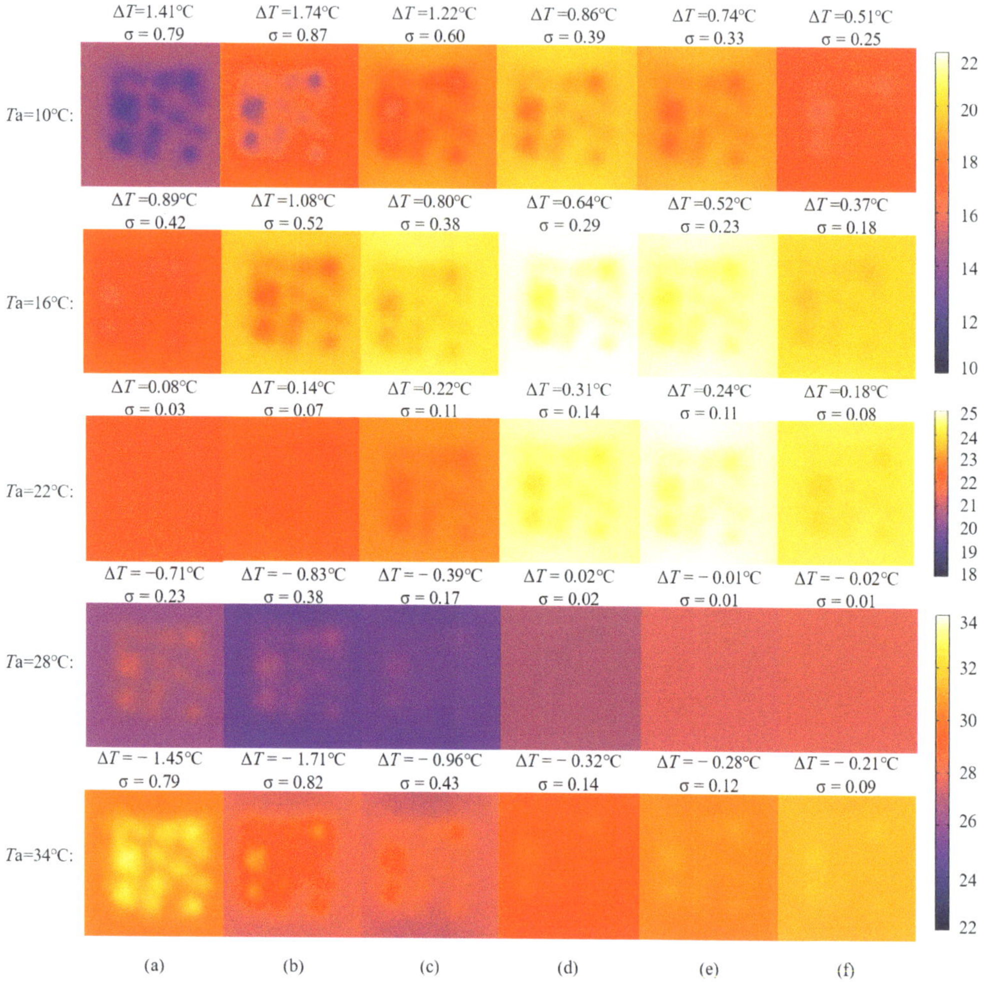

Figure 13. Evolution of infrared images of the honeycomb defect under different ambient temperatures: (**a**) t = 20 min; (**b**) t = 50 min; (**c**) t = 100 min; (**d**) t = 300 min; (**e**) t = 600 min; (**f**) t = 1200 min.

Simultaneously, statistical analysis of temperature data is conducted for each infrared image, with the standard deviation (σ) calculated for each image. As Figure 13 demonstrates, a larger σ value indicates a more uneven distribution of temperature data across the image. Furthermore, the standard deviation exhibits a highly synchronized relationship with the absolute value of thermal contrast ($|\Delta T|$), revealing their synergistic role in defect detection. These findings validate the accuracy and reliability of the selected evaluation metric (ΔT) proposed in this study. Consistent conclusions are confirmed in subsequent analyses presented in Sections 4.2 and 4.3, reinforcing the robustness of the methodology.

As evidenced by the temporal evolution characteristics in Figure 12, the study reveals that 50 min post-casting constitutes a critical threshold for thermal contrast development, during which the thermal contrast (ΔT) approaches its peak values across varying ambient temperature conditions. Figure 14 further elucidates the quantitative heat transfer within this 50 min window, demonstrating that the temperature gradient between the formwork surface and ambient environment ($Ts - Ta$) serves as the governing parameter for ΔT evolution. A statistically significant linear correlation emerges between these variables ($y = 0.347x + 0.0075$, $R^2 = 0.99$). Specifically, as $Ts - Ta$ increases from −6 °C to 6 °C, ΔT transitions linearly from negative to positive thermal contrast regimes, achieving peak magnitudes of approximately ±2 °C. The positive temperature gradient domain ($Ts - Ta > 0$) exhibits a maximum ΔT of +2 °C, while the negative gradient regime ($Ts - Ta < 0$) reaches a contrasting minimum of −2 °C.

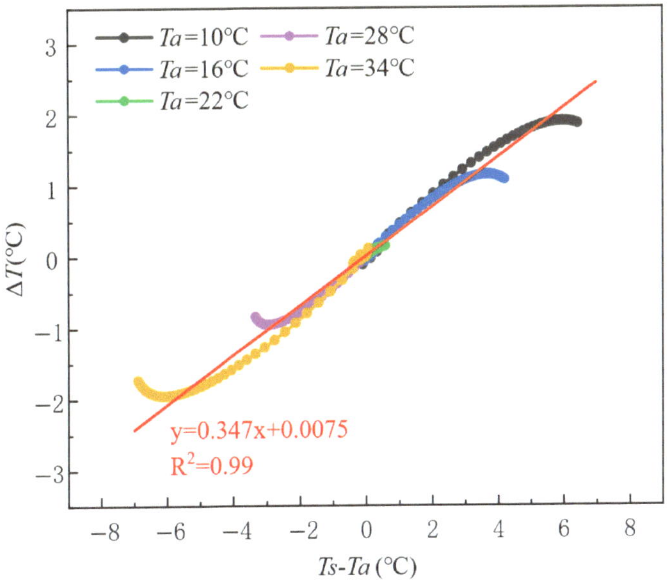

Figure 14. Response characteristics of thermal contrast (ΔT) to surface-to-ambient temperature gradient ($Ts - Ta$) under varying ambient temperatures during the initial casting phase (Phase I).

4.2. *Influence of Honeycomb Particle Size on Thermal Contrast*

This section discusses the relationship between the thermal contrast of the area of the honeycomb defect and the non-defect area under the conditions of an initial concrete pouring temperature of 22.5 °C, a formwork thickness of 14 mm, and an ambient temperature of 16 °C. The honeycomb defect sizes considered are 4–10 mm, 10–30 mm, 30–60 mm (with a thickness of 30 mm), 30–60 mm (with a thickness of 60 mm), and a 200 × 200 mm void (with a thickness of 60 mm).

Figure 15 shows the variation trend of thermal contrast (ΔT) over time for honeycomb defects of different sizes. From the figure, it can be observed that, in the early stages after concrete pouring, the thermal contrast of honeycomb defects of different sizes initially increases sharply, then reaches a peak (Phase I) and rapidly decreases (Phase II). After 200 min, the thermal contrast of the honeycomb defect starts to decrease slowly and eventually levels off (Phase III). At the same time point, the larger the size of the honeycomb defect, the higher its thermal contrast. Among them, void defects should be observed using the data on the right y-axis. It can be seen that the thermal contrast of the void defect is significantly higher than that of other types of honeycomb defects, indicating that the surface temperature of the formwork in the void defect areas reflects the differences in heat conduction more noticeably than in other defect areas.

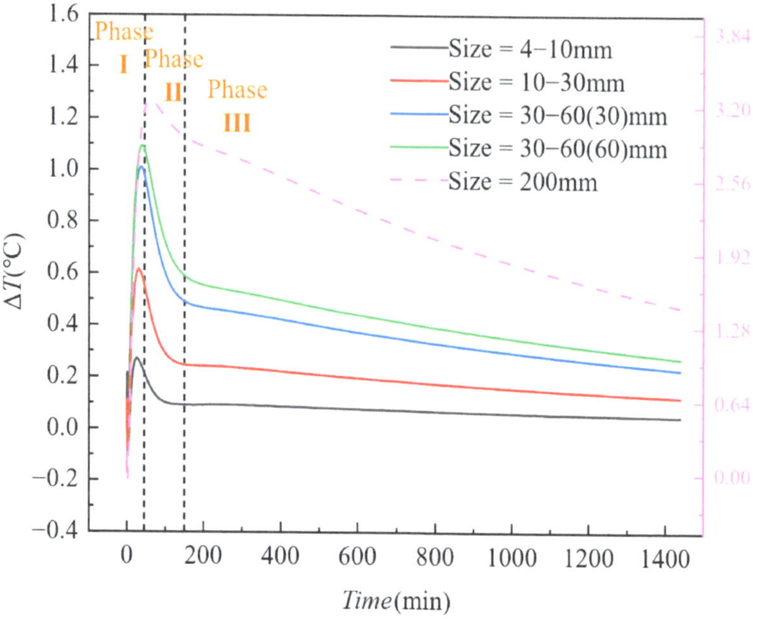

Figure 15. Thermal contrast trends of honeycomb defects in concrete columns under different honeycomb defect sizes.

Figure 16 illustrates the evolution pattern of infrared images of honeycomb defects with different sizes. Overall, the colors in the figure transition gradually from deep purple to yellow, reflecting the gradual increase in the formwork surface temperature, followed by a subsequent decrease. This temperature variation is primarily influenced by the differences in heat transfer between the defect and non-defect areas, as the presence of a honeycomb defect significantly alters the temperature distribution characteristics of the formwork surface. Additionally, honeycomb defects of different sizes exhibit noticeable differences in their thermal behavior. For smaller honeycomb defects (4–10 mm, 10–30 mm), during the hydration process, the color contrast in the defect areas of the infrared images is minimal, especially for defects in the 4–10 mm range, where only a slight color difference is visible in the early stages. For defects in the 10–30 mm range, the color contrast in the infrared images is slightly more noticeable compared to the 4–10 mm defects. For medium-sized honeycomb defects (30–60 mm), throughout the entire hydration stage, the color contrast in the defect areas of the infrared images is higher than that of small-sized defects, indicating a greater temperature variation between the defect and non-defect areas. Additionally, the defect remains visible in the infrared images for a longer period. For void defects (200 mm), throughout the entire hydration process, the color contrast in the defect areas of the infrared images is consistently significantly higher than that of defects of other sizes. This indicates that void defects strongly impede heat transfer, resulting in a pronounced increase in color contrast between defect and non-defect areas, with the thermal contrast being the most prominent. Finally, at the same point in time, the larger the size of the honeycomb defect, the more pronounced the thermal contrast between the defect area and the non-defect area, and the clearer the characteristics of the defect area appear in the infrared images.

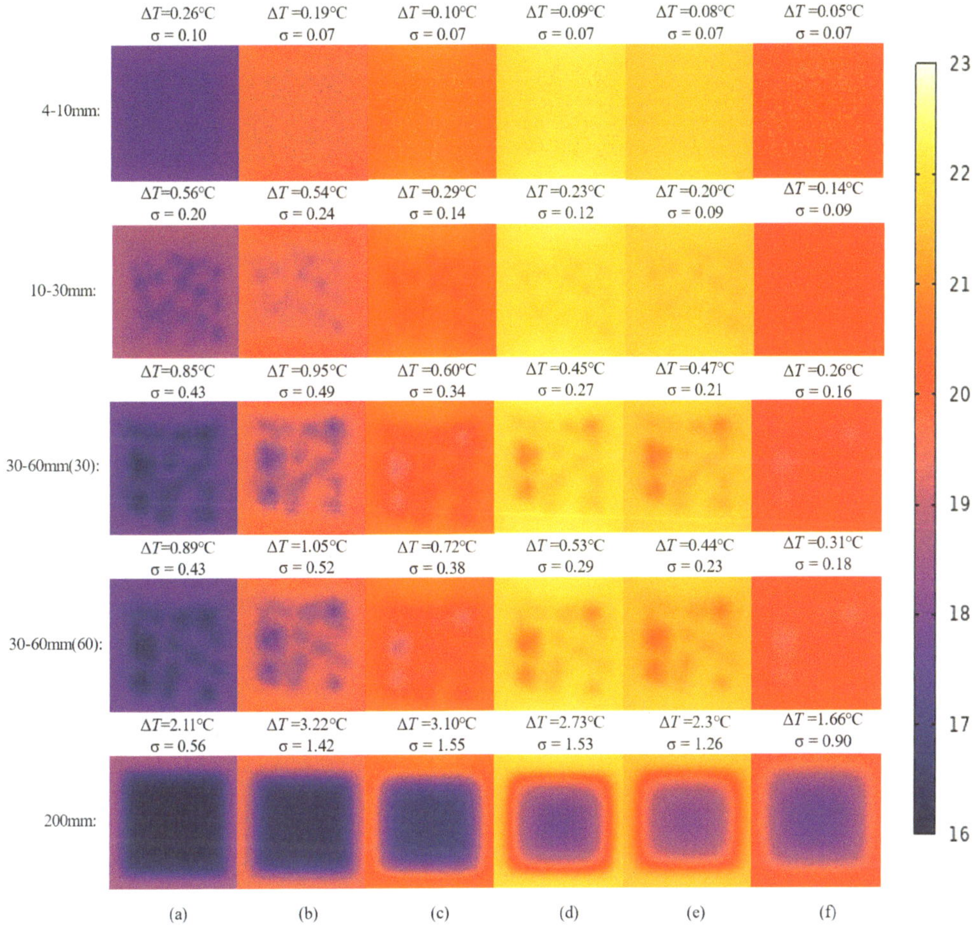

Figure 16. Infrared image evolution of the honeycomb defect with different honeycomb defect sizes: (**a**) t = 20 min; (**b**) t = 50 min; (**c**) t = 100 min; (**d**) t = 300 min; (**e**) t = 600 min; (**f**) t = 1200 min.

As evidenced by the temporal evolution characteristics in Figure 15, the critical time threshold for thermal contrast development is identified at 45 min post-casting, during which the thermal contrast ΔT reaches peak responses across varying honeycomb defect sizes. Figure 17 further elucidates the coupled size-gradient mechanism governing heat transfer: When the formwork surface-to-ambient temperature gradient ($Ts - Ta$) exceeds 2 °C, the fitted slope increases ninefold from 0.0913 to 0.8287 as defect size expands from 4 to 10 mm to 200 mm, confirming enhanced linear sensitivity in large-scale defects. Notably, sublinear growth (R^2 = 0.74–0.82) dominates small defects (<30 mm) under $Ts - Ta \in [0, 3]$ °C, whereas large defects (>30 mm) exhibit robust linearity ($R^2 > 0.96$) within the same gradient range. Dual-axis scaling analysis reveals a 12-fold amplification in peak ΔT (3.2 °C for 200 mm defects vs. 0.26 °C for 4–10 mm defects), demonstrating that defect size governs thermal contrast amplification effects.

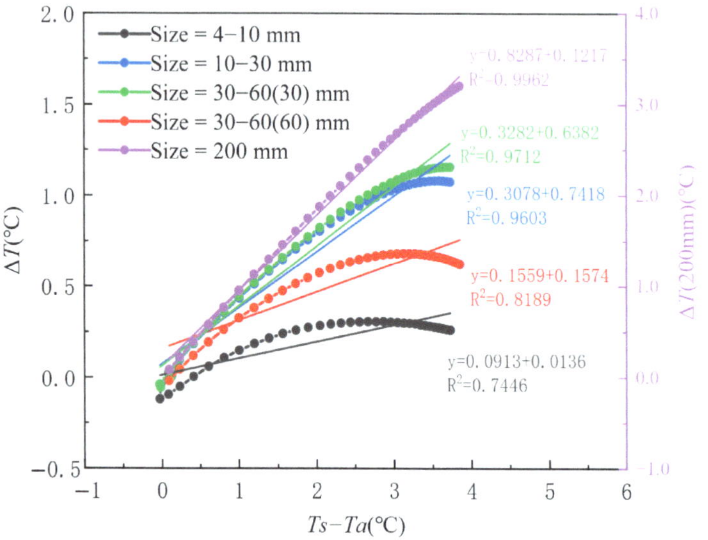

Figure 17. Response characteristics of thermal contrast (ΔT) to surface-to-ambient temperature gradient ($Ts - Ta$) under varying defect sizes during the initial casting phase (Phase I).

4.3. Influence of Plate Thickness on Thermal Contrast

This section discusses the relationship between the thermal contrast of honeycomb defect areas and non-defect areas under different formwork thicknesses. The study was conducted under the conditions of an initial concrete pouring temperature of 22.5 °C, an ambient temperature of 16 °C, and honeycomb particle sizes of 30–60 mm (with a thickness of 60 mm).

Figure 18 illustrates the trend of the thermal contrast (ΔT) of honeycomb defects over time under different formwork thicknesses. As shown in Figure 18a, after concrete pouring, the thermal contrast (ΔT) for formwork of different thicknesses initially increases sharply and reaches a peak (Phase I), followed by a rapid decrease (Phase II). After approximately 200 min, all curves begin to decrease gradually (Phase III), with the trends of the ΔT curves for formworks of different thicknesses becoming essentially similar. Furthermore, it is observed that the smaller the formwork thickness, the higher the peak value and the greater the final thermal contrast. Conversely, the larger the formwork thickness, the lower the peak value and the smaller the final thermal contrast. In Figure 18b, bar charts representing the maximum thermal contrast and the thermal contrast at t = 1200 min for different formwork thicknesses are presented. The figure demonstrates an inverse relationship between thermal contrast and formwork thickness. By fitting the relationship between maximum thermal contrast and formwork thickness, the following equation is obtained:

$$y = -0.0483 + 1.7717$$
$$R^2 = 0.9986$$

Figure 19 illustrates the simulated infrared temperature images at different times under various formwork thickness conditions. Overall, the colors in the figure transition from deep purple to light yellow and then back to a deeper shade, reflecting the gradual increase in the formwork surface temperature initially, followed by a subsequent decrease. The yellow areas correspond to higher surface temperatures of the formwork, while the deep purple areas correspond to lower temperatures. Secondly, for formworks with smaller thicknesses, the color contrast in the defect areas of the infrared images is greater, indicating that thinner formworks pose less resistance to heat transfer, resulting in more pronounced thermal contrast and making the defect areas easier to observe. In contrast, for thicker formworks, the color contrast in the defect areas of the infrared images is smaller, suggesting that thicker formworks provide stronger resistance to heat transfer, resulting in less pronounced thermal contrast and making the defect areas harder to detect.

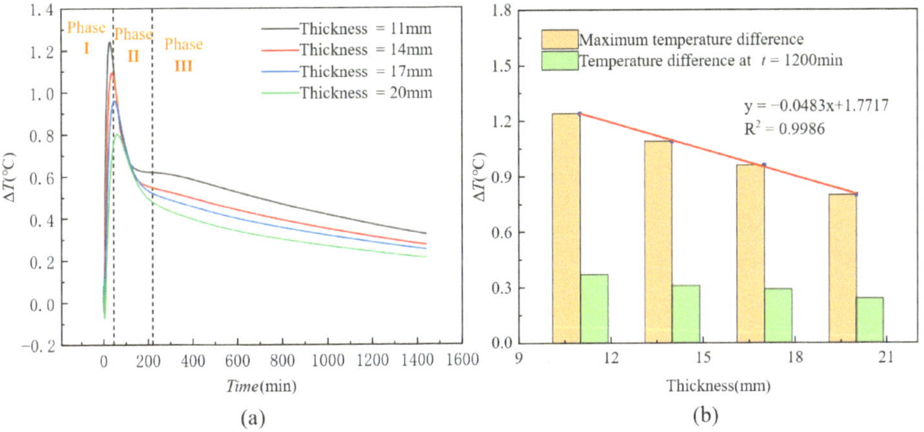

Figure 18. Influence of formwork thicknesses on thermal contrast: (**a**) variation in thermal contrast with the hydration process; (**b**) relationship between thermal contrast and formwork thickness.

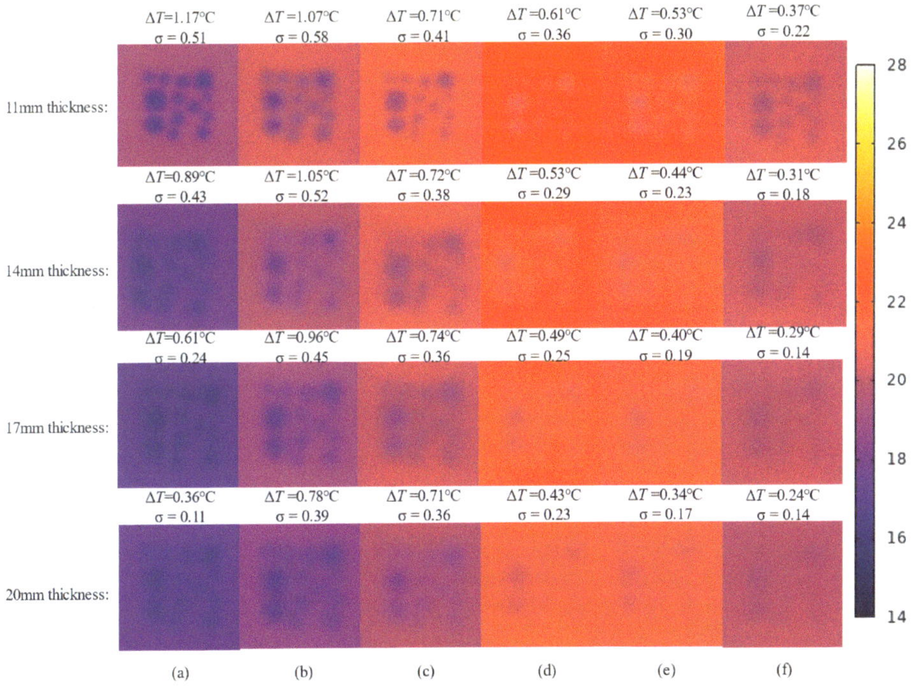

Figure 19. Evolution of infrared images of honeycomb defects under different formwork thicknesses: (**a**) t = 20 min; (**b**) t = 50 min; (**c**) t = 100 min; (**d**) t = 300 min; (**e**) t = 600 min; (**f**) t = 1200 min.

Temporal evolution analysis in Figure 18 reveals that 40 min post-concrete casting marks the critical time threshold for thermal contrast development, during which the thermal contrast ΔT reaches peak responses across varying formwork thickness conditions. Figure 20 further quantifies the inverse correlation between formwork thickness and heat transfer efficiency: Under a constant surface-to-ambient temperature gradient ($Ts - Ta$) of 3 °C, increasing formwork thickness from 11 mm to 20 mm reduces the thermal contrast sensitivity coefficient (slope of linear fit) by 21.4% (from 0.3552 to 0.2794), demonstrating that thickened formwork significantly impedes thermal flux transmission and thereby diminishes defect detection sensitivity. Furthermore, all thickness formworks exhibit robust linear responses within $Ts - Ta \in [0, 3.5]$ °C ($R^2 = 0.9566 - 0.9777$).

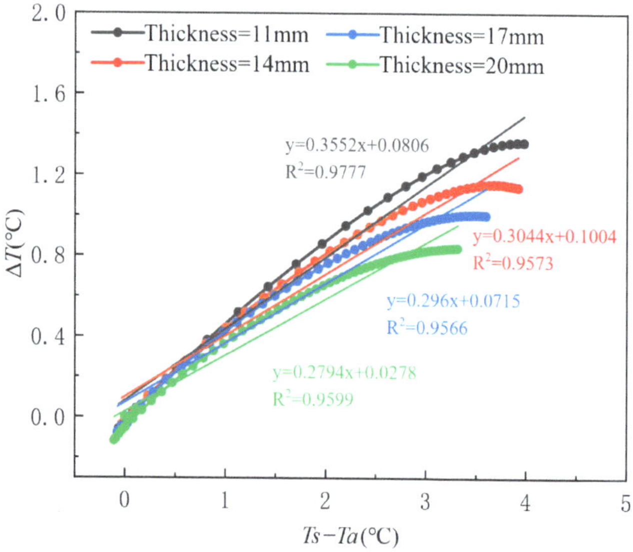

Figure 20. Response characteristics of thermal contrast (ΔT) to surface-to-ambient temperature gradient ($Ts - Ta$) under varying formwork thicknesses during the initial casting phase (Phase I).

4.4. Comprehensive Analysis and Technical Validation

4.4.1. Influence of Key Parameters on Thermal Contrast

Based on the experimental data analysis, the control mechanisms of thermal contrast (ΔT) by ambient temperature (Ta), defect size (D), and formwork thickness (L) can be summarized as follows:

1. Ambient Temperature: The extreme response of ΔT is strictly controlled by the temperature gradient between the formwork surface and the environment ($Ts - Ta$). When the absolute value of the temperature gradient exceeds 4 °C ($|Ts - Ta| > 4$ °C), the peak value of ΔT can reach ± 2 °C ($R^2 = 0.99$). Under low-temperature conditions ($Ta < Tc$), ΔT is primarily dominated by heat dissipation from the environment ($Tc - Ta > 0$), whereas under high-temperature conditions ($Ta > Tc$), the accumulation of hydration heat takes over ($Tc - Ta < 0$).
2. Defect Size: The sensitivity of ΔT exhibits a nonlinear increase with defect size. When $D \geq 30$ mm, the sensitivity coefficient (k) increases sharply from 0.0913 to 0.8287 (a 9-fold increase), with the peak value of ΔT reaching 3.2 °C (12 times higher than for D = 4–10 mm). Large defects ($D \geq 30$ mm) show a strong linear response within

the temperature gradient range of $Ts - Ta \in [1, 4]$ °C ($R^2 > 0.96$), while small defects (D < 30 mm) exhibit a significantly reduced linear response ($R^2 = 0.74$–0.82).

3. Formwork Thickness: The detection sensitivity of ΔT is negatively correlated with formwork thickness. As the thickness increases from 11 mm to 20 mm, the sensitivity coefficient (k) decreases from 0.3552 to 0.2794 (a 21.4% reduction).

   In summary, the temperature gradient between the formwork surface and the environment is the primary driving force of ΔT. The defect size determines the thermal signal strength of both the defect and non-defect areas, while the formwork thickness governs the attenuation of detection sensitivity. Together, these three factors define the technical boundary for infrared detection of concrete honeycomb defects during the initial casting phase.

4.4.2. Analysis of the Optimal Detection Window Based on Temporal Evolution

The time-varying characteristics of thermal contrast (ΔT) reveal the key time constraints for infrared detection. As shown in Figures 12a and 15, and Figure 18, the evolution of ΔT under different conditions exhibits a three-stage pattern: "rapid rise—peak maintenance—slow decay". The time threshold analysis is as follows:

1. Effect of Ambient Temperature: Under low-temperature ($Ta \leq 16$ °C) and high-temperature ($Ta \geq 28$ °C) conditions, ΔT reaches positive (+2.1 °C) and negative (−1.8 °C) peak values 50 ± 5 min after pouring. When the ambient temperature is close to the initial concrete temperature (Tc), the peak value is delayed to around 200 min.
2. Effect of Defect Size: For small defects (D < 30 mm), the ΔT peak occurs earlier, between 30 and 35 min, while for large defects (D ≥ 30 mm), the peak is delayed to 50 to 60 min.
3. Effect of Formwork Thickness: As the formwork thickness increases, the time to reach the ΔT peak is delayed. For example, with a formwork thickness of 11 mm, the peak occurs at 35 min, while with a thickness of 20 mm, the peak is delayed to 65 min.

In conclusion, considering the combined temporal characteristics of various factors, the optimal window for infrared detection of concrete honeycomb defects is between 35 and 60 min after pouring. During this period, the thermal contrast is close to its peak across different conditions, making it the most favorable phase for defect detection.

4.4.3. Error and Uncertainty Analysis and Future Research Directions

This study demonstrates the theoretical feasibility of using infrared detection technology for identifying concrete defects through idealized models. However, its practical application still faces the following key limitations:

1. Incomplete Decoupling of Environmental Coupling Effects: In actual engineering sites, environmental conditions are complex and variable. High wind speeds can cause significant convective heat loss, which in turn affects the accuracy of temperature measurements by infrared devices and reduces the thermal contrast (ΔT). In addition, fluctuations in solar radiation may induce extra temperature variations, further impacting both the precision of infrared temperature measurements and the subsequent defect identification.
2. Idealization Deviations of Material and Model Parameters: In this study, both experimental and simulation models assumed that material property parameters are uniformly distributed. However, in practice, discrepancies exist between the actual thermal conductivities—such as the uniformity of heat conduction in formwork materials and concrete—and those assumed in the models. These differences can lead to errors in predicting the thermal contrast (ΔT). In particular, the presence of air gaps

between the formwork and concrete may introduce additional thermal resistance, resulting in deviations between the predicted ΔT peak and the actual condition.

3. Limitations in Dynamic Equipment Detection: The thermal sensitivity of infrared cameras poses challenges for detecting small-scale defects within thicker formworks. Moreover, the operational complexity of infrared cameras limits their practical application on concrete pouring sites.

To address these challenges, future research should focus on the following directions:

1. Multi-Physical Field Coupled Modeling: Future studies should integrate complex experimental conditions—including ambient temperature, humidity, wind speed, and solar radiation intensity—to conduct experiments and simulations under multiple coupled conditions. The primary focus should be on investigating the mechanisms and performance of infrared detection of honeycomb defects when influenced by multiple factors simultaneously.
2. Refined Characterization of Materials and Interfaces: Based on practical engineering requirements, further research should be conducted on the material parameters and interfacial heat transfer characteristics. The goal is to elucidate the influence of material properties and interfacial heat transfer on the infrared detection of concrete honeycomb defects, thereby achieving more accurate prediction outcomes.
3. Upgrading Intelligent Detection Systems: Develop an unmanned aerial vehicle (UAV) platform equipped with a high-frame-rate infrared module. By integrating image enhancement techniques and deep learning-based image fusion methods, the signal-to-noise ratio of the thermal contrast (ΔT) can be improved, thereby enhancing the precision and reliability of infrared detection.

5. Conclusions

This study investigated honeycomb defects in concrete using infrared thermography during the hydration heat process. Finite element modeling (FEM) was employed to simulate the thermal conduction of concrete containing honeycomb defects, exploring the feasibility of defect detection via the heat generated during hydration. Infrared testing verified the accuracy and reliability of the model, ensuring the applicability of the simulation results. Furthermore, the study analyzed the effects of ambient temperature, formwork thickness, and defect size on defect detection. The experimental results show that ambient temperature, defect size, and formwork thickness have significant impacts on thermal contrast, with different environmental conditions and material parameter combinations leading to varying thermal contrast responses. Specifically, when the temperature difference between the ambient temperature and the initial concrete temperature is large, the peak value of thermal contrast is more pronounced. Large defects generate significantly higher thermal contrast than small defects, and thinner formworks provide higher thermal contrast, which is beneficial for defect detection. By analyzing the temporal evolution characteristics of thermal contrast, the optimal time window for infrared detection was determined to be between 35 and 60 min after concrete pouring. Although this study demonstrates the theoretical feasibility of infrared thermography, practical applications still face challenges such as environmental coupling effects, idealization deviations of material parameters, and limitations in equipment detection capabilities. Future research should focus on multi-physical field coupled modeling, refined characterization of material and interface heat transfer properties, and the upgrading of intelligent detection systems to enhance the accuracy and reliability of infrared detection.

Author Contributions: F.Y.: Project formal analysis, funding acquisition, writing—original draft, review and editing. X.Z.: Methodology, data curation, formal analysis, writing—original draft, writing—review and editing. Q.X.: Validation, writing—review and editing. L.Y.: Funding acquisition, writing—review and editing. H.C.: Data curation, writing—review and editing. C.C.: Conceptualization, project administration, funding acquisition, writing—review and editing. All authors have read and agreed to the published version of the manuscript.

Funding: This work was partially supported by the National Natural Science Foundation of China (52478301), the Chongqing Natural Science Foundation of China (CSTB2022NSCQ-MSX1379), the Special Support Program of Chongqing Postdoctoral Research (2021XMT007), and the China Postdoctoral Science Foundation under Grant Number 2024T171100.

Institutional Review Board Statement: Not applicable.

Informed Consent Statement: Not applicable.

Data Availability Statement: The raw data supporting the conclusions of this article will be made available by the authors on request.

Acknowledgments: The authors would like to thank the State Key Laboratory of Mountain Bridge and Tunnel Engineering and the No. 3 Construction Co., Ltd., of Chongqing Construction Engineering Group for its invaluable support in the experimental phase of this research.

Conflicts of Interest: Authors Fan Yang and Qilong Xia are employed by the No.3 Construction Co., Ltd., of Chongqing Construction Engineering Group. The remaining authors declare that the research was conducted in the absence of any commercial or financial relationships that could be construed as a potential conflict of interest.

References

1. Liew, M.S.; Nguyen-Tri, P.; Nguyen, T.A.; Kakooei, S. (Eds.) *Smart Nanoconcretes and Cement-Based Materials: Properties, Modelling and Applications*; Elsevier: Amsterdam, The Netherlands, 2019. [CrossRef]
2. de Brito, J.; Poon, C.S.; Zhan, B. New trends in recycled aggregate concrete. *Appl. Sci.* **2019**, *9*, 2324. [CrossRef]
3. Maruyama, I.; Lura, P. Properties of early-age concrete relevant to cracking in massive concrete. *Cem. Concr. Res.* **2019**, *123*, 105770. [CrossRef]
4. Nguyen, T.T.; Waldmann, D.; Bui, T.Q. Computational chemo-thermo-mechanical coupling phase-field model for complex fracture induced by early-age shrinkage and hydration heat in cement-based materials. *Comput. Methods Appl. Mech. Eng.* **2019**, *348*, 1–28. [CrossRef]
5. Jain, H.; Patankar, V.H.; Kulkarni, M.S. *Ultrasonic Testing of Concrete Structures: A Review*; Bhabha Atomic Research Centre: Mumbai, India, 2021.
6. Fischer, O.; Thoma, S.; Crepaz, S. Distributed fiber optic sensing for crack detection in concrete structures. *Civ. Eng. Des.* **2019**, *1*, 97–105. [CrossRef]
7. Sirca, G.F., Jr.; Adeli, H. Infrared thermography for detecting defects in concrete structures. *J. Civ. Eng. Manag.* **2018**, *24*, 508–515. [CrossRef]
8. Monsberger, C.M.; Lienhart, W. Distributed fiber optic shape sensing of concrete structures. *Sensors* **2021**, *21*, 6098. [CrossRef]
9. Hashimoto, K.; Shiotani, T.; Ohtsu, M. Application of impact-echo method to 3D SIBIE procedure for damage detection in concrete. *Appl. Sci.* **2020**, *10*, 2729. [CrossRef]
10. Zhao, B.; Lei, D.; Fu, J.; Yang, L.; Xu, W. Experimental study on micro-damage identification in reinforced concrete beam with wavelet packet and DIC method. *Constr. Build. Mater.* **2019**, *210*, 338–346. [CrossRef]
11. Janků, M.; Cikrle, P.; Grošek, J.; Anton, O.; Stryk, J. Comparison of infrared thermography, ground-penetrating radar and ultrasonic pulse echo for detecting delaminations in concrete bridges. *Constr. Build. Mater.* **2019**, *225*, 1098–1111. [CrossRef]
12. Jiao, L.; Ye, Q.; Cao, X.; Huston, D.; Xia, T. Identifying concrete structure defects in GPR image. *Measurement* **2020**, *160*, 107839. [CrossRef]
13. McCabe, T.; Erdogmus, E.; Kodsy, A.; Morcous, G. Early detection of honeycombs in concrete pavement using GPR. *J. Perform. Constr. Facil.* **2021**, *35*, 04020138. [CrossRef]
14. Chow, J.K.; Liu, K.F.; Tan, P.S.; Su, Z.; Wu, J.; Li, Z.; Wang, Y.H. Automated defect inspection of concrete structures. *Autom. Constr.* **2021**, *132*, 103959. [CrossRef]

15. Völker, C.; Shokouhi, P. Multi-sensor data fusion approach for automatic honeycomb detection in concrete. *NDT E Int.* **2015**, *71*, 54–60. [CrossRef]
16. Xu, J.; Yu, X. Detection of concrete structural defects using impact echo based on deep networks. *J. Test. Eval.* **2021**, *49*, 109–120. [CrossRef]
17. Rocha, J.A.; Póvoas, Y.V. Infrared thermography as a non-destructive test for the inspection of reinforced concrete bridges: A review of the state of the art. *Rev. Alconpat* **2017**, *7*, 200–214. [CrossRef]
18. Cheng, C.C.; Cheng, T.M.; Chiang, C.H. Defect detection of concrete structures using both infrared thermography and elastic waves. *Autom. Constr.* **2008**, *18*, 87–92. [CrossRef]
19. Vemuri, S.H.; Atadero, R.A. Case study on rapid scanning techniques for concrete bridge decks with asphalt overlay: Ground-penetrating radar and infrared thermography. *Pract. Period. Struct. Des. Constr.* **2017**, *22*, 05016004. [CrossRef]
20. Cheng, C.; Shen, Z. Semi real-time detection of subsurface consolidation defects during concrete curing stage. *Constr. Build. Mater.* **2021**, *270*, 121489. [CrossRef]
21. Hong, M.; Lei, D.; Hu, F.; Chen, Z. Assessment of void and crack defects in early-age concrete. *J. Build. Eng.* **2023**, *70*, 106372. [CrossRef]
22. Cheng, C.; Chen, D.; Shao, S.; Na, R.; Cai, H.; Zhou, H.; Wu, B. Revealing the Impact of Depth and Surface Property Variations on Infrared Detection of Delamination in Concrete Structures Under Natural Environmental Conditions. *Buildings* **2024**, *15*, 10. [CrossRef]
23. Chen, H.L.R.; Mardmomen, S.; Leon, G. On-site measurement of heat of hydration of delivered mass concrete. *Constr. Build. Mater.* **2021**, *269*, 121246. [CrossRef]
24. Garg, S.; Misra, S. Efficiency of NDT techniques to detect voids in grouted post-tensioned concrete ducts. *Nondestruct. Test. Eval.* **2021**, *36*, 366–387. [CrossRef]
25. Li, S.; Han, S.; Wang, J.; Han, X.; Zheng, P.; Cui, C.; Sun, S. Infrared thermography detection of grouting defects in external post-tensioned tendon ducts under construction hydration heat excitation. *NDT E Int.* **2023**, *134*, 102785. [CrossRef]
26. Wan, R.; Zhao, H.; Zhang, Q.; Zheng, P.; Zhong, Z.; Xue, C.; Huang, L. Semi-real-time infrared thermography for detecting layering defects in plasters solidification within indoor environments. *Front. Mater.* **2024**, *11*, 1405745. [CrossRef]
27. Cai, H.; Cheng, C.; Na, R.; Zhang, H.; Zhou, J.; Jing, S.; Miao, C. Cooling-excited infrared thermography for enhancing the detection of concrete filled steel tube interfacial debonding at concrete hydration. *Case Stud. Constr. Mater.* **2024**, *20*, e02995. [CrossRef]
28. Cheng, C.; Cheng, X.; Zhang, H.; Cai, H.; Zhou, J.; Na, R.; Wu, B. Experimental study on infrared detection of debonding in concrete-filled steel tubular structure under acceleratory period of hydration heat action. *Case Stud. Constr. Mater.* **2024**, *21*, e03928. [CrossRef]
29. Zhu, B. Temperature stress and temperature control of mass concrete. *J. Civ. Eng.* **1999**, *4*, 15.
30. Liu, Z. Study on the Temperature Field of Concrete-Filled Steel Tube Arch Rib Cross-Section. Ph.D. Thesis, Fuzhou University, Fuzhou, China, 2006.
31. Lin, C.; Zheng, J.; Huang, H. Experimental study on the heat of hydration temperature field of concrete-filled steel tube arch with circular cross-section. *Concrete* **2009**, *10*, 13–15.
32. Lin, C.; Zheng, J.; Qin, R. Analysis of the influence of hydration heat on the forming process of concrete-filled steel tube arch ribs. *J. Guangxi Univ. Nat. Sci. Ed.* **2007**, *32*, 186–188.
33. Zhu, B. *Temperature Stress and Temperature Control of Mass Concrete*; China Water & Power Press: Beijing, China, 2012.
34. *JGJ162*; Technical Code for Safety of Forms in Construction. Ministry of Housing and Urban-Rural Development of the People's Republic of China: Beijing, China, 2008.

Disclaimer/Publisher's Note: The statements, opinions and data contained in all publications are solely those of the individual author(s) and contributor(s) and not of MDPI and/or the editor(s). MDPI and/or the editor(s) disclaim responsibility for any injury to people or property resulting from any ideas, methods, instructions or products referred to in the content.

materials

MDPI

Article

Accelerating Multi-Objective Optimization of Composite Structures Using Multi-Fidelity Surrogate Models and Curriculum Learning

Bartosz Miller † and Leonard Ziemiański *,†

Faculty of Civil and Environmental Engineering and Architecture, Rzeszów University of Technology, Al. Powstancow Warszawy 12, 35-959 Rzeszow, Poland; bartosz.miller@prz.edu.pl

* Correspondence: ziele@prz.edu.pl; Tel.: +48-17-865-1353

† These authors contributed equally to this work.

Abstract: The optimization of multilayer composite structures requires balancing mechanical performance, economic efficiency, and computational feasibility. This study introduces an innovative approach that integrates Curriculum Learning (CL) with a multi-fidelity surrogate model to enhance computational efficiency in engineering design. A multi-fidelity strategy is introduced to generate training data efficiently, leveraging a high-fidelity finite element model for accurate simulations and a low-fidelity model to provide a larger dataset at reduced computational cost. Unlike conventional surrogate modeling approaches, the proposed method applies CL to iteratively refine the surrogate model, enabling step-by-step learning of complex structural patterns and improving prediction accuracy. Genetic algorithms (GAs) are then applied to optimize structural parameters while minimizing computational expense. The integration of CL and multi-fidelity modeling allows for a reduction in computational burden while preserving accuracy, demonstrating practical applicability in real-world structural design problems. The effectiveness of this methodology is validated by evaluating Pareto front quality using selected performance indicators. Results demonstrate that the proposed approach reduces optimization burden while achieving accurate predictions, highlighting the benefits of integrating surrogate modeling, multi-fidelity analysis, CL, and GAs for efficient composite structure optimization. This work contributes to the advancement of optimization methodologies by providing a scalable framework applicable to complex engineering problems requiring high computational efficiency.

Keywords: multi-objective optimization; composite; multi-fidelity models; surrogate models; deep neural networks; genetic algorithms; curriculum learning

Academic Editors: Patryk Rozylo, Katarzyna Falkowicz and Pawel Wysmulski

Received: 27 February 2025
Revised: 20 March 2025
Accepted: 24 March 2025
Published: 26 March 2025

Citation: Miller, B.; Ziemiański, L. Accelerating Multi-Objective Optimization of Composite Structures Using Multi-Fidelity Surrogate Models and Curriculum Learning. *Materials* **2025**, *18*, 1469. https://doi.org/10.3390/ma18071469

Copyright: © 2025 by the authors. Licensee MDPI, Basel, Switzerland. This article is an open access article distributed under the terms and conditions of the Creative Commons Attribution (CC BY) license (https://creativecommons.org/licenses/by/4.0/).

1. Introduction

Multilayer composite structures are widely used in various industries, including aerospace, automotive, and construction, where the optimization of parameters such as load-bearing capacity, fatigue resistance, and vibration damping is essential [1–4]. The design of multilayer composite structures poses a significant engineering challenge, particularly in the selection of materials with different mechanical properties and economic factors [5]. This process requires precise selection of materials for individual layers so that the resulting structure meets specified strength, technological, and operational requirements while minimizing costs. The complexity of this issue makes classical single-optimization methods insufficient, leading to the necessity of applying a multi-criteria approach that allows the simultaneous analysis of multiple design aspects, including both mechanical and economic parameters.

One of the effective tools for solving complex optimization problems is the application of genetic algorithms (GAs) [6–9]. These methods enable the efficient exploration of the solution space to minimize the objective function while considering both mechanical and economic constraints. Genetic algorithms, inspired by evolutionary processes, allow for the iterative improvement of solutions through selection, crossover, and mutation of a population of solutions. However, the use of GAs involves repeated objective function evaluations, which, in the case of advanced analyses based on the Finite Element Method (FEM), leads to high computational costs.

To reduce computational time, surrogate models are used to quickly approximate FEM analysis results [10–12]. A key challenge is building a sufficient dataset to train the surrogate model. A multi-fidelity (MF) approach is often applied in this process, utilizing FEM models with varying levels of accuracy [13]. Lower-resolution numerical models (low-fidelity, LF) enable fast data generation but require correction techniques to improve their reliability. Various methods exist for integrating results obtained from models of different fidelity levels, including statistical methods and machine learning-based algorithms. The choice of an appropriate strategy affects both the training time and the accuracy of the surrogate model.

Different methods, such as Kriging and co-Kriging, statistical models, and deep neural networks, are commonly used to develop surrogate models [11,12,14–20]. Each of these approaches offers varying levels of accuracy and computational complexity, making it crucial not only to assess the effectiveness of the surrogate model itself but also its impact on optimization results. Kriging, as one of the widely used interpolation methods, enables the precise modeling of nonlinear relationships between input parameters and FEM analysis results. On the other hand, deep neural networks can model highly complex nonlinear dependencies but require substantial computational resources and large training datasets [21–23].

The concept of Curriculum Learning (CL) originates from human cognition, where learning involves acquiring knowledge through exposure to successive samples in a structured sequence—progressing from the simplest to the most complex examples. This idea was first introduced into machine learning by Bengio et al. [24] in 2009, who suggested that such an approach not only accelerates the training process but also improves the quality of the local minima obtained. In subsequent years, the concept of progressive learning was applied across various domains with numerous modifications and extensions. Despite these differences, a common feature of these approaches is the emphasis on refining models by focusing on increasingly challenging or problematic examples. For instance, Hsiao and Chang [25] utilized CL for constructing surrogate models to describe chemical processes (namely, an amine scrubbing process), demonstrating its effectiveness in improving model accuracy.

The concept has also been extended to continual learning, where models are incrementally trained on new data while retaining previously acquired knowledge, as explored by Fayek et al. [26]. Another advanced adaptation, referred to as adaptive continual learning, was proposed by Kong et al. [27], in which each learning step was dynamically adjusted based on results obtained in preceding steps, further enhancing model performance. A notable application of CL to highly complex systems includes its use in modeling unsteady hypersonic flows in chemical nonequilibrium, as demonstrated by Scherding et al. [28]. This study highlights the potential of curriculum-based approaches in computationally demanding simulations, reinforcing the broader utility of this learning paradigm across various scientific and engineering disciplines.

The effectiveness of CL has been demonstrated across various complex applications, where structured, progressive training significantly enhances model performance. In many challenging problems, the approach described in this study as CL provides substantial improvements in result quality while simultaneously reducing computational effort, making it a valuable tool for optimizing surrogate modeling and numerical simulations.

The verification of solution quality cannot rely solely on comparing the surrogate model with FEM results but must also consider the correctness of the optimization at a global level. This necessitates evaluating the influence of the surrogate model on results obtained using GA-based optimization. Furthermore, the optimization process must account for modeling uncertainties, which may require applying probabilistic methods for result analysis.

Due to the multi-criteria nature of the problem and the need for a continuous comparison of optimization results, appropriate indicators are used to assess solution quality [29,30]. One of the widely used tools is the analysis of Pareto fronts, enabling the evaluation of trade-offs between different optimization criteria. The Pareto front identifies a set of non-dominated solutions, where each represents an optimal compromise between multiple optimization objectives. This allows for determining optimal material configurations and evaluating their effectiveness concerning predefined design criteria.

Addressing these challenges is a key aspect of effective composite structure optimization, enabling the development of design methodologies that provide optimal solutions both in technical and economic terms. By integrating computational methods, optimization algorithms, and machine learning techniques, it is possible to create more efficient tools to support the design process of multilayer composites. Modern optimization approaches also consider sustainability and environmental constraints, which can serve as additional factors in the design analysis.

Queipo et al. [31] explored surrogate-based methods for analysis and optimization, addressing key aspects such as loss function selection, regularization criteria, experimental design strategies, sensitivity analysis, and convergence assessment. Their study also illustrated state-of-the-art applications through a multi-objective optimization case involving a liquid rocket injector.

In their comprehensive review, Forrester and Keane [12] examined the latest advancements in surrogate model construction and their integration into optimization strategies. Their work provided a detailed evaluation of the advantages and limitations of various surrogate modeling techniques, offering practical guidelines for their implementation. Additionally, Hwang and Martins [32] analyzed the behavior of several popular surrogate modeling approaches when applied to problems requiring thousands of training samples.

The optimization of the dynamic behavior of shell structures has been widely studied, with numerous algorithms proposed in recent research to tackle this challenge. For example, Jing et al. [33] introduced a sequential permutation search algorithm aimed at optimizing the stacking sequence of doubly curved laminated composite shallow shells to maximize the fundamental frequency. Similarly, Chen et al. [34] developed a multivariate improved sparrow search algorithm to enhance the fundamental frequency of composite laminated cylindrical shells while minimizing vibrational resonance. Chaudhuri et al. [35] performed a numerical investigation into the free vibration response of composite stiffened hypar shells with cutouts, utilizing an FE analysis. Their optimization relied on parametric tuning based on the Taguchi approach to achieve the desired frequency response. Another study by Serhat [36] focused on optimizing the eigenfrequencies of circular cylindrical laminates by examining the influence of parameters such as cylinder length, radius, thickness, and boundary conditions. Likewise, Alshabatat [37] explored the optimization of natural frequencies in circular cylindrical shells using axially functionally graded materials. Collectively, these studies contribute to advancing optimization methodologies for improving the dynamic performance of composite structures.

This study aims to address the challenges associated with optimizing the dynamic properties of multilayer composite structures while minimizing computational costs. The proposed methodology integrates MF FE models with deep neural network-based surrogate modeling, enabling efficient and accurate multi-objective optimization.

The novelty of this research lies in the systematic use of surrogate models within a CL framework specifically tailored for multi-objective optimization. Unlike traditional surrogate modeling approaches, where training is performed using a fixed dataset, the proposed method dynamically improves the surrogate model by incorporating new high-fidelity samples in successive CL iterations. This iterative refinement enhances the predictive accuracy of the surrogate model while ensuring better convergence of the optimization process. By progressively increasing the quality of the surrogate model, the CL-based approach enables a more reliable identification of the Pareto front, leading to improved trade-off solutions between competing objectives while maintaining computational efficiency.

Furthermore, this study explores different architectures for the surrogate model, comparing three distinct configurations. The effectiveness of these variants is assessed using Pareto front quality indicators, providing a comprehensive evaluation of their impact on optimization performance.

By incorporating these innovations, the proposed methodology offers a robust and scalable solution for optimizing composite structures, demonstrating its applicability to engineering problems requiring a balance between accuracy and computational feasibility.

2. Materials and Methods

2.1. Vibration Problem

In dynamic structural analysis, an essential issue is determining the system's natural frequencies and mode shapes. The equation of motion describing the system's dynamics can be written as:

$$\mathbf{M}\ddot{\mathbf{x}} + \mathbf{C}\dot{\mathbf{x}} + \mathbf{K}\mathbf{x} = \mathbf{P} \tag{1}$$

where

$\mathbf{M}$ is a $n \times n$ mass matrix;
$\mathbf{C}$ is a $n \times n$ damping matrix;
$\mathbf{K}$ is a $n \times n$ stiffness matrix;
$\mathbf{x}$ is a n-element vector of nodal displacements;
$\mathbf{P}$ is a n-element vector of external forces at nodes;
n is the number of dynamic degrees of freedom.

For the free-vibration analysis, when external forces are absent and damping is neglected, the equation simplifies to:

$$\mathbf{M}\ddot{\mathbf{x}} + \mathbf{K}\mathbf{x} = \mathbf{0} \tag{2}$$

Solving this system leads to the so-called eigenvalue problem:

$$\mathbf{K}\boldsymbol{\Phi} = \mathbf{M}\boldsymbol{\Phi}\boldsymbol{\Omega}^2 \tag{3}$$

where $\boldsymbol{\Phi}$ represents the matrix of mode shapes ϕ_i (stored in successive columns of matrix $\boldsymbol{\Phi}$), and $\boldsymbol{\Omega}$ is the matrix of eigenvalues ω_i. The angular frequencies ω_i divided by 2π yield the natural frequencies f_i corresponding to the vibration shapes ϕ_i. Determining the system's eigenvalues and eigenvectors allows for the analysis of the dynamic properties of the structure, which is crucial for designing and optimizing structures subjected to dynamic excitation.

2.2. Analysis of Dynamic Parameters to Avoid the Resonance Phenomenon

In the analysis of structures subjected to dynamic loads, a key aspect is optimizing their dynamic properties to prevent resonance, which can lead to catastrophic consequences. Resonance occurs when the excitation frequency coincides with or is very close to one of the system's natural frequencies, resulting in a rapid increase in vibration amplitude, which may lead to structural failure. To avoid this, it is necessary to shape the system's natural frequency spectrum appropriately at the design stage.

If the excitation frequency is known, optimizing the natural frequency spectrum involves maximizing the separation of natural frequencies from this value, creating a frequency gap around the excitation frequency. This approach significantly reduces the risk of resonance. It is also crucial for low-stress structures, where even minor vibrations can cause premature damage or degradation of functional properties.

Shaping the natural frequency spectrum can be performed as part of an optimization procedure with a properly defined objective function. In its basic form, the objective function can be formulated to maximize the distance between the natural frequencies and the excitation frequency:

$$g_f(\boldsymbol{p}) = -\min|\boldsymbol{f}(\boldsymbol{p}) - f_{\text{exc}}|, \tag{4}$$

where the vector $\boldsymbol{f}(\boldsymbol{p})$ gathers the natural frequencies of the investigated model obtained for specific values of design parameters collected in vector $\boldsymbol{p}$, and f_{exc} stands for the considered excitation frequency. In this paper, $f_{\text{exc}} = 80$ Hz.

If an additional criterion, such as minimizing the structure's cost, is considered, the optimization problem becomes multi-objective. In this case, the second objective function can be expressed as:

$$g_c(\boldsymbol{p}) = \sum_{i=1}^{8}\left(\frac{V(\boldsymbol{p})}{8} \cdot c_i\right) \tag{5}$$

where $V(\boldsymbol{p})$ is the total volume of the structure, and c_i is the cost per unit volume of the material used in layer i.

In this case, the multi-objective optimization aims to minimize both objective functions simultaneously. The standard formulation of the multi-objective optimization problem—finding the values of the arguments collected in an m-element structure's control parameters vector $\boldsymbol{p}$ for which two considered objective functions yield the lowest possible values—is given by:

$$\boldsymbol{p}^{\text{opt}} = \underset{\boldsymbol{p}\in\mathbb{P}^m}{\arg\min}\left\{g_f(\boldsymbol{p}), g_c(\boldsymbol{p})\right\}, \tag{6}$$

where $\boldsymbol{p}$ is a vector of structure parameters, and $\mathbb{P}^m$ is the m-dimensional space of the decision parameters gathered in vector $\boldsymbol{p}$.

The solution to the multi-objective optimization problem is the so-called Pareto front. The Pareto front represents the set of non-dominated solutions in a multi-objective optimization problem. A solution is considered non-dominated if no other solution exists that improves one objective without worsening at least one other. In practical applications, the Pareto front provides decision-makers with a range of optimal trade-offs between competing objectives, allowing for a more informed selection of the most suitable design configuration.

To compare results obtained from different optimization approaches, numerical measures of Pareto front quality must be introduced. Pareto front indicators assess the distribution and diversity of solutions. One example is the hypervolume indicator, which measures the volume of space enclosed by the Pareto front concerning a reference point. The greater the value of this indicator, the better the quality of the obtained solutions regarding the distribution of trade-offs among objective functions. Another commonly used metric is the distance of generated solutions from the theoretically optimal Pareto solution, which helps evaluate the accuracy of the optimization process.

Considering these aspects in the design process allows us to obtain a system with optimized dynamic properties while simultaneously minimizing production costs and reducing the risk of damage due to uncontrolled dynamic excitations.

2.3. The Analyzed Structure

The axisymmetric structure analyzed in this study was generated by rotating a flat hyperbola (marked with blue in Figure 1) around a fixed axis. This hyperbola had predefined fixed start point A and end point C (namely, $r_1 = 61.03$ cm, $r_2 = 73.236$ cm, and the overall length equaled 600 cm), while its middle point B could change its position (given by d parameter) along the axis perpendicular to the axis of rotation, allowing control over the shape of the generated shell. This geometry enabled a broad range of structural configurations with varying dynamic and mechanical properties.

The shell was asymmetrically supported—one end was fixed, meaning all degrees of freedom are constrained, while the other end remained free. These boundary conditions led to specific dynamic properties of the structure, directly affecting its natural frequency spectrum and susceptibility to resonance phenomena. The structure is shown in Figure 1.

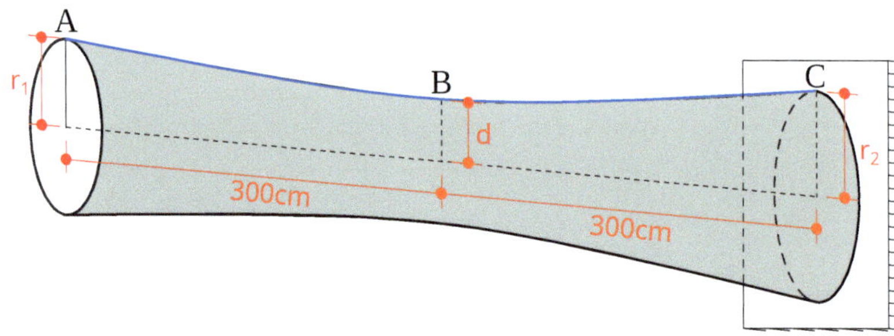

Figure 1. The analyzed structure: cantilever axisymmetric hyperboloid with varying middle-length diameter d.

The analyzed structure was made of a composite material with a constant thickness of 16 mm and consisted of eight layers. Each layer had the same thickness but could be made from one of three available composite materials. Additionally, each layer had a unique fiber orientation, meaning that the orientation of fibers in each layer differed, affecting the mechanical and dynamic properties of the shell.

The complete configuration of the structure was described by the parameter vector $\boldsymbol{p}$, which consisted of $m = 17$ variables: eight fiber orientation angles λ_i, material selections for each of the eight layers μ_i, and one coordinate defining the position of the middle point of the base hyperbola d; see Equation (7). This set of parameters allowed for a precise modeling of the shell and its optimization concerning various criteria, including structural dynamics, stiffness, and material and manufacturing costs.

$$\underset{17\times 1}{\boldsymbol{p}} = \{\lambda_1, \lambda_2, \ldots, \lambda_8\, \mu_1, \mu_2, \ldots, \mu_8, d\}'. \tag{7}$$

The materials used to construct the shell included two real composite materials: Carbon Fiber-Reinforced Polymer (CFRP) and Glass Fiber-Reinforced Polymer (GFRP). Additionally, a theoretical material, the theoretical Fiber-Reinforced Polymer (*t*FRP), was introduced for optimization purposes. The parameters of this material were calculated as the average values of the properties of the CFRP and GFRP. The introduction of this material increased the complexity of the optimization task by introducing an additional value for one of the decision variables.

Table 1 presents a summary of the mechanical and physical properties of the available materials.

Table 1. The properties of three fiber-reinforced composite materials: CFRP, GFRP and *t*FRP (see [38]).

Material Label	μ [-]	E_a [GPa]	E_b [GPa]	E_c [GPa]	ν_{ab} [-]	ν_{ac} [-]	ν_{bc} [-]	G_{ab} [GPa]	G_{ac} [GPa]	G_{bc} [GPa]	Mass Density [kg/m^3]	Cost [-]
CFRP	1	120	8	8	0.014	0.028	0.028	5	5	3	1536	10.20
*t*FRP	2	80	6	6	0.020	0.036	0.036	4	4	3	1428	5.78
GFRP	3	40	4	4	0.026	0.044	0.028	3	3	3	1320	1.36

2.4. Finite Element Models

This study employed two FE models that differed only in their FE size, which effectively means variations in mesh density. Each model consisted of four-node MITC4 multilayered shell elements, which are based on the first-order shear deformation theory [39].

Each layer of the shell structure corresponded to a single composite layer, with potentially different material properties and fiber orientation angles. The maximum side length of an approximately square finite element, denoted as h, for the primary FE model (referred to here as M1), was selected to be approximately $h = 1.25$ cm. However, slight variations existed in both the circumferential and longitudinal directions, and also at different locations along the shell's axis. In addition to the M1 model, one coarse model, labeled as M5, was introduced, featuring element sizes of $h = 5$ cm.

The high-fidelity M1 model served as the basis for constructing a pseudo-experimental model. Meanwhile, the lower-fidelity M5 model contributed to expanding the dataset for training the surrogate model. Given that the element size in the coarser model M5 was four times larger than that in M1, the computational cost was reduced by a factor 4^2. However, this efficiency gain came at the expense of accuracy—errors in M5 increased by factors of 4^4. While this loss of precision was substantial, the proposed methodology was designed to account for and mitigate this issue.

The pseudo-experimental model was derived from the M1 FE model, where the computed natural frequencies underwent the following nonlinear transformation:

$$f^{\mathrm{Me}} = f^{\mathrm{M1}} + 20 \cdot \sin\left(\frac{1}{60} \cdot f^{\mathrm{M1}} - 5\right) = \mathrm{Me}\left(f^{\mathrm{M1}}\right), \tag{8}$$

where f^{M1} represents the vector of natural frequencies (in Hz) obtained from the M1 model, corresponding to specific mode shapes ϕ_i within the mode shape matrix $\mathbf{\Phi}$ (see [40]). Unlike a typical approach where the frequency vector contains the lowest natural frequencies in sequential order, in this study, it included only frequencies corresponding to selected mode shapes. To enable such selection, the mode shapes obtained from numerical simulations first had to be identified and subsequently filtered to retain only the eleven most relevant ones [40].

This strategy enhanced the accuracy of the surrogate model by focusing on the most meaningful vibrational modes and eliminating unnecessary information that could introduce noise into the learning process. As a result, the optimization procedure benefited from improved convergence and solution quality, as demonstrated in the authors' previous studies [40].

The transformed vector f^{Me} represents the pseudo-experimental model's natural frequencies, and the function $\mathrm{Me}(\cdot)$ mimics experimental testing procedures. The neural network-based approximation—surrogate model application—of f^{Me} is denoted as $f^{\mathrm{Me}}_{\mathrm{SM}}$.

It is important to note that the function Me(·) does not stem from actual experimental research but is instead an attempt to model discrepancies between numerical simulations and laboratory experiments. The authors' previous studies relied entirely on numerical analyses; thus, incorporating the pseudo-experimental model into the optimization framework enables the consideration of potential deviations encountered in experimental studies. Furthermore, this approach helps address practical limitations related to the number of feasible experimental tests.

2.5. Optimization Strategy Using Genetic Algorithms, Surrogate Models, and Curriculum Learning

The optimization problems given by Equation (6) were herein solved using the Non-dominated Sorting Genetic Algorithm II (NSGAII) [41], a GA-based multi-objective search method that is not derivative-based. Genetic algorithms are widely used in complex engineering problems, particularly where traditional optimization methods prove insufficient [6,40]. They work on a population of possible solutions and use deterministic computations and random number generators. The GA's advantage, crucial from the point of view of the problem to be solved, is the ability to search the entire solution space when trying to find a global minimum. However, this requires repeated evaluations of the objective function, which is computationally expensive when the FEM is applied. In the proposed optimization procedure, the objective function was solved using a surrogate model instead of FEM calculations. Therefore, the GA procedure worked extremely fast.

However, one of the key challenges associated with GAs is the need to repeatedly evaluate the objective functions. This process can be computationally expensive, especially when the objective functions require time-consuming numerical analyses, such as FEM simulations. To significantly mitigate this issue, the present approach employed surrogate models based on deep neural networks (DNNs).

The use of DNNs as surrogate models enables the rapid approximation of analysis results, replacing costly simulations with near-instantaneous predictions. This allows for large-scale optimization while drastically reducing computation time. The effectiveness of this approach was confirmed in the authors' previous studies, where it was demonstrated that using a DNN for objective function prediction led to a significant reduction in computational burden compared to conventional methods [42].

The process of selecting DNN parameters required a thorough evaluation of network errors, taking into account the following aspects:

- The number of input variables, denoted as I;
- The number of layers, represented by N_L;
- The number of neurons H within each hidden layer (expressed as $H^{(\cdot)}$, maintaining consistency across all hidden layers within a specific network);
- The number of output nodes, denoted as O;
- The choice of learning algorithms and regularization techniques, along with other contributing factors;
- The choice of activation and loss functions.

A summary of the different network parameter values considered is presented in Table 2. It is worth noting that the architecture 17-50-50-50-11, in combination with the Tanh activation function and the RMSProp learning algorithm, provided optimal performance in over 80% of the evaluated DNNs. This configuration was frequently used alongside Batch Normalization (BN) for regularization and Early Stopping strategies. Also, the best results were achieved using the MAE as the loss function.

Preparing surrogate models in the form of a DNN requires generating an appropriate dataset for training. To achieve this, an MF approach was introduced to limit the number of calls to the high-fidelity M1 model during data generation. The less accurate M5 model was employed, allowing the acquisition of a large number of training samples at the cost of reduced accuracy. In the authors' previous study [43], it was demonstrated that increasing the FE size by a factor of h resulted in an approximately h^2 reduction in computational time. However, this simplification came at the expense of accuracy, as the numerical error increased by a factor of h^4. This trade-off underscores the necessity of incorporating correction mechanisms, such as auxiliary neural networks, to mitigate the errors introduced by the lower-fidelity M5 model. The number of cases computed using the M1 model (which also provided pseudo-experimental data samples) was an order of magnitude smaller than the number of cases evaluated with the M5 model. To further enhance the accuracy of the surrogate model, auxiliary neural networks were incorporated to compensate for the errors introduced by the lower-fidelity M5 model. The number of M1 model evaluations was denoted as n_{M1}, while the number of M5 model evaluations was denoted as n_{M5}.

Table 2. Architecture, algorithms, function, and methods used in DNN simulations [44].

DNN architecture $I - H^{(\cdot)} - O$	$I = \{17, 28\}$ $N_L = \{4, \mathbf{5}, 6, 7, 8\}$ $H^{(\cdot)} = \{20, 30, 40, \mathbf{50}, 75, 100\}$ $O = 11$
Learning algorithms	ADAM * RMSProp SGD
Regularization methods	* Early Stopping L_2 Regularization Dropout * Batch Normalization
Activation functions	SoftMax * Tanh ReLu Sigmoid
Loss functions	MSE * MAE ArcSin

* Option selected based on preliminary testing.

Within this framework, two FEM models of varying accuracy were utilized: a high-fidelity model (M1) and a low-fidelity model (M5). The M1 model served as a reference and was used to generate pseudo-experimental data by introducing a nonlinear perturbation function $\text{Me}(\cdot)$. This function aimed to account for potential discrepancies between numerical results and real experimental data, thereby enabling optimization under conditions closer to real-world scenarios. This introduced an additional verification step, allowing for the assessment of the robustness of the applied optimization methods against inevitable errors and uncertainties present in experimental data. The low-fidelity model M5, on the other hand, facilitated the rapid estimation of preliminary values while significantly reducing computational costs.

The integration of MF modeling with deep neural networks enhanced the efficiency of the surrogate model training process, allowing for more precise representation of dependencies in the design space, while maintaining an acceptable computation time. The following sections provide a detailed discussion of three different approaches, each varying in the construction of surrogate models and their integration with the optimization procedure.

Regardless of the applied variant, the primary objective of the surrogate model remained unchanged. Its purpose was to predict the pseudo-experimental frequency values f^{Me} based on a given vector of model parameters $\boldsymbol{p}$. Ultimately, regardless of the methodology adopted for constructing and training the surrogate model, its operation can be symbolically depicted as in Figure 2.

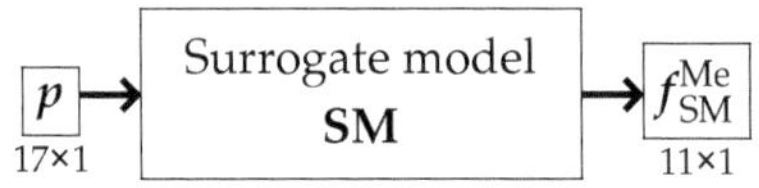

Figure 2. The surrogate model.

The optimization procedure, whose concept is presented in Figure 3, was based on an iterative approach involving multiple refinements of the surrogate model within the framework of CL.

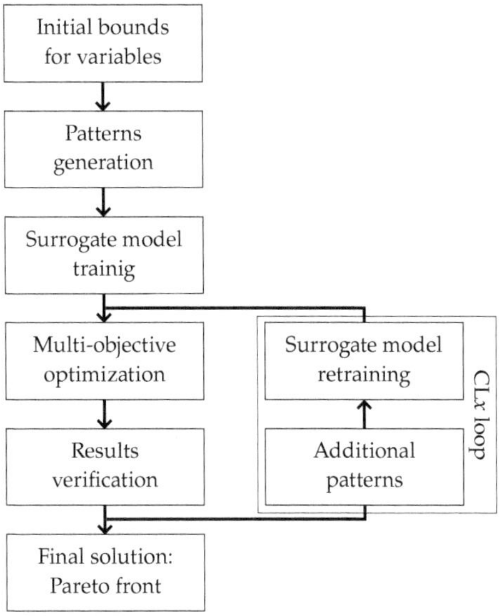

Figure 3. The optimization scheme, including CLx loops.

The process begins with data generation, which includes a large number of samples obtained using the simplified M5 model (n_{M5}) and a significantly smaller number of samples derived from the pseudo-experimental Me(M1) model (n_{M1}). This approach allows for the collection of a comprehensive dataset while simultaneously limiting the computational cost associated with the high-fidelity M1 model.

Based on the generated dataset, a surrogate model in the form of a deep neural network is constructed and appropriately trained. Once the training process is completed and the surrogate model is prepared, the optimization procedure is initiated using a GA. At that stage, the surrogate model plays a crucial role in enabling the efficient and rapid estimation of the objective function values.

After completing the first optimization cycle, the obtained results are validated and subsequently used to build an additional dataset. The new samples focus on regions of the design space located in the vicinity of the optimal solution, facilitating the further refinement of the surrogate model.

In the subsequent steps, the surrogate model is retrained based on the newly generated data, and the optimization process is restarted, this time utilizing the improved surrogate model. The iterative refinement cycles of the surrogate model form the core of the CL approach, where x represents the number of performed iterations.

The procedure terminates after reaching a predefined number of CL cycles, ensuring a systematic improvement in the quality of the surrogate model and yielding the final optimized solution.

2.5.1. Variant I

In the first approach variant, an auxiliary surrogate model was first developed to generate training data for the primary surrogate model. The purpose of the auxiliary model was to refine the results obtained from the low-fidelity M5 model so that they would closely match the values derived from the pseudo-experimental model $\mathrm{Me}\left(f^{\mathrm{M1}}\right)$. To achieve this, FEM calculations were performed for a limited number of cases using both the high-fidelity M1 model and the low-fidelity M5 model. Based on the collected data, an auxiliary model was trained. Its inputs consisted of the structural design parameters, gathered in the vector $\boldsymbol{p}$, along with a vector f^{M5} of eleven selected natural frequencies obtained from the M5 model. The neural network was trained to accurately estimate the pseudo-experimental frequencies $f^{\mathrm{Me}} = \mathrm{Me}\left(f^{\mathrm{M1}}\right)$, which served as approximations of real experimental measurements (see Figure 4a).

Upon completion of the training process, the trained auxiliary surrogate model was used to predict pseudo-experimental frequency values $f^{\mathrm{Me}}_{\mathrm{aux}}$ based on the results from rapid calculations using the M5 model only (see Figure 4b).

This approach enabled the generation of a large dataset, which was subsequently used to train the primary surrogate model (see Figure 5a). The role of this final surrogate model was to predict pseudo-experimental frequency values $f^{\mathrm{Me}}_{\mathrm{SM}}$ solely based on the design parameter vector $\boldsymbol{p}$, eliminating the need for any additional numerical simulations (see Figure 5b).

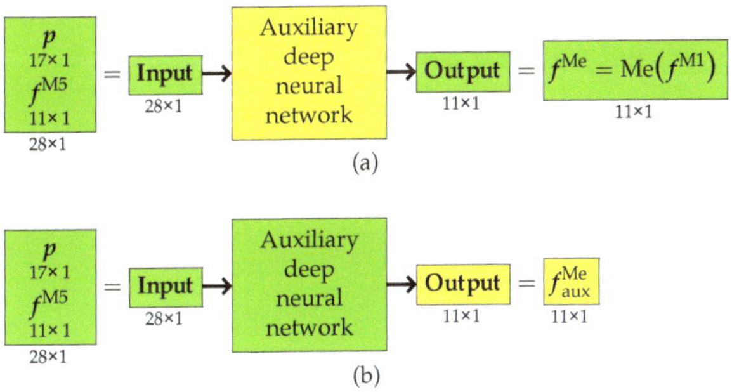

Figure 4. Variant I: (**a**) training and (**b**) application phase of the auxiliary surrogate model.

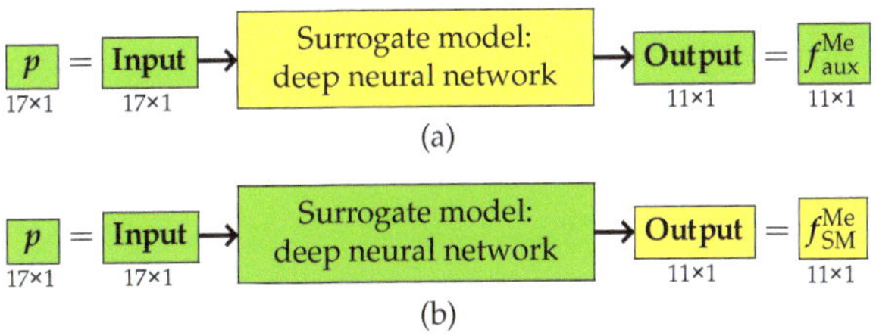

Figure 5. Variant I: (**a**) training and (**b**) application phase of the primary surrogate model.

This methodology significantly reduced the necessity of repeatedly utilizing the computationally expensive M1 model (as well as the pseudo-experimental model). Moreover, it facilitated the development of an accurate and efficient primary surrogate model. The large number of training samples generated by the auxiliary model allowed for precise predictions while maintaining a low computational cost.

2.5.2. Variant II

In the second approach, a different architecture was employed for the auxiliary surrogate model, while the primary surrogate model remained unchanged from the first variant. The key modification introduced in this version was the division of the auxiliary neural network structure into two distinct modules: one dedicated exclusively to processing linear dependencies and the other responsible for capturing nonlinear components of the mapping. Despite its more complex architecture, the auxiliary surrogate model remained a single neural network.

This architectural choice for the auxiliary model was based on the assumption that for functions that can be decomposed into linear and nonlinear components, processing these elements separately should yield more accurate approximation results [45–47]. By structuring the auxiliary model in this manner, it was possible to better align its design with the characteristics of the data, thereby improving its ability to capture the relationships between structural parameters and the resulting pseudo-experimental frequencies.

The training procedure of the auxiliary surrogate model (see Figure 6a), its application phase (see Figure 6b), and its objective remained identical to those in the first variant. The precomputed values from the simplified M5 model were still utilized and subsequently corrected using the trained network to best match the values obtained from the pseudo-experimental model. The refined data were then used to construct the main surrogate model, whose purpose was to estimate the pseudo-experimental frequency values f_{SM}^{Me} based solely on the design parameter vector $\boldsymbol{p}$, eliminating the need for multiple costly numerical computations (see Figure 5b).

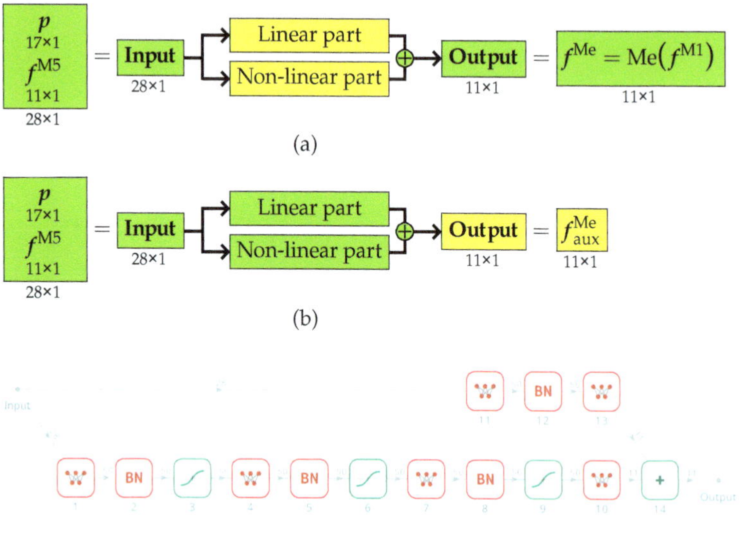

Figure 6. Variant II: (**a**) training and (**b**) application phase of the auxiliary surrogate model; (**c**) the structure of the applied two-module DNN.

A similar modular architecture to the one described above for the auxiliary surrogate model (see Figure 6c) was also tested for the primary surrogate model. The goal was to examine whether separating linear and nonlinear processing could enhance the accuracy of pseudo-experimental frequency predictions. However, the results obtained with this configuration did not show significant improvements over the standard approach, and in some cases, even led to increased approximation errors in the surrogate model. Consequently, this approach was abandoned.

2.5.3. Variant III

The third variant of the approach differed significantly from the two previous methods. It still utilized two surrogate models; however, their role and application underwent substantial modifications. Unlike variants I and II, where the auxiliary surrogate model was used solely for preparing training data for the primary surrogate model, in this approach, both models were employed simultaneously and actively participated in the entire optimization process.

The first surrogate model was designed to replace computations performed using the simplified M5 model. Its primary function was to directly estimate the selected natural frequencies $\boldsymbol{f}^{M5}$ obtained originally from the M5 model based on the vector of design parameters $\boldsymbol{p}$. This eliminated the need for the repeated use of the M5 model during the optimization process.

The second surrogate model, in turn, was responsible for estimating the pseudo-experimental frequencies $\boldsymbol{f}^{Me}$, which are essential for optimization. Its input consisted of an extended input vector comprising both the design parameter vector $\boldsymbol{p}$ and the vector of frequencies $\boldsymbol{f}^{M5}_{aux}$ obtained from the first surrogate model. As a result, this model accounted for both the structural characteristics and the dynamic properties derived from the analysis of the M5 model (or, more precisely, from the first surrogate model). The training and application of both surrogate models is presented in Figures 7 and 8.

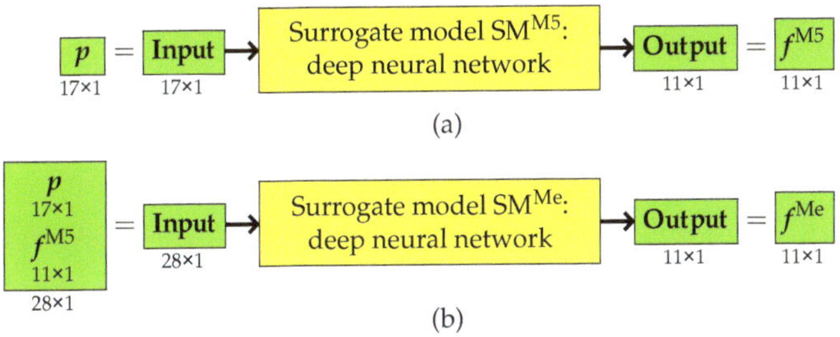

Figure 7. Variant III: training phase of (**a**) the first surrogate model and (**b**) the second surrogate model.

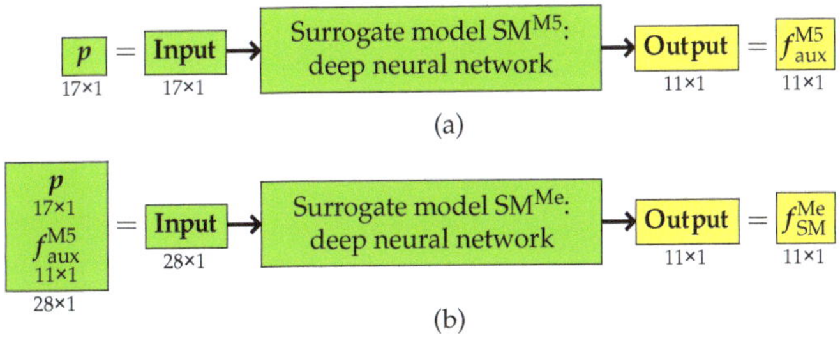

Figure 8. Variant III: application phase of (**a**) the first surrogate model and (**b**) the second surrogate model.

With this configuration, both surrogate models were utilized at every stage of the optimization process.

2.6. Indicators: Pareto Front Quality Metrics

The results of the multi-objective optimization problem analyzed in this study, which involved two objective functions, resulted in a two-dimensional Pareto front.

For an objective assessment of the quality of solutions obtained through multi-objective optimization, appropriate evaluation metrics must be introduced. While visual inspection of several Pareto fronts is effective for distinguishing qualitative differences, it becomes insufficient when variations between the compared fronts are merely quantitative. In such cases, the repeated comparison of Pareto fronts necessitates the definition of numerical quality metrics. These indicators allow for an objective evaluation of various characteristics of the analyzed fronts. Audet et al. [30] reviewed a total of 57 performance indicators and categorized them based on the evaluated parameters into four groups: (i) cardinality indicators, (ii) convergence indicators, (iii) distribution and spread indicators, and (iv) convergence and distribution indicators. Alternatively, Tian et al. [48] proposed a more simplified classification, distinguishing only between (i) diversity indicators (assessing the evenness and spread of the Pareto front) and (ii) convergence indicators.

In this study, four indicators were selected. The first was the hypervolume indicator, denoted as I_H, and the second was the relative hypervolume indicator, denoted as I^r_H. The hypervolume indicators are classified as convergence and distribution indicators in [30] or as convergence and diversity indicators in [48]. The hypervolume indicator I_H is recognized as the most widely used metric [29]. The third metric utilized was the Epsilon ϵ-indicator [49], referred to as I_ϵ. It is classified as a convergence indicator in [30] and ranks as the third most frequently used indicator according to [29]. The second most common metric, the Generational Distance indicator, was applied in this study as the fourth indicator.

Originally introduced by Zitzler [50], the hypervolume indicator measures the area covered by the examined Pareto front A relative to a suitably chosen reference point. When comparing two fronts, A and B, this indicator can be adapted as the difference $I_H(A) - I_H(B)$. If one of the compared fronts represents the true Pareto front (TPF), meaning the optimal front sought during the optimization process, the indicator can be redefined as a unary metric: $I_{H2}(A) = I_H(TPF) - I_H(A)$. The relative hypervolume indicator used in this study is given by:

$$I^r_H = \frac{I_H(TPF) - I_H(A)}{I_H(TPF)}, \tag{9}$$

where $I_H(TPF)$ and $I_H(A)$ denote the areas covered by the TPF and the examined Pareto front A, respectively. The true Pareto front was defined in this study as the envelope of the results obtained from all examined approaches and variants considered in the analysis. Therefore, it did not represent a fully legitimate TPF, which should ideally be derived analytically. Instead, it served as the most accurate possible approximation of the true optimal front within the scope of this study.

The third selected indicator, $I_\epsilon(A, B)$, represents the smallest scalar ϵ that scales Pareto front B so that every point in $\epsilon \cdot B$ is dominated by at least one point in A. If the second Pareto front corresponds to the TPF, this metric can be treated as a unary indicator, denoted as $I_{\epsilon 1}(A)$, which was applied in this form in the present study.

The fourth selected indicator, the Generational Distance indicator I_{GD} [51] measures the average distance of the obtained Pareto front solutions from the TPF and is defined as:

$$I_{GD} = \left(\frac{1}{N} \sum_{i=1}^{N} d_i^2 \right)^{\frac{1}{2}} \tag{10}$$

where

- N is the number of points in the approximated Pareto front;
- d_i is the distance of each solution from the nearest point in the TPF.

3. Results

3.1. Evaluation of High-Fidelity Sample Size and Training Strategies for Surrogate Models

The initial analyses aimed to determine the optimal number of M1 samples in the training dataset for the surrogate model. Table 3 and Figure 9 present the first set of results, where the table provides numerical values, and the figures offer a graphical representation with the vertical axes representing the values of four selected indicators and the horizontal axes indicating the number of M1 samples used in the applied datasets. For the reader's convenience, the desired trend for each indicator is repeated in parentheses (higher or lower values preferred):

- Hipervolume indicator I_H: see Figure 9a (the higher the better),
- ϵ indicator I_ϵ: see Figure 9b (the lower the better),
- relative Hipervolume indicator I_H^r: see Figure 9c (the lower the better),
- Generational Distance indicator I_{GD}: see Figure 9d (the lower the better).

Table 3. The optimization outcomes obtained using a surrogate model trained with a single-step training Variant II (no CLx) or CL iterative approach on Variant II CL0–CL2.

		Me(M1) Samples			
		125	250	500	750
I_H	Variant II	14.8924	15.1581	14.8327	14.8594
	Variant II CL0–CL2	—	15.1581	15.3016	15.3572
	"Variant II CL0–CL2" improvement vs. "Variant II"			3%	3%
I_ϵ	Variant II	1.3082	1.2241	1.3453	1.3258
	Variant II CL0–CL2	—	1.2241	1.2004	1.2004
	"Variant II CL0–CL2" improvement vs. "Variant II"			11%	9%
I_H^r	Variant II	0.0464	0.0294	0.0502	0.0485
	Variant II CL0–CL2	—	0.0294	0.0202	0.0166
	"Variant II CL0–CL2" improvement vs. "Variant II"			60%	66%
I_{GD}	Variant II	0.0391	0.0230	0.0289	0.0223
	Variant II CL0–CL2	—	0.0230	0.0187	0.0089
	"Variant II CL0–CL2" improvement vs. "Variant II"			35%	60%

The analysis of these indicators enabled the evaluation of the effectiveness of successive iterations in improving the surrogate model and their impact on the final quality of the optimization outcomes.

This analysis focused on comparing the optimization outcomes obtained using a surrogate model trained with a nearly identical number of computationally expensive Me(M1) samples but employing two distinct training strategies: (i) a single-step training approach utilizing all randomly generated samples at once, and (ii) an iterative refinement approach based on CL.

The tests presented in the table and figures were conducted exclusively for the second variant of the surrogate model, in which the auxiliary neural network consisted of separate modules dedicated to processing linear and nonlinear dependencies. It is important to note that Figure 9 does not encompass the entire range of simulations performed within the CL procedure. The results displayed in the figures correspond only to those cases for which calculations were also performed using a surrogate model trained without CL iterations, allowing for a direct comparison.

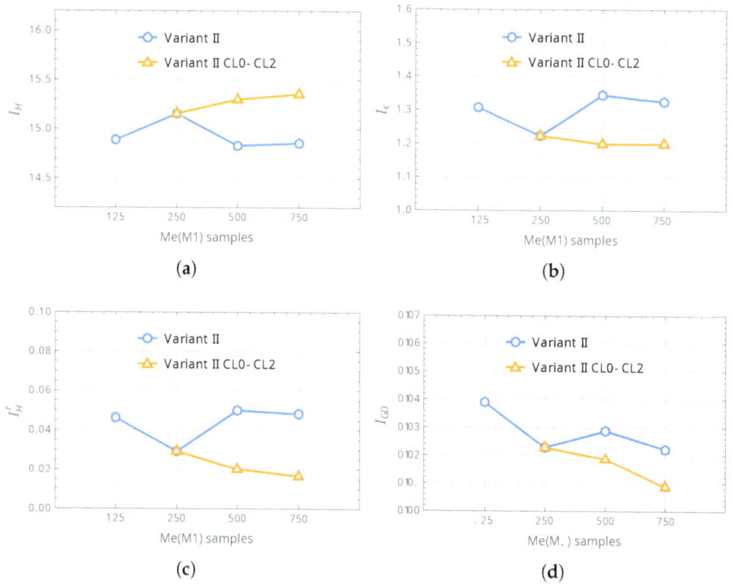

Figure 9. The optimization outcomes obtained using a surrogate model trained with a single-step training (no CLx) or CL iterative approach: (**a**) I_H, (**b**) I_ϵ, (**c**) I_H^r, (**d**) I_{GD}.

The results obtained for the variant that did not utilize CL are additionally presented not only in the form of Pareto front quality indicators but also through the resulting Pareto fronts themselves. A graphical representation of these fronts is provided in Figure 10, allowing for a direct comparison of their shape and distribution.

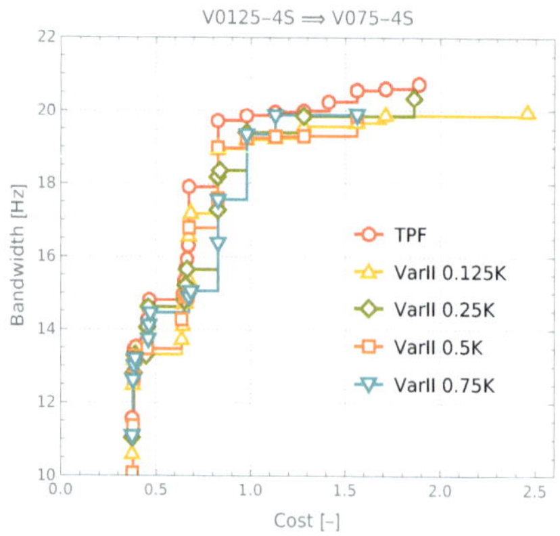

Figure 10. Comparison of Pareto fronts obtained without CL iterations, using different numbers of high-quality samples for surrogate model training.

The analysis of the results shown in Figure 9a–d leads to several key conclusions. First and foremost, there is no need to use more than 250 high-fidelity M1 samples in the initial stage of optimization (CL0), as increasing their number did not improve results and could even degrade performance in some cases. Comparing Pareto fronts obtained from different approaches that did not incorporate CLx loops (see Figure 10) was more complex; however, even in this case, the benefits of using 250 M1 samples could be observed. Another important finding is the impact of successive CLx loops on optimization quality. The conducted analyses demonstrated that the iterative refinement of the surrogate model led to a noticeable improvement in all applied performance indicators, as confirmed for CL1 and CL2 iterations. These findings highlight the effectiveness of the proposed approach and confirm that the key factors influencing the quality of the surrogate model are the proper management of high-fidelity samples and the application of iterative learning.

3.2. Evaluation of Different Surrogate Model Configurations at CL0 Stage

In the next step, a preliminary comparison (see Figure 11) was conducted at the CL0 stage (i.e., without refinement loops improving the model's accuracy) to evaluate the results obtained from three different surrogate model construction approaches (Variants I through III). Additionally, two alternative surrogate models were examined, where the auxiliary neural network (see Variant I, Figure 4a) was replaced by either Gradient Boosted Trees (GBTs) or Kriging inference (Krig). In this comparison, 250 high-fidelity M1 samples and 4000 low-fidelity, corrected M5 samples, were used. The only exception was the case labeled as "VarI: 5S", where the number of M5 samples was increased to 5000.

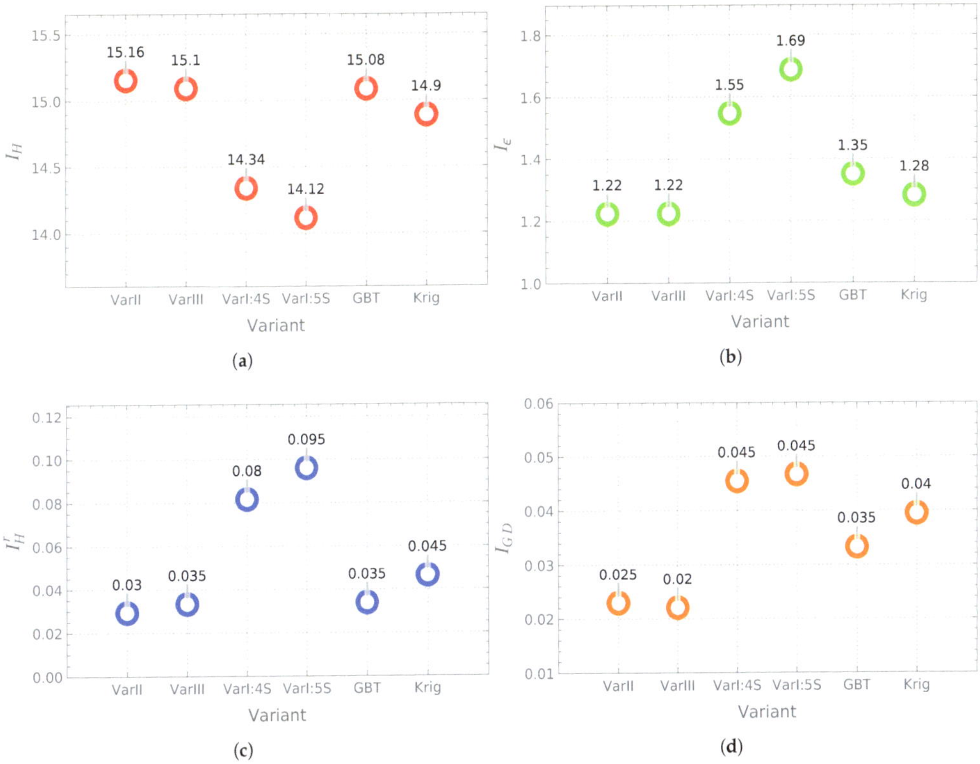

Figure 11. Comparison of performance indicators for different surrogate model variants at the CL0 stage: **(a)** I_H, **(b)** I_ϵ, **(c)** I_H^r, **(d)** I_{GD}.

The analysis of the obtained results indicates that the best optimization outcomes were achieved using Variant II and Variant III surrogate models.

A comparison of the results obtained for the Variant I surrogate model (denoted in the figures as "VarI: 4S" and "VarI: 5S") suggests that employing 4000 M5 samples was justified. A further analysis of these cases demonstrated that increasing the number of M5 samples to 5000 (as in "VarI: 5S") did not provide significant improvements in result quality.

Additionally, the results obtained using Gradient Boosted Trees (GBTs) and Kriging (Krig) indicated that in the Variant I surrogate model, during the construction phase of the auxiliary surrogate model (Figure 4a), alternative machine learning methods could be applied. This was feasible because the number of training samples in this phase was relatively small (250), allowing the effective use of techniques such as GBTs and Kriging. However, in the second phase (where the number of training samples increased to 4000), the results obtained using GBTs and Krig significantly deteriorated and were unsuitable for optimization purposes. Consequently, a DNN (deep neural network) was selected as the final surrogate model.

This finding suggests the potential for a hybrid approach, where GBTs or Kriging could be applied in the initial phase, followed by a DNN as the final surrogate model (GBT→DNN or Krig→DNN). However, this hybrid methodology was not further explored in the present study.

3.3. *Analysis of Effectiveness of Optimization Utilizing Curriculum Learning*

This section presents the results of the analysis of three surrogate model variants, each subjected to an iterative refinement process within the framework of CL; see Figure 12. The plots in the subsequent subfigures show the values of the applied performance indicators (vertical axes) as a function of the number of applied CL loops (horizontal axes). A detailed analysis was conducted for the case where the number of M1 samples (and consequently Me samples) was 250, while the number of M5 samples was 1000. This configuration is symbolically referred to as V025-4S. In each case of different surrogate models variants, five optimization cycles were conducted, denoted as CL0, CL1, CL2, CL3, and CL4. The results are presented in figures summarizing the tendencies of four selected performance indicators.

Each CL iteration required verification of results, allowing for the preparation of a new batch of training samples. In the applied approach, each CL iteration introduced approximately 250 additional training samples. Consequently, iterations from CL0 to CL3 resulted in around 1000 new samples generated using the M1 model, which was utilized for verification.

The surrogate models applied in the CL0 iteration were trained on an initial dataset containing 250 Me(M1) samples. After completing the CL3 stage, these models were further refined through four additional training phases, each incorporating approximately 250 new samples. With reasonable approximation, it can be stated that the surrogate models used in the CL4 iteration were trained on a dataset containing a total of $250 + 4 \times 250 = 1250$ samples of M1 quality.

Additionally, this study examined an alternative approach in which a surrogate model, built according to Variant I, was trained using all 1250 M1 simulation samples from the outset, without iterative refinement through CL cycles. In Figure 12, the results of this approach are marked with a red dot labeled V125-4S (1250 high-quality M1 samples and 4000 lower-quality auxiliary network-refined M5 samples). Due to the comparable number of high-fidelity samples, these results are presented in the CL4 column. However, it should be noted that formally, these correspond to CL0 since no iterative learning process was applied. Nevertheless, the number of high-fidelity samples used in this approach closely matched that of the V025-4S configuration at the CL4 stage.

For the reader's convenience, the numerical results presented graphically in Figure 12 are additionally provided in Table 4. To clarify the interpretation of the data presented in the table, several explanations are necessary. The columns labeled CL1, CL2, CL3, and CL4 display the values obtained in successive iterations of the CL process. The values in parentheses within these columns indicate the improvement achieved compared to the previous step, allowing for an assessment of the effectiveness of each iterative refinement in the surrogate model training process. The second-to-last column represents the overall improvement between CL4 and CL0. Additionally, the last column presents the overall improvement in results obtained using Variant I or Variant II relative to Variant III. The values in this column provide insight into which surrogate model approach yielded superior final results and quantify the extent to which Variant I or Variant II outperformed Variant III.

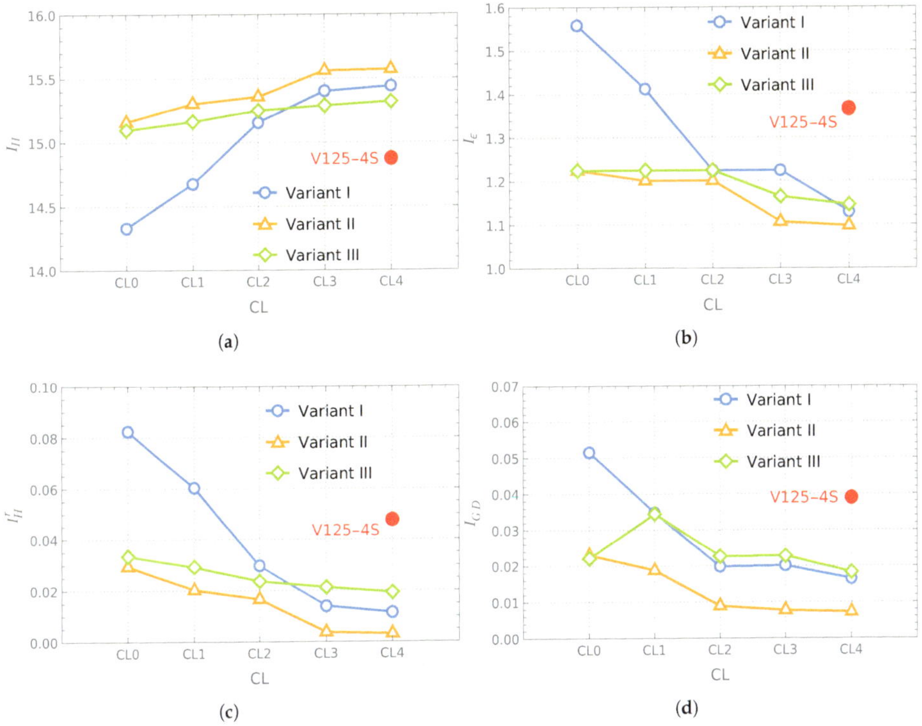

Figure 12. Surrogate model variants and Pareto front indicators obtained for subsequent CL loops: (**a**) I_H, (**b**) I_ϵ, (**c**) I^r_H, (**d**) I_{GD}.

Table 4. Surrogate model variants and Pareto front indicators obtained for subsequent CL loops.

		CL0	CL1	CL2	CL3	CL4	CL4 vs. CL0	Improvement vs. Variant III
I_H	Variant I	14.331	14.676 (2%)	15.153 (3%)	15.400 (2%)	15.439 (0%)	8%	1%
	Variant II	15.158	15.302 (1%)	15.357 (0%)	15.559 (1%)	15.569 (0%)	3%	2%
	Variant III	15.095	15.161 (0%)	15.248 (1%)	15.286 (0%)	15.318 (0%)	1%	—
I_ϵ	Variant I	1.5580	1.4119 (9%)	1.2241 (13%)	1.2241 (0%)	1.1288 (8%)	28%	1%
	Variant II	1.2241	1.2004 (2%)	1.2004 (0%)	1.1057 (8%)	1.0968 (1%)	10%	4%
	Variant III	1.2241	1.2241 (0%)	1.2241 (0%)	1.1640 (5%)	1.1449 (2%)	6%	—
I^r_H	Variant I	0.0824	0.0603 (27%)	0.0297 (51%)	0.0139 (53%)	0.0114 (18%)	86%	41%
	Variant II	0.0294	0.0202 (31%)	0.0166 (18%)	0.0037 (78%)	0.0031 (16%)	89%	84%
	Variant III	0.0334	0.0292 (13%)	0.0236 (19%)	0.0212 (10%)	0.0192 (9%)	43%	—
I_{GD}	Variant I	0.0515	0.0347 (33%)	0.0198 (43%)	0.0201 (−2%)	0.0164 (18%)	68%	10%
	Variant II	0.0230	0.0187 (19%)	0.0089 (52%)	0.0076 (15%)	0.0072 (5%)	69%	60%
	Variant III	0.0221	0.0344 (−56%)	0.0226 (34%)	0.0227 (0%)	0.0182 (20%)	18%	—

The Pareto fronts (not their indicators, as previously) obtained from the optimization process using different surrogate model variants and varying numbers of CL loops are presented in Figure 13a–c. These plots also include the TPF, which serves as a reference (benchmark) for assessing the quality of the obtained solutions.

An alternative visualization of the same results (displayed only for every second CL loop) is provided in Figure 13d–f. The magenta color is used to indicate the region enclosed by the Pareto front obtained for CL0. The green-shaded region corresponds to the area bounded by the Pareto front obtained in the CL2 iteration. However, only those portions of this region where the CL2 front dominated over the CL0 front are visible in the figure. Notably, the front obtained in CL2 was never inferior to the one from CL0. Similarly, regions where the Pareto front from CL4 outperformed the front from CL2 are highlighted in yellow. Finally, the red color marks areas where the True Pareto Front (TPF) provided superior results compared to the front obtained in CL4.

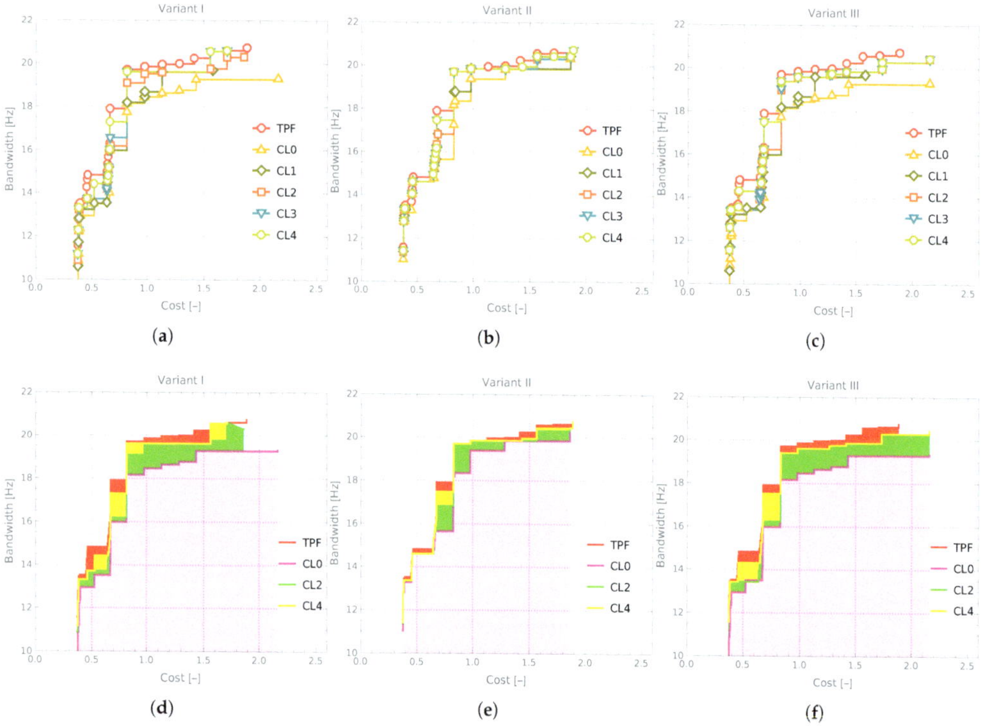

Figure 13. Pareto fronts obtained for subsequent CL loops: (**a**,**d**) Variant I, (**b**,**e**) Variant II, (**c**,**f**) Variant III.

The next figure, namely, Figure 14, presents the Pareto fronts obtained after the CL4 iteration for each of the considered surrogate model variants, as well as the Pareto front corresponding to the V125-4S case, previously described in the context of Figure 12. In the four cases depicted in the figure (supplemented by the TPF), the number of Me samples, which required computationally expensive evaluations, was very similar, amounting to approximately 1250.

This figure provides a clear assessment of the quality of the CL0 approach (where no CL iterations were applied, as in the case of V125-4S) compared to the iterative improvement strategy, in which the surrogate model was refined through successive CL loops.

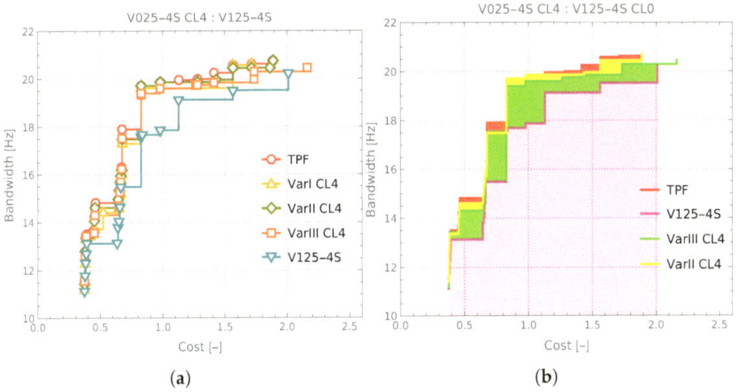

Figure 14. Pareto fronts obtained for CL4 loops and CL0 case V125-4S: (**a**) line-based representation of Pareto fronts, (**b**) surface-based representation of Pareto fronts.

4. Discussion

The analytical mode shape identification procedure employed in this study did not achieve the same level of precision for the analyzed geometries as it did for the originally considered cylindrical structure, on which it was initially developed (see [40,52]). This method involved identifying the node with the highest displacement for a given mode shape and determining the corresponding vibration mode based on the displacement direction and comparison with selected reference points. The application of concave or convex hyperboloid geometries introduced additional challenges in the identification process, as the curvature variation affected the displacement patterns and complicated the interpretation of mode shapes.

Although the identification accuracy could potentially be improved by fine-tuning selected parameters and coefficients within the identification procedure, the authors recognized this as an opportunity to assess the robustness of the proposed optimization framework in the presence of identification errors. While for a cylindrical structure, the identification method proved highly effective, with an estimated error rate of only about 1%, the more complex hyperboloid of revolution, which could exhibit both concave and convex configurations, resulted in a significantly higher error rate, increasing by several times.

Despite this increased error rate, the optimization process remained stable and effective, demonstrating its resilience to imperfections in mode shape identification. The observed identification inaccuracies did not introduce critical disruptions in the algorithm's performance, further validating the applicability of the proposed approach to optimizing complex geometries.

5. Conclusions

Based on the presented results, the following key conclusions can be drawn:

1. Effectiveness of CL iterations: The introduction of CL loops significantly enhanced the optimization outcomes. The results demonstrated that iterative refinement of the surrogate model through CL1 and CL4 led to a noticeable improvement in all applied performance indicators. This confirmed the effectiveness of CL in refining surrogate models and improving optimization performance.
2. Pareto front analysis: The visualization of Pareto fronts obtained for different surrogate models and CL iterations confirmed the positive impact of iterative learning on optimization quality. Moreover, the comparison of Pareto fronts from CL4 iterations with the V125-4S approach provided insights into the advantages of an iterative model refinement strategy over direct surrogate model training with a large dataset.

3. Comparison of different surrogate model variants: The best optimization results were achieved using the surrogate models from Variant II, then Variant I. These models consistently outperformed Variant III in terms of Pareto front quality indicators. Additionally, the comparison of Variant I results confirmed that increasing the number of low-fidelity samples above 4000 did not yield significant benefits.
4. Optimal number of high-fidelity samples: The analyses indicated that using more than 250 high-fidelity M1 samples in the initial optimization stage (CL0) did not improve results and, in some cases, could even degrade performance. This suggests that the selection of an appropriate number of high-fidelity samples is crucial for balancing computational cost and optimization accuracy.
5. Evaluation of alternative machine learning approaches: the findings suggest that, in the auxiliary surrogate model construction phase, alternative machine learning methods such as GBTs and Kriging can be effectively used.

These findings underscore the importance of iterative refinement in surrogate-based optimization and suggest that a carefully structured training approach, incorporating CL, can significantly enhance optimization performance while maintaining computational efficiency.

Author Contributions: Conceptualization, B.M. and L.Z.; methodology, B.M. and L.Z.; software, B.M. and L.Z.; investigation, B.M.; writing—original draft preparation, B.M.; writing—review and editing, L.Z. All authors have read and agreed to the published version of the manuscript.

Funding: Financed by the Minister of Science and Higher Education Republic of Poland within the program "Regional Excellence Initiative".

Institutional Review Board Statement: Not applicable.

Informed Consent Statement: Not applicable.

Data Availability Statement: The original contributions presented in this study are included in the article. Further inquiries can be directed to the corresponding author.

Conflicts of Interest: The authors declare no conflicts of interest.

Abbreviations

The following abbreviations are used in this manuscript:

GA	Genetic algorithm
FEM	Finite Element Method
MF	Multi-fidelity
LF	Low-fidelity
HF	High-fidelity
CFRP	Carbon Fiber-Reinforced Polymer
GFRP	Glass Fiber-Reinforced Polymer
tFRP	Theoretical Fiber-Reinforced Polymer
NSGAII	Non-dominated Sorting Genetic Algorithm
DNN	Deep neural network
CL	Curriculum Learning
TPF	True Pareto Front

References

1. Nikbakt, S.; Kamarian, S.; Shakeri, M. A review on optimization of composite structures Part I: Laminated composites. *Compos. Struct.* **2018**, *195*, 158–185. [CrossRef]
2. Ghiasi, H.; Pasini, D.; Lessard, L. Optimum stacking sequence design of composite materials Part I: Constant stiffness design. *Compos. Struct.* **2009**, *90*, 1–11. [CrossRef]

3. Setoodeh, S.; Abdalla, M.M.; Gürdal, Z. Design of variable–stiffness laminates using lamination parameters. *Compos. Part Eng.* **2006**, *37*, 301–309. [CrossRef]
4. Chiachio, M.; Chiachio, J.; Rus, G. Reliability in composites—A selective review and survey of current development. *Compos. Part Eng.* **2012**, *43*, 902–913. [CrossRef]
5. Reddy, J.N. *Mechanics of Laminated Composite Plates and Shells: Theory and Analysis*; CRC Press: Boca Raton, FL, USA, 2004.
6. Goldberg, D.E. *Genetic Algorithms in Search, Optimization and Machine Learning*, 1st ed.; Addison-Wesley Longman Publishing Co., Inc.: Boston, MA, USA, 1989.
7. Callahan, K.J.; Weeks, G.E. Optimum design of composite laminates using genetic algorithms. *Compos. Eng.* **1992**, *2*, 149–160. [CrossRef]
8. Riche, R.L.; Haftka, R.T. Optimization of laminate stacking sequence for buckling load maximization by genetic algorithm. *AIAA J.* **1993**, *31*, 951–956. [CrossRef]
9. Sivanandam, S.; Deepa, S.N. *Introduction to Genetic Algorithms*; Springer: Berlin/Heidelberg, Germany, 2008. [CrossRef]
10. Wang, G.G.; Shan, S. Review of Metamodeling Techniques in Support of Engineering Design Optimization. *J. Mech. Des.* **2006**, *129*, 370–380. [CrossRef]
11. Forrester, A.I.J.; Sóbester, A.; Keane, A.J. *Engineering Design via Surrogate Modelling*; John Wiley & Sons, Ltd.: Chichester, West Sussex, UK, 2008. [CrossRef]
12. Forrester, A.I.; Keane, A.J. Recent advances in surrogate-based optimization. *Prog. Aerosp. Sci.* **2009**, *45*, 50–79. [CrossRef]
13. Peherstorfer, B.; Willcox, K.; Gunzburger, M. Survey of Multifidelity Methods in Uncertainty Propagation, Inference, and Optimization. *SIAM Rev.* **2018**, *60*, 550–591. [CrossRef]
14. Kleijnen, J.P. Kriging metamodeling in simulation: A review. *Eur. J. Oper. Res.* **2009**, *192*, 707–716. [CrossRef]
15. Toal, D.J.J. Some considerations regarding the use of multi-fidelity Kriging in the construction of surrogate models. *Struct. Multidiscip. Optim.* **2015**, *51*, 1223–1245. [CrossRef]
16. Dadras Eslamlou, A.; Huang, S. Artificial-Neural-Network-Based Surrogate Models for Structural Health Monitoring of Civil Structures: A Literature Review. *Buildings* **2022**, *12*, 2067. [CrossRef]
17. Alizadeh, R.; Allen, J.K.; Mistree, F. Managing computational complexity using surrogate models: a critical review. *Res. Eng. Des.* **2020**, *31*, 275–298. [CrossRef]
18. Zhang, L.; Choi, S.K.; Xie, T.; Jiang, P.; Hu, J.; Koo, J. Multi-fidelity surrogate model-assisted fatigue analysis of welded joints. *Struct. Multidiscip. Optim.* **2021**, *63*, 2771–2787. [CrossRef]
19. Zhang, X.; Xie, F.; Ji, T.; Zhu, Z.; Zheng, Y. Multi-fidelity deep neural network surrogate model for aerodynamic shape optimization. *Comput. Methods Appl. Mech. Eng.* **2021**, *373*, 113485. [CrossRef]
20. Liao, P.; Song, W.; Du, P.; Zhao, H. Multi-fidelity convolutional neural network surrogate model for aerodynamic optimization based on transfer learning. *Phys. Fluids* **2021**, *33*, 127121. [CrossRef]
21. Waszczyszyn, Z.; Ziemiański, L. Neural Networks in Mechanics of Structures and Materials-New Results and Prospects of Applications. *Comput. Struct.* **2001**, *79*, 2261–2276. [CrossRef]
22. LeCun, Y.; Bengio, Y.; Hinton, G. Deep learning. *Nature* **2015**, *521*, 436–444. [CrossRef]
23. Goodfellow, I.; Bengio, Y.; Courville, A. *Deep Learning*; MIT Press: Cambridge, MA, USA, 2016.
24. Bengio, Y.; Louradour, J.; Collobert, R.; Weston, J. Curriculum learning. In Proceedings of the 26th Annual International Conference on Machine Learning, New York, NY, USA, 14–18 June 2009; ICML '09, pp. 41–48. [CrossRef]
25. Hsiao, Y.D.; Chang, C.T. Progressive learning for surrogate modeling of amine scrubbing CO2 capture processes. *Chem. Eng. Res. Des.* **2023**, *194*, 653–665. [CrossRef]
26. Fayek, H.M.; Cavedon, L.; Wu, H.R. Progressive learning: A deep learning framework for continual learning. *Neural Netw.* **2020**, *128*, 345–357. [CrossRef]
27. Kong, Y.; Liu, L.; Wang, J.; Tao, D. Adaptive Curriculum Learning. In Proceedings of the 2021 IEEE/CVF International Conference on Computer Vision (ICCV), Virtual Event, 11–17 October 2021; pp. 5047–5056. [CrossRef]
28. Scherding, C.; Rigas, G.; Sipp, D.; Schmid, P.J.; Sayadi, T. An adaptive learning strategy for surrogate modeling of high-dimensional functions—Application to unsteady hypersonic flows in chemical nonequilibrium. *Comput. Phys. Commun.* **2025**, *307*, 109404. [CrossRef]
29. Riquelme, N.; Lückenand, C.V.; Baran, B. Performance metrics in multi-objective optimization. In Proceedings of the 2015 Latin American Computing Conference (CLEI), Arequipa, Peru, 19–23 October 2015, pp. 1–11. [CrossRef]
30. Audet, C.; Bigeon, J.; Cartier, D.; Le Digabel, S.; Salomon, L. Performance indicators in multiobjective optimization. *Eur. J. Oper. Res.* **2021**, *292*, 397–422. [CrossRef]
31. Queipo, N.V.; Haftka, R.T.; Shyy, W.; Goel, T.; Vaidyanathan, R.; Kevin Tucker, P. Surrogate-based analysis and optimization. *Prog. Aerosp. Sci.* **2005**, *41*, 1–28. [CrossRef]
32. Hwang, J.T.; Martins, J.R. A fast-prediction surrogate model for large datasets. *Aerosp. Sci. Technol.* **2018**, *75*, 74–87. [CrossRef]

33. Jing, Z.; Sun, Q.; Zhang, Y.; Liang, K.; Li, X. Stacking sequence optimization of doubly-curved laminated composite shallow shells for maximum fundamental frequency by sequential permutation search algorithm. *Comput. Struct.* **2021**, *252*, 106560. [CrossRef]
34. Chen, Y.; Wang, Q.; Zhong, R.; Shi, X.; Qin, B. Fiber orientation and boundary stiffness optimization of laminated cylindrical shells with elastic boundary for maximum the fundamental frequency by an improved sparrow search algorithm. *Thin-Walled Struct.* **2023**, *193*, 111299. [CrossRef]
35. Chaudhuri, P.B.; Mitra, A.; Sahoo, S. Maximization of Fundamental Frequency of Composite Stiffened Hypar Shell with Cutout by Taguchi Method. *Mech. Adv. Compos. Struct.* **2023**, *10*, 43–54. [CrossRef]
36. Serhat, G. Design of Circular Composite Cylinders for Optimal Natural Frequencies. *Materials* **2021**, *14*, 3203. [CrossRef]
37. Alshabatat, N.T. Natural Frequencies Optimization of Thin-Walled Circular Cylindrical Shells Using Axially Functionally Graded Materials. *Materials* **2022**, *15*. [CrossRef]
38. Chen, D.; Sun, G.; Meng, M.; Jin, X.; Li, Q. Flexural performance and cost efficiency of carbon/basalt/glass hybrid FRP composite laminates. *Thin-Walled Struct.* **2019**, *142*, 516–531. [CrossRef]
39. Bathe, K.J. *ADINA: Theory and Modeling Guide Volume I: ADINA Solids & Structures*; ADINA R&D, Inc.: Watertown, NY, USA, 2016.
40. Miller, B.; Ziemiański, L. Optimization of dynamic behavior of thin-walled laminated cylindrical shells by genetic algorithms and deep neural networks supported by modal shape identification. *Adv. Eng. Softw.* **2020**, *147*, 102830. [CrossRef]
41. Deb, K.; Pratap, A.; Agarwal, S.; Meyarivan, T. A fast and elitist multiobjective genetic algorithm: NSGA-II. *IEEE Trans. Evol. Comput.* **2002**, *6*, 182–197. [CrossRef]
42. Miller, B.; Ziemiański, L. Multi-Objective Optimization of Thin-Walled Composite Axisymmetric Structures Using Neural Surrogate Models and Genetic Algorithms. *Materials* **2023**, *16*, 6794. [CrossRef] [PubMed]
43. Miller, B.; Ziemiański, L. Optimizing composite shell with neural network surrogate models and genetic algorithms: Balancing efficiency and fidelity. *Adv. Eng. Softw.* **2024**, *197*, 103740. [CrossRef]
44. *Mathematica*, Version 12.0; Wolfram Research, Inc.: Champaign, IL, USA, 2019.
45. Ahn, J.G.; Yang, H.I.; Kim, J.G. Multi-fidelity meta modeling using composite neural network with online adaptive basis technique. *Comput. Methods Appl. Mech. Eng.* **2022**, *388*, 114258. [CrossRef]
46. Guo, M.; Manzoni, A.; Amendt, M.; Conti, P.; Hesthaven, J.S. Multi-fidelity regression using artificial neural networks: Efficient approximation of parameter-dependent output quantities. *Comput. Methods Appl. Mech. Eng.* **2022**, *389*, 114378. [CrossRef]
47. Tan, J.; Shao, Y.; Zhang, J.; Zhang, J. Efficient Antenna Modeling and Optimization Using Multifidelity Stacked Neural Network. *IEEE Trans. Antennas Propag.* **2024**, *72*, 4658–4663. [CrossRef]
48. Tian, Y.; Cheng, R.; Zhang, X.; Li, M.; Jin, Y. Diversity Assessment of Multi-Objective Evolutionary Algorithms: Performance Metric and Benchmark Problems [Research Frontier]. *IEEE Comput. Intell. Mag.* **2019**, *14*, 61–74. [CrossRef]
49. Zitzler, E.; Thiele, L.; Laumanns, M.; Fonseca, C.M.; da Fonseca, V.G. Performance assessment of multiobjective optimizers: an analysis and review. *IEEE Trans. Evol. Comput.* **2003**, *7*, 117–132. [CrossRef]
50. Zitzler, E.; Deb, K.; Thiele, L. Comparison of Multiobjective Evolutionary Algorithms: Empirical Results. *Evol. Comput.* **2000**, *8*, 173–195. [CrossRef]
51. Vargas, A.; Bogoya, J. A Generalization of the Averaged Hausdorff Distance. *Comput. Sist.* **2018**, *22*, 331–345. [CrossRef]
52. Miller, B.; Ziemiański, L. Identification of Mode Shapes of a Composite Cylinder Using Convolutional Neural Networks. *Materials* **2021**, *14*, 2801. [CrossRef] [PubMed]

Disclaimer/Publisher's Note: The statements, opinions and data contained in all publications are solely those of the individual author(s) and contributor(s) and not of MDPI and/or the editor(s). MDPI and/or the editor(s) disclaim responsibility for any injury to people or property resulting from any ideas, methods, instructions or products referred to in the content.

materials

MDPI

Article

Dynamic Load Identification of Thin-Walled Cabin Based on CNN-LSTM-SA Neural Network

Jun Wang [1], Shaowei Song [1,2], Chang Liu [1] and Yali Zhao [3,*]

[1] National Key Laboratory of Aerospace Liquid Propulsion, Xi'an Aerospace Propulsion Institute, Xi'an 710100, China; junw83@163.com (J.W.); songshw82@163.com (S.S.); 13720753215@163.com (C.L.)
[2] School of Aerospace Engineering, Xi'an Jiaotong University, Xi'an 710049, China
[3] State Key Laboratory for Strength and Vibration of Mechanical Structures, School of Mechanical Engineering, Xi'an Jiaotong University, Xi'an 710049, China
* Correspondence: zhaoyali2000@163.com; Tel.: +86-15803997952

Abstract: Spacecraft are subjected to various external loads during flight, and these loads have a direct impact on the structural safety and functional stability of the spacecraft. Obtaining external load information can provide reliable support for spacecraft health detection and fault warning, so accurate load identification is very important for spacecraft. Compared with the traditional time-domain load identification method, the neural network-based time-domain load identification method can avoid the establishment of the inverse model and realize the response-load time-sequence mapping, which has a broad application prospect. In this paper, a CNN-LSTM-SA neural network-based load identification method is proposed for load acquisition of a thin-walled spacecraft model. Simulation results show that the method has higher identification accuracy and robustness (RMSE and MAE of 8.47 and 10.83, respectively, at a 20% noise level) in the load identification task compared to other network structures. The experimental results show that the coefficients of determination (R^2) of the proposed neural network load recognition model for time-domain identification tasks of sinusoidal and random loads are 0.98 and 0.93, respectively, indicating excellent fitting performance. This study provides a reliable new method for load identification in thin-walled spacecraft cabin structures.

Keywords: load identification; CNN-LSTM-SA; thin-walled compartment; time sequence mapping

Academic Editor: Valentino Paolo Berardi

Received: 17 January 2025
Revised: 24 February 2025
Accepted: 28 February 2025
Published: 12 March 2025

Citation: Wang, J.; Song, S.; Liu, C.; Zhao, Y. Dynamic Load Identification of Thin-Walled Cabin Based on CNN-LSTM-SA Neural Network. *Materials* **2025**, *18*, 1255. https://doi.org/10.3390/ma18061255

Copyright: © 2025 by the authors. Licensee MDPI, Basel, Switzerland. This article is an open access article distributed under the terms and conditions of the Creative Commons Attribution (CC BY) license (https://creativecommons.org/licenses/by/4.0/).

1. Introduction

In the field of aerospace, the acquisition of external load information is directly related to the stability and reliability of spacecraft. However, due to the harsh external environment and complex dynamic loads during spacecraft flight, it is difficult to measure external loads directly with sensors, so it is necessary to use load identification techniques to estimate external loads from the response inside the spacecraft [1]. The concept of load identification originated in the 1970s. At that time, load identification techniques were used to determine actual aircraft loads in order to improve aircraft performance [2]. It has since been widely applied in various engineering fields due to its ability to improve product reliability and durability. Dynamic load identification is divided into two categories: time-domain load identification methods and frequency-domain load identification methods. Time-domain load identification technology does not need to consider the frequency characteristics of the structure and is suitable for dynamic problems with significant nonlinearity and uncertainty. At the same time, the time-domain method can realize the on-line monitoring of structural loads, so the research on time-domain load identification methods has received more and more attention in recent years [3].

Time-domain load identification usually relies on the Green's kernel function deconvolution method (GKFM) [4], which is mainly for linear systems. Meanwhile, in order to solve the matrix ill-posedness problem, the regularization method is needed in the solution process. However, the selection of the regularization parameters is a more complicated task [5]. With the development of deep learning, neural networks are gradually being applied to the task of load identification. Compared with the traditional time-domain load identification methods, the neural network-based method has the ability of automatic feature extraction and powerful nonlinear modeling capability, as well as better robustness [6]. Luis et al. [7] used a neural network to achieve the identification of horizontal and vertical forces on automobile wheel hubs and discussed the potential of the method for online monitoring of loads applied to automobiles. Tang et al. [8] proposed a method combining stochastic response power spectral density and deep convolutional neural network (CNN) to accurately recognize vehicle load information. Zhang et al. [9]. proposed a CNN network based on a transform domain approach, using decomposed signals from wavelet transforms of multiple vibration signals as input; combined with a CNN network to achieve load identification in the time domain. Zhou et al. [10] performed impact load identification for nonlinear structures using a deep recurrent neural network consisting of two long short-term memory network (LSTM) layers and one bi-directional long short-term memory network (BiLSTM) layer; the network trained on numerous dynamic responses and impact loads demonstrated the ability to identify complex impact loads, even when the impact location was unknown. Yang et al. [11] proposed a neural network model with a bi-directional LSTM layer, an LSTM layer, and two fully connected layers to identify typical dynamic loads (sinusoidal, impact, and random loads) for simply supported beams. Furthermore, Yang et al. [12] developed a depth-expanded convolutional neural network, which directly constructs a transfer model between the structural acceleration response and excitation for data-driven dynamic load identification. Wang et al. [13] proposed a deep regression adaptive network based on model migration that learns to improve the accuracy and efficiency of neural networks for load identification.

Due to the different characteristics of different neural networks, recent research in time-series data prediction has focused on combining multiple networks, thereby leveraging their strengths and building more accurate and efficient models. Lu et al. [14] proposed a CNN-BiLSTM-AM (attention mechanism) neural network for predicting stock data, which confirms the significant advantages of this method over other neural network methods. Huy et al. [15] used a CNN-LSTM model combined with an attention mechanism to predict short-term power loads, addressing issues with input–output relationships, mitigating information loss from long input time-series data and improving prediction accuracy. Marios [16] examined dynamic structural loads of gated cyclic units, using an LSTM and CNN trained on small datasets, and compared the results with a physically based residual Kalman filter (RKF). Zhang et al. [17] proposed a method to predict wave height based on a CNN-LSTM neural network, and the results show that the method can effectively improve the prediction accuracy as well as robustness. Hu et al. [18] proposed a fall detection method based on a CNN-LSTM neural network and compared the CNN network and LSTM network alone, and the results showed that the network can significantly improve the accuracy of detection. Divya et al. [19] proposed a new hybrid deep learning method SSA (Singular Spectrum Analysis)-CNN-LSTM, which demonstrated the superior performance of this network for solar power generation prediction over a long time period in the future. Numerous studies have shown that higher prediction accuracy as well as robustness can usually be achieved by employing hybrid neural networks in time-series tasks. Whereas most of the related studies are currently focused on time-series prediction tasks, there are fewer related studies on time-series mapping tasks similar to load identification.

In order to further improve the identification accuracy and applicability of neural networks in load identification tasks, and to promote the engineering application of neural network load identification models, this paper proposes a load identification model based on a CNN-LSTM-SA (self-attention mechanism) hybrid neural network. The CNN layer performs feature extraction on the time-series data, the LSTM layer captures the long-term dependencies in the time series, and the SA layer establishes the global dependencies of the time series and assigns different weights. In this paper, the initial time-series data are segmented and used as inputs to the neural network for network training, the load identification accuracy of this neural network is compared to other network models, and its noise immunity is discussed. Finally, experimental studies are conducted on the identification of sinusoidal and random load time-domain information using the method to prove its value for engineering applications.

2. CNN-LSTM-SA Neural Network Load Identification Model Building

2.1. Load–Response Relationship in the Time Domain

For a linear system, the response induced by the load in the time domain satisfies the principle of linear superposition, and the time-domain response data $R(t)$ and the time-domain load $F(t)$ at the response measurement point satisfy the following Equation [20]:

$$R(t) = \int_0^t F(t)h(t-\tau)d\tau \tag{1}$$

where $R(t)$ denotes the dynamic response of the system measurement points, e.g., strain, stress, displacement, velocity, acceleration, etc. $h(t)$ is the response point–excitation point impulse response function. By discretizing the continuous time domain into a number of discrete time periods Δt, Equation (1) can be expressed as follows:

$$\begin{bmatrix} R_1 \\ R_2 \\ \vdots \\ R_n \end{bmatrix} = \begin{bmatrix} h_1 & 0 & \ldots & 0 \\ h_2 & h_1 & \ldots & 0 \\ \vdots & \vdots & \ddots & \vdots \\ h_n & h_{n-1} & \ldots & h_1 \end{bmatrix} \begin{bmatrix} F_0 \\ F_1 \\ \vdots \\ F_{n-1} \end{bmatrix} \Delta t \tag{2}$$

That is to say, for time-series data with time sampling length, the relationship between response and loading satisfies Equation (3). From Equation (3), it can be seen that the response $r(t)$ of the structure at the moment t is not only affected by the current moment $f(t)$ but also by all previous moments $f(0)$~$f(t-1)$. In other words, the response at any moment t contains all the information about the loads between the moments 0~t. For the inverse of the above process, the load $f(t)$ at the moment t can be written as Equation (4).

$$\begin{Bmatrix} r(1) \\ r(2) \\ \vdots \\ r(k) \end{Bmatrix} = \begin{bmatrix} h(1)\Delta t & & & \\ h(2)\Delta t & h(1)\Delta t & & \\ \vdots & \vdots & \ddots & \\ h(k)\Delta t & h(k-1)\Delta t & \cdots & h(1)\Delta t \end{bmatrix} \begin{Bmatrix} f(0) \\ f(1) \\ \vdots \\ f(k-1) \end{Bmatrix} \tag{3}$$

$$f(t) = G(r(t), r(t+\Delta t), \cdots, r(t+n\Delta t)) \tag{4}$$

2.2. CNN-LSTM-SA Network Architecture

The CNN model has excellent feature extraction ability and is widely used in various classification problems. The LSTM model has a special gate structure and weight sharing mechanism, which can avoid losing the long-term features of time-series data and is widely used in time-series analysis. The SA model has the importance of adding the past feature state of the time-series data to the output results and is widely used in adjusting the prediction results from an LSTM model. In this paper, according to the characteristics of the CNN, LSTM, and SA models, a load identification model based on a CNN-LSTM-SA architecture is established, and the structure of the model is shown in Figure 1. The main structure of the model consists of an input layer, CNN layer (one-dimensional convolutional layer and maximum pooling layer), LSTM layer, SA layer, fully connected layer, and output layer.

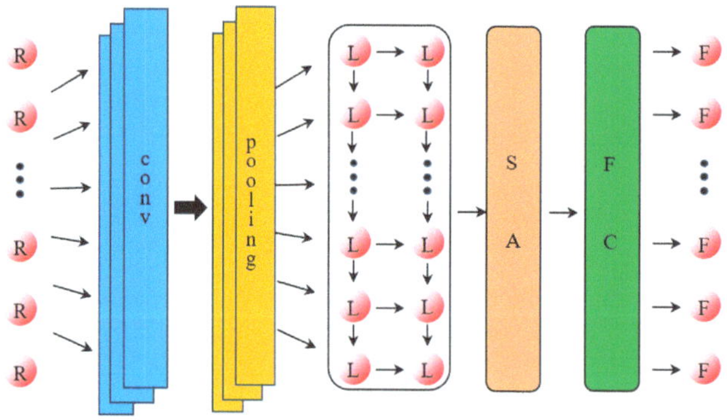

Figure 1. CNN-LSTM-SA load identification model.

CNN is a feed-forward neural network with good performance in digital image processing, image denoising [21], and classification problems [22]. Its main structure includes a convolutional layer, pooling layer, fully connected layer, and output layer, as shown in Figure 2. The convolutional layer is the core of the CNN, which performs convolutional operations on input data by convolutional kernels of specific sizes. Each convolutional layer contains multiple convolutional kernels, which can extract high-dimensional features from the data and establish the relationship between input and output units through sparse connections. Its unique weight sharing feature allows all output units to share a common set of parameters to connect with the input units, which greatly reduces the training parameters required for the model and saves training time. The pooling layer is used to downsize the data, reducing redundant information while retaining key information and reducing the number of parameters required by the network. The calculational procedure of the convolutional layer is shown in Equation (5):

$$y_t = f(x_t * k_t + b_t) \tag{5}$$

where x_t is the input, k_t is the weight of the convolution kernel, b_t is the bias, f is the activation function, and y_t is the output.

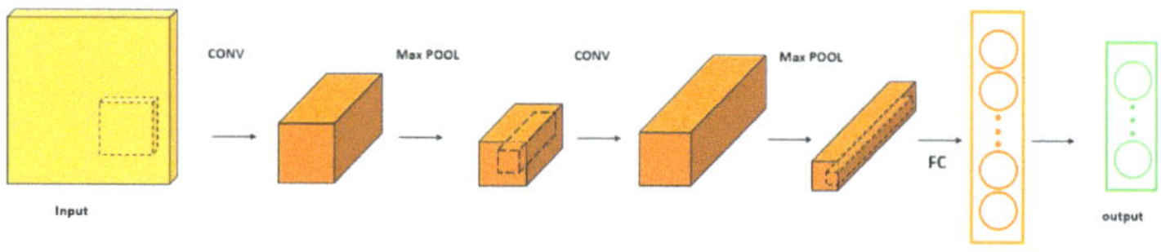

Figure 2. CNN network model (The dashed squares are convolution kernels and the circles indicate the flattened data).

The LSTM neural network is a special type of recurrent neural network which is especially used for processing and predicting temporal data [23]. Compared with a recurrent neural network (RNN), a LSTM solves the problems of gradient vanishing and gradient explosion through a special network structure, which is especially suitable for tasks involving long-term dependencies. The memory cell of an LSTM is shown in Figure 3, which controls the flow of data through a "gating mechanism" to selectively remember or forget information to preserve the long-term dependencies of the temporal data. The memory cell of the LSTM consists of three gates (forget gate, input gate, and output gate), each consisting of a sigmoid activation function and a point multiplication operation with values between 0 and 1, which are used to control the flow of data. In Figure 3, C_{t-1} is the state of the cell at the previous moment, h_{t-1} is the final output value of the LSTM cell at the previous moment, f_t is the output value of the forget gate, i_t is the input of the current input gate, $\widehat{C_t}$ is the temporary state of the current cell, o_t is the output value of the output gate, C_t is the state of the cell at the current moment, and h_t is the final output value of the current cell. The computational procedure of an LSTM is as follows:

(1) The forget gate generates an output value f_t between 0 and 1 by reading the final output h_{t-1} of the previous moment and the input x_t of the current moment and using Equation (6), where 1 represents the complete retention of information and 0 represents the complete discarding of information.

$$f_t = \sigma\left(W_f \cdot [h_{t-1}, x_t] + b_f\right) \tag{6}$$

where W_f is the weight matrix of the forget gate, b_f is the bias term, and σ is the Sigmoid activation function;

(2) The input gate generates the temporary state of the cell at the current moment $\widehat{C_t}$ by reading the final output of the previous moment h_{t-1} and the input of the current moment x_t and then updates the cell state in conjunction with the output of the forget gate to obtain the new cell state C_t which taking the values from 0 to 1.

$$i_t = \sigma(W_i \cdot [h_{t-1}, x_t] + b_i) \tag{7}$$

$$\widehat{C_t} = \tanh(W_c \cdot [h_{t-1}, x_t] + b_c) \tag{8}$$

$$C_t = f_t * C_{t-1} + i_t * \widehat{C_t} \tag{9}$$

where i_t is the output value of the input gate at the current moment, which determines the extent to which the current input x_t affects the update of the unit state C_t. W_c and b_c are the weight matrix and bias terms for computing the temporary unit state $\widehat{C_t}$. tanh is the hyperbolic tangent activation function and $*$ denotes the matrix element-by-element multiplication;

(3) The output gate extracts and outputs key information from the current unit state. It also reads the final output value h_{t-1} of the cell at the previous moment and the input value x_t of the cell at the current moment and calculates o_t through Equation (10).

$$o_t = \sigma(W_o[h_{t-1}, x_t] + b_o) \tag{10}$$

where o_t takes values from 0 to 1, W_o is the weight of the output gate, and b_o is the bias of the output gate;

(4) Finally, the final output of the unit at the current moment is calculated from the output value of the output gate o_t and the current state of the unit C_t by Equation (11).

$$h_t = o_t * \tanh(C_t) \tag{11}$$

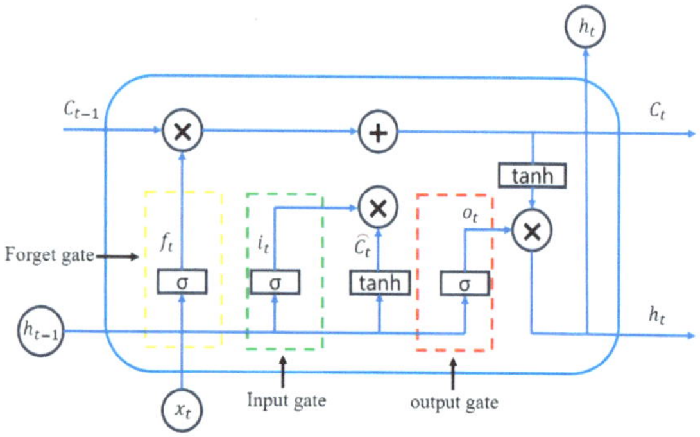

Figure 3. LSTM network memory cell.

SA (self-attention mechanism) is a technique that can dynamically capture the dependencies between different positions in a sequence and is widely used in natural language processing, computer vision, and time-series prediction. Its core idea is to decide how to summarize information by calculating the correlation weights of each element in a sequence with other elements. This mechanism allows the model to model the backward and forward relationships of the input sequences on a global scale, effectively capturing long-term dependencies. The main computational procedure for SA is as follows:

The input sequence X is mapped to the Query (Q), Key (K), and Value (V) for representing the characteristics of the current element, the importance of other elements, and the information to be summarized, respectively, through three sets of learnable linear transformations as in Equation (12).

$$Q = XW_Q,\ K = XW_K,\ V = XW_V \tag{12}$$

where W_Q, W_K, and W_V are trainable weight matrices.

The similarity score is obtained by computing the dot product of the Query and Key as in Equation (13), which is usually divided by $\sqrt{d_k}$ in order to balance the numerical stability, where d_k is the Key dimension:

$$Similarity_{i,j} = \frac{Q_i \cdot K_j^T}{\sqrt{d_k}} \tag{13}$$

Use the softmax function to normalize the scores to a *probability* distribution:

$$\alpha_{i,j} = \text{softmax}\left(Similarity_{i,j}\right) = \frac{\exp\left(Similarity_{i,j}\right)}{\sum_{k=1}^{n} \exp\left(Similarity_{i,k}\right)} \tag{14}$$

A new representation Z is obtained by weighted summation of V using the attention weights:

$$Z_i = \sum_{j=1}^{n} \alpha_{i,j} V_j \tag{15}$$

2.3. Constructing the CNN-LSTM-SA Load Identification Model

Since the load identification problem is a time-series mapping problem rather than a direct time-series prediction, therefore, it is necessary to construct the obtained response data and load data into a time series that the network can handle. The method used here is to construct the time-series mapping dataset by cutting the time-series data $X(t)$ with a sampling length of n into m segments with a length of k short time-series data $X_1(t), X_2(t) \ldots X_m(t)$, as shown in Equation (16). The time series constructed by this method can fully utilize the information of the original time domain data, ensure that the length of each time sample is consistent, and can also perform load identification with an arbitrary time-series length. At the same time, the data were normalized to the interval [−1,1] to prevent too large a gap in the order of magnitude of the data from causing the model to converge slowly or fail to converge.

$$\begin{bmatrix} X_1(t) \\ X_2(t) \\ \vdots \\ X_m(t) \end{bmatrix} = \begin{bmatrix} X_1(1) \cdots X_1(k) \\ X_2(1) \cdots X_2(k) \\ \vdots \\ X_m(1) \cdots X_m(k) \end{bmatrix} = \begin{bmatrix} x(1) & x(2) & \cdots & x(k) \\ x(2) & x(3) & \cdots & x(k+1) \\ \vdots & \vdots & & \vdots \\ \vdots & \vdots & & \vdots \\ x(m) & x(m+1) & \cdots & x(n) \end{bmatrix} \tag{16}$$

The structural diagram of the constructed CNN-LSTM-SA network is shown in Figure 4, and the main steps of the model applied to the load identification work are as follows:

(1) Segment the load and response data according to the method described above and perform data normalization as well as division of the training and test sets;
(2) After initialization of the network, the response data first pass through a one-dimensional convolutional CNN network for convolutional operation and average pooling operation to extract the high-dimensional features of the data;
(3) In order to make the highly dimensional data adapt to the LSTM neural network, the data output from the CNN network need to be flattened. The flattening process consists of pulling the feature maps of each channel into one-dimensional vectors in order and then connecting the vectors of all channels;
(4) The flattened data go sequentially through the LSTM neural network, the SA network, and the fully connected network, and finally the predicted load is generated;
(5) Judge whether to terminate network training based on the error between the network's predicted load and the actual load;
(6) The trained network performs load identification on the test set to verify the recognition effect.

The hyperparameters of this network are as follows:

(1) CNN layer: 1D convolutional kernel size, number of convolutional kernels, number of convolutional layers, and pooling size;
(2) LSTM layer: number of LSTM units, number of LSTM layers, and dropout rate;
(3) SA layer: number of attention heads, attention head dimension, and dropout rate;
(4) Hyperparameters of the network: optimizer type, learning rate, learning rate decay, batch size, max epochs, etc.

The hyperparameters of the network structure described above can be adjusted according to the complexity of the task and whether it is overfitted or not. The network hyperparameters can be determined by grid search, Bayesian optimization, and empirically. Due to the limitation of the arithmetic power, the key hyperparameters of the neural network are determined in this study by small batch grid search.

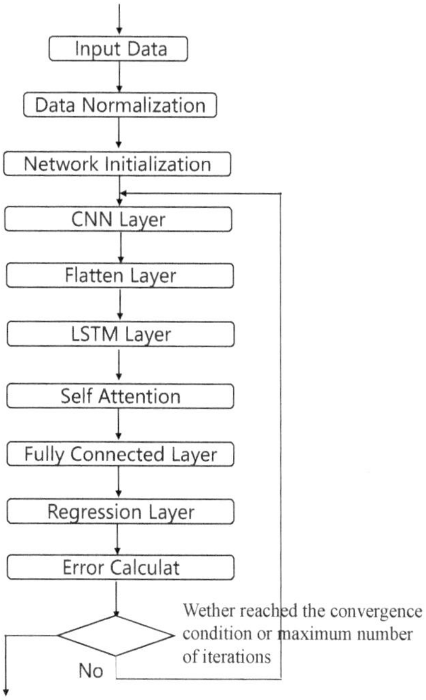

Figure 4. CNN-LSTM-SA network structure diagram.

It is worth noting that the optimal value of the segmentation sequence length is affected by a variety of factors, which will not be investigated in detail here, and the length of the segmentation time sequence set in the subsequent simulation and experimental studies in this paper is 200.

3. Numerical Simulation Study

Thin-walled cabins are efficient structural designs widely used in aerospace, industrial production, and scientific research, and have attracted considerable attention due to their light weight, high strength, and excellent material utilization. This paper presents both numerical simulations and experiments using a thin-walled cabin as the research object, applying the proposed method for load identification.

The simulation model is shown in Figure 5. The bottom face of the model is constrained with a fixed constraint and a simple harmonic excitation is applied at a point on the bulkhead wall, and the response is measured at the bulkhead bracket. Both the wall thickness of the bulkhead model and the bracket thickness inside the bulkhead are 5 mm. The model is made of aluminum alloy with a density of 2770 kg/m^3, Young's modulus of 71 GPa, and Poisson's ratio of 0.33. The simulation was performed by transient analysis. The transient simulation module of ANSYS(2024R2) Mechanical software is used to calculate the simulation results. The simulation was performed using transient analysis with a time step of 0.001 s and a damping ratio of 0.03. Considering the calculation time and the accuracy of the simulation results, the mesh size is set to 2 mm.

Fifty sets of transients were analyzed for sinusoidal excitations of different frequencies and amplitudes lasting 1 s. The acceleration response at the measurement points was recorded. The sinusoidal excitation frequency ranges from 5 to 7 Hz, and a simulation calculation frequency is selected every 0.2 Hz (except 7 Hz), and transient analyses of sinusoidal excitations with amplitudes of 80 N, 110 N, 140 N, 170 N, and 200 N are performed at each frequency, respectively. The applied load time-domain data and the acquired acceleration time-domain data were divided into training (80%) and test (20%) sets for network training and load identification. To evaluate the effectiveness of the model, LSTM, CNN-LSTM, and CNN-LSTM-SA networks were selected for load identification, and their performance in load identification and ability to handle noisy data were compared.

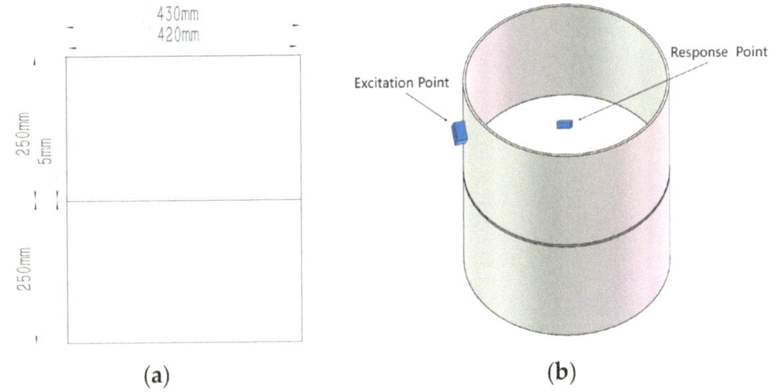

Figure 5. (**a**) Simulation object model; (**b**) simple harmonic excitation application position and response pickup position.

This section uses test set error metrics to evaluate the effectiveness of the neural network load identification. The root mean square error (RMSE) and mean absolute error (MAE) are chosen as the error metrics, as defined by the following formulas:

$$\mathrm{RMSE} = \sqrt{\frac{1}{n}\sum_{i=1}^{n}(\hat{y}_i - y_i)^2} \tag{17}$$

$$\mathrm{MAE} = \frac{1}{n}\sum_{i=1}^{n}(\hat{y}_i - y_i) \times 100\% \tag{18}$$

where $\hat{y}_i$ is the neural network predictive value load, y_i is the true load of the test set, and n is the number of samples in the test set; the smaller the value of RMSE and MAE, the higher the prediction accuracy of the model.

The three neural networks are trained separately, and the load identification results of each network under 0% noise data are shown in Table 1. It can be seen that the load identification accuracy of CNN-LSTM-SA has a significant advantage over the other two neural networks. Figure 6 shows the recognition effect and absolute error of the three neural networks for load identification under the data without the influence of noise. From the figure, it can be seen that the CNN-LSTM-SA network can not only recognize the peak of sinusoidal excitation well, but also the absolute error is smaller than the other two networks in the whole sampling time, indicating that the CNN-LSTM-SA load identification method proposed in this paper has obvious advantages.

Table 1. Neural network load recognition effect (0% noise).

	RMSE	**MAE**
LSTM	1.10	0.61
CNN-LSTM	0.74	0.55
CNN-LSTM-SA	0.47	0.53

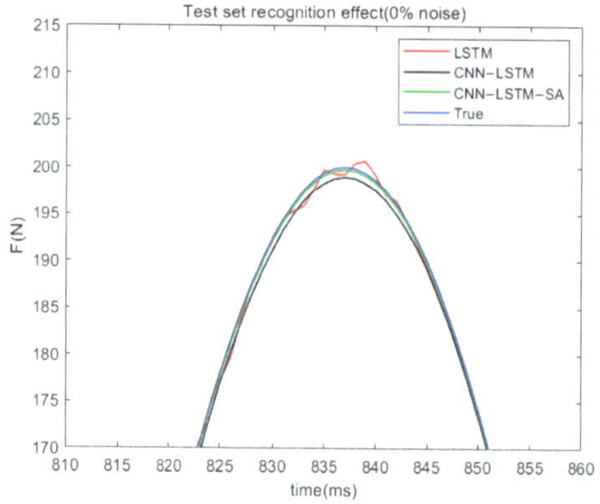

(a)

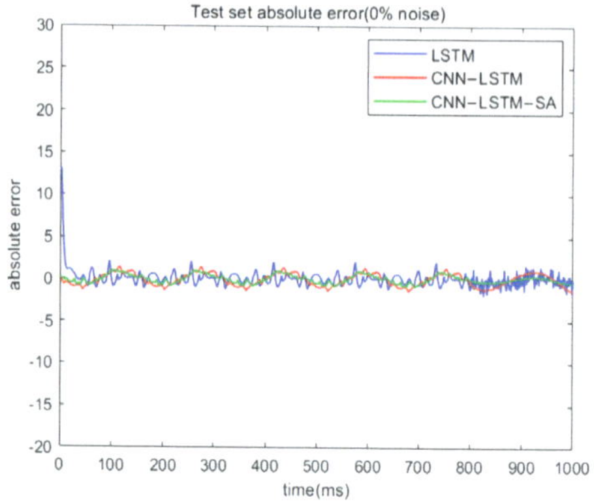

(b)

Figure 6. (**a**) Effectiveness of three neural network load recognition (0% noise); (**b**) absolute error of three neural network load recognition (0% noise).

In practical engineering problems, there is often noise interference during signal data acquisition using sensing devices, so it is necessary to verify the effectiveness of the CNN-LSTM-SA network for load recognition tasks in the presence of noisy signals. Researchers have confirmed in previous studies that CNN networks have filtering capabilities [23]; in this paper, we use the CNN-LSTM neural network as a reference to evaluate the adaptability of the CNN-LSTM-SA for load recognition tasks with noisy signals. White noise signals of 2%, 5%, 10%, and 20% are added to the simulated collected dataset, and the load recognition task is performed using the above two neural networks. Here, RMSE and MAE are still used as the indexes to evaluate the model load recognition effect, and the specific results are shown in Table 2. It can be seen that the CNN-LSTM-SA neural network has higher recognition accuracy and better noise immunity than the CNN-LSTM neural network in the load recognition work in the case of containing noise. Figure 7 shows the effect and absolute error of load recognition of both networks under different noise levels.

Table 2. Neural network load recognition effect (with noise).

Noise Level	2% Noise		5% Noise		10% Noise		20% Noise	
	RMSE	MAE	RMSE	MAE	RMSE	MAE	RMSE	MAE
CNN-LSTM	1.62	1.27	3.55	3.18	6.43	5.10	13.89	12.73
CNN-LSTM-SA	1.29	1.13	2.58	2.86	4.76	4.45	8.47	10.83

Based on the results of the analysis of the graphical data, it can be clearly seen that with the gradual increase in the noise level, the performance of both neural networks in the load recognition task shows a significant decrease, which is manifested by the trend of increasing their absolute error and root mean square error (RMSE). However, it is worth noting that under the influence of noise interference, the CNN-LSTM-SA neural network is still able to maintain the accuracy of load recognition well, with a relatively small increase in its error and a more stable recognition performance. In contrast, the performance of the CNN-LSTM neural network decreases more significantly, and the recognition accuracy is much lower than that of the CNN-LSTM-SA network. This indicates that the CNN-LSTM-SA network has obvious advantages in robustness and accuracy in the noisy environment.

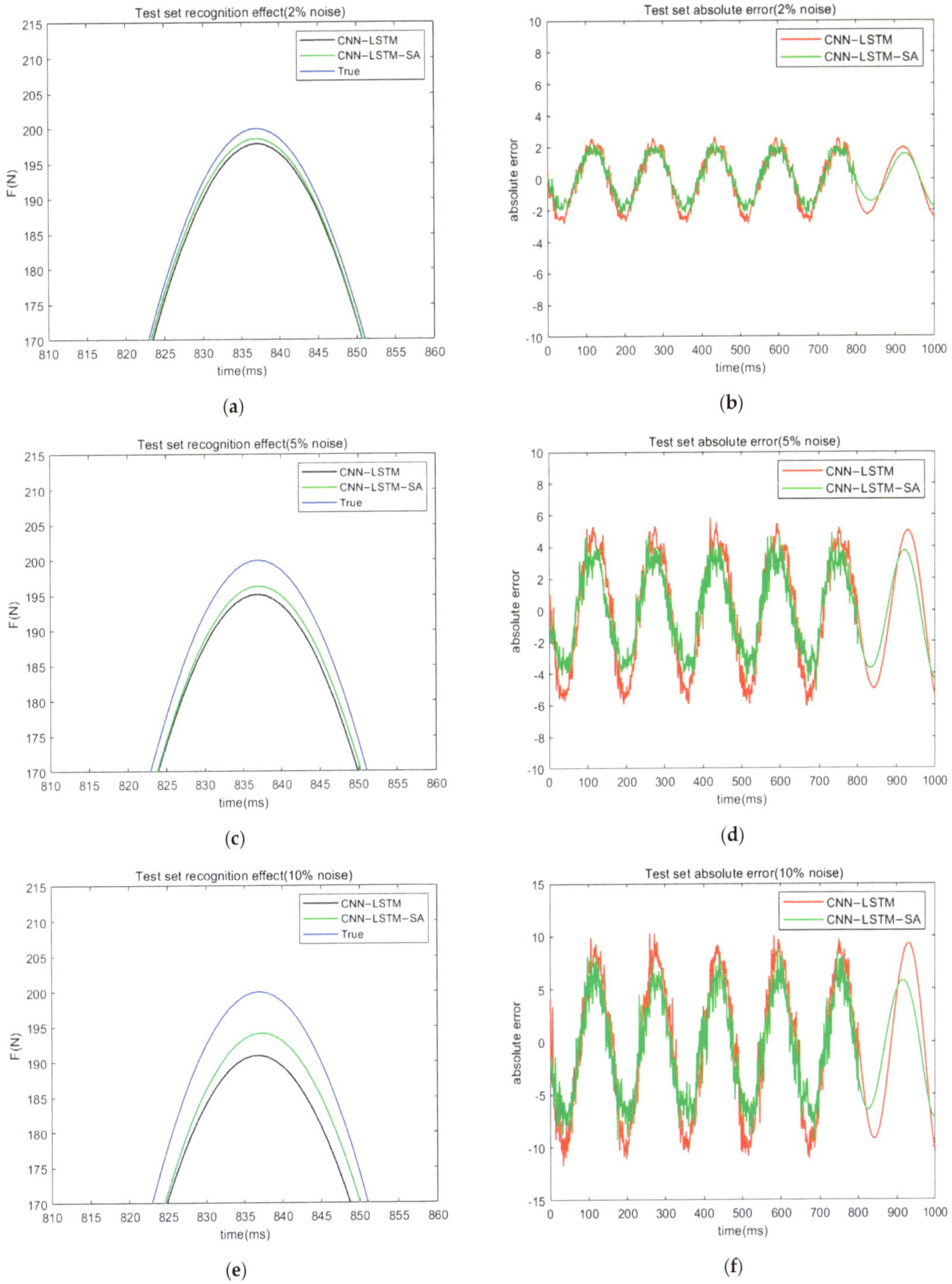

Figure 7. *Cont.*

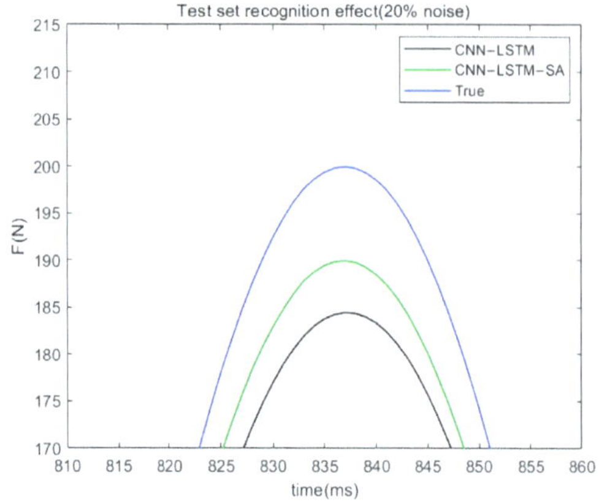

(**g**)

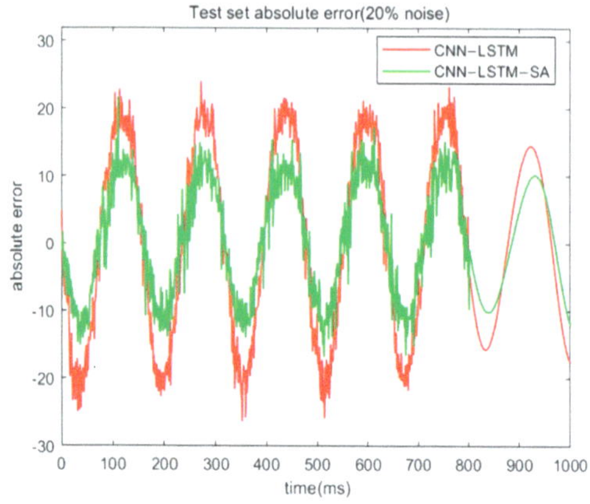

(**h**)

Figure 7. Effectiveness and absolute error (with noise) of two neural network's load recognition. (**a**) effectiveness of two neural network's load recognition (2% noise); (**b**) absolute error of two neural network's load recognition (2% noise); (**c**) effectiveness of two neural network's load recognition (5% noise); (**d**) absolute error of two neural network's load recognition (5% noise); (**e**) effectiveness of two neural network's load recognition (10% noise); (**f**) absolute error of two neural network's load recognition (10% noise); (**g**) effectiveness of two neural network's load recognition (20% noise); (**h**) absolute error of two neural network's load recognition (20% noise).

4. Experimental Study

4.1. Test Object and Test System

The bottom of the test object is a fixed constraint, as shown in Figure 8, which is used to simulate the thin-walled segment of the spacecraft; the material is aluminum alloy, and the wall thickness of the model is 5 mm. Excitation is applied by the shaker to simulate the external load of the thin-walled cabin, and the acceleration sensor is used to measure the structural response. The excitation position of the shaker and the mounting position of the acceleration sensor are shown in Figure 9a, and the test system consists of a personal computer, a power amplifier, and a data acquisition system, as shown in Figure 9b. In order to reflect the superiority of the load recognition method proposed in this paper, two load recognition methods, an LSTM neural network and a CNN-LSTM neural network, are still used for comparison, and the evaluation indexes are the RMSE, MAE, and coefficient of determination R2, where R2 is used to evaluate the fitting effect of the regression model, and the closer the value is to 1, the better the model fitting effect.

Figure 8. Test object.

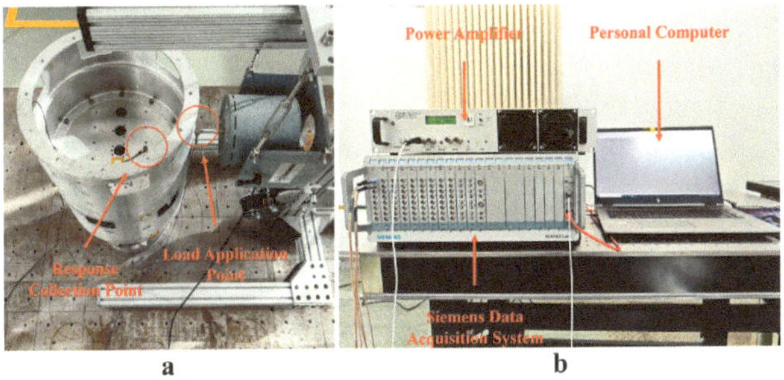

Figure 9. (**a**) Data acquisition location; (**b**) test system.

4.2. Time-Domain Identification of Sinusoidal Load

The experimental study of sinusoidal load identification is first carried out, and the sampling frequency of the time-series data is 6400 Hz. In this paper, 30 sets of sinusoidal excitation information with different frequencies and amplitudes and the corresponding structural acceleration responses are collected. A total of 25 sets were used to train the neural network load recognition model, and 5 test sets were used to evaluate the load recognition effectiveness of the model. The sinusoidal excitation frequency ranges from 50 to 100 Hz, and an experimental frequency (including 50 Hz) is selected every 10 Hz; data acquisition is performed at each frequency with amplitudes of 1 N, 1.5 N, 2 N, 2.5 N, and 3 N, respectively.

The sinusoidal time-domain load recognition effect of the model is shown in Figure 10a–c. From the figure, it can be observed that all three neural networks exhibit high accuracy for sinusoidal load identification. Specifically, Figure 10d shows the absolute errors of the three networks, where the CNN-LSTM-SA neural network has significantly lower errors than the other two networks in the time domain of the test set. This indicates that this network outperforms the other two models in the sinusoidal load identification task. Therefore, the CNN-LSTM-SA neural network used in this study has more superior accuracy and robustness in the sinusoidal load identification task. The model evaluation indices are shown in Table 3; compared with the other two neural networks, the neural network load identification method used in this paper has a smaller RMSE and MAE. This indicates that the method has excellent load recognition performance. Meanwhile, the R2 value of the proposed network in this paper is closest to 1, indicating that the method used in this paper has the best fitting effect among the three neural networks.

Table 3. Effect of time-domain identification of sinusoidal loads.

	RMSE	MAE	R^2
LSTM	0.12	0.14	0.97
CNN-LSTM	0.09	0.11	0.97
CNN-LSTM-SA	0.08	0.10	0.98

4.3. Time-Domain Identification of Random Load

Since the structure is often subjected to not only stable sinusoidal loads but also random loads under real working conditions, this paper also applies the proposed method to the time-domain load identification of random vibration tests to verify the generalizability of the method. Random signals are commonly described by power spectral density (PSD). In the experimental part of this study, 30 sets of time-domain excitation force and acceleration response information are collected under the same PSD with a sampling frequency of 12,800 Hz. A total of 25 of these sets are used as a training set to train the neural network load identification model, and 5 sets are used as a test set to evaluate the load identification effect of the model. The evaluation metrics used here are the same as those used for sinusoidal excitation identification. The PSD for the experimental setup is shown in Figure 11.

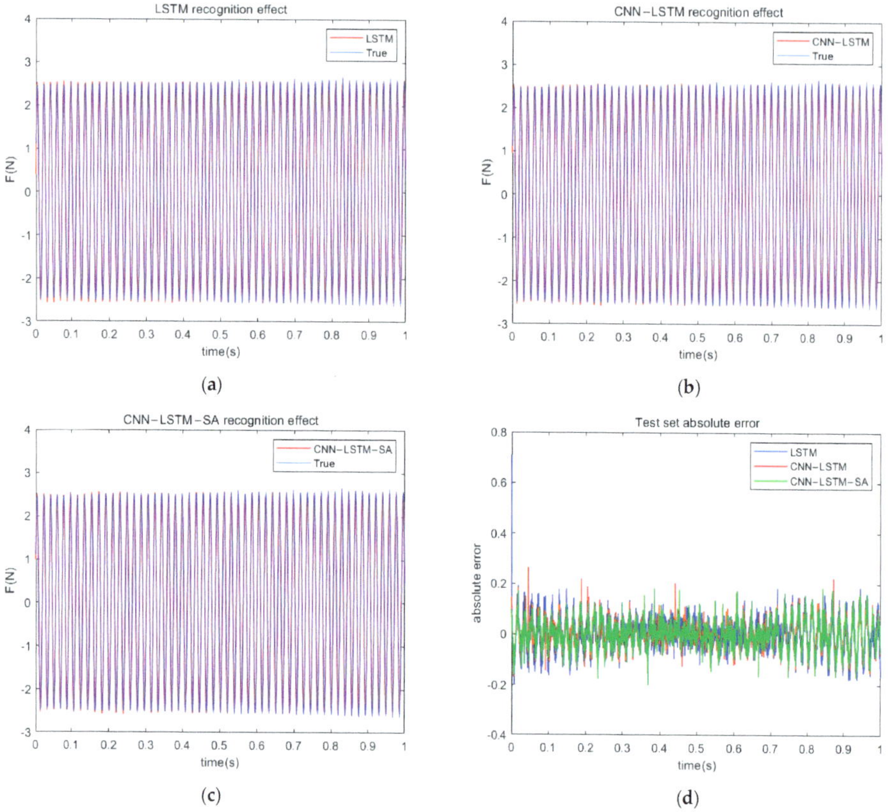

Figure 10. (**a**) LSTM sinusoidal load recognition effect; (**b**) CNN-LSTM sinusoidal load recognition effect; (**c**) CNN-LSTM-SA sinusoidal load recognition effect; (**d**) absolute errors in sinusoidal load recognition by three neural networks.

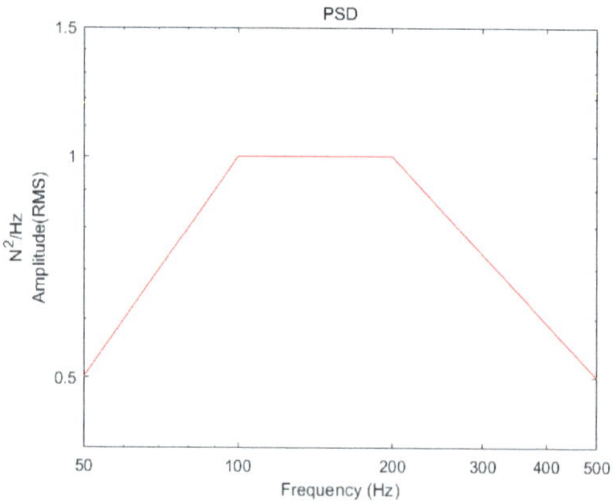

Figure 11. PSD for random vibration experiments.

The random time-domain load recognition effect of the model is shown in Figure 12a–c, from which it can be seen that the three kinds of neural networks for random time-domain load have this good recognition effect. But the CNN-LSTM-SA network still has this certain advantage, which can be proved from the absolute error diagram; the absolute error diagram of the three kinds of neural networks is shown in Figure 12d. The model evaluation indices are shown in Table 4. It can be seen that, compared with the other two neural networks, the RMSE and MAE of the neural network load identification method used in this paper are smaller, which indicates that this method has an excellent load identification effect. At the same time, the R2 value of this network is closest to 1, which indicates that the method used in this paper has the best fitting effect among the three neural networks, and it also indicates that the neural network model used in this paper has the best result for recognizing the time-domain information of random loads. In summary, the neural network model used in this paper has good applicability for the time-domain information recognition of random loads.

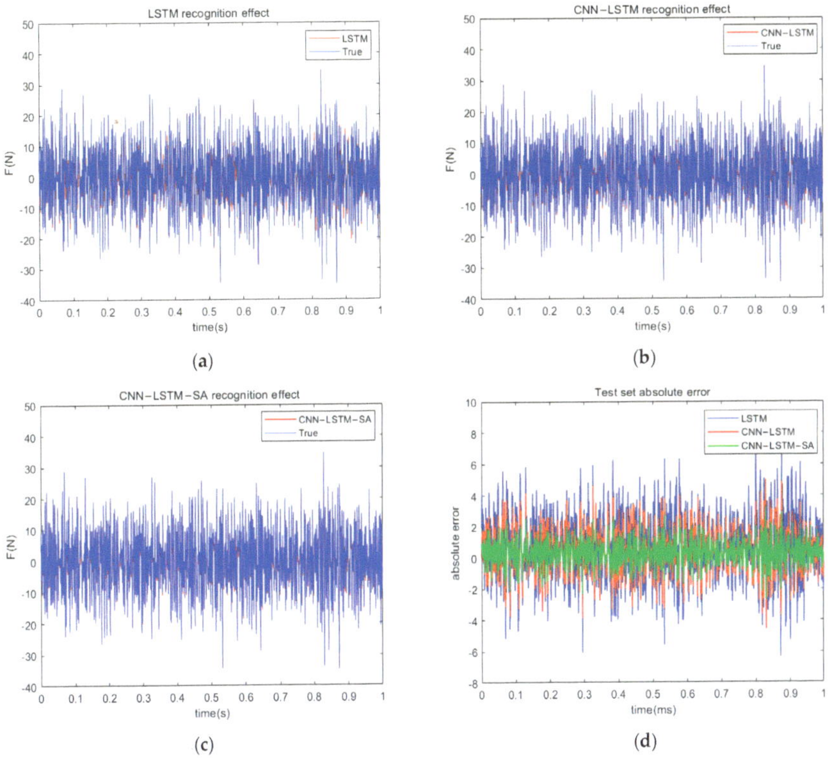

Figure 12. (**a**) LSTM random load recognition effect; (**b**) CNN-LSTM random load recognition effect; (**c**) CNN-LSTM-SA random load recognition effect; (**d**) absolute errors in random load recognition by three neural networks.

Table 4. Random load time-domain recognition effect.

	RMSE	MAE	R^2
LSTM	2.07	3.42	0.88
CNN-LSTM	1.39	2.57	0.90
CNN-LSTM-SA	0.83	1.69	0.93

It can be seen from the above study that under the experimental data, the load identification effect of the neural network decreases compared to the simulation data. This is mainly due to the fact that the experimental data are not only disturbed by white noise but also by abnormal time-series data and other uncertainties. Nevertheless, the neural network model proposed in this paper still shows good identification ability, which fully proves its adaptability and effectiveness in complex environments. On the other hand, the number of training parameters for the model increases significantly due to the mixing of multiple network structures. In addition, under high-frequency excitation, a higher sampling frequency is required to ensure sufficient accuracy, which also leads to an increase in the size of the dataset. The combination of these factors increases the cost of network training.

5. Conclusions

This paper proposes a time-domain load identification method based on a CNN-LSTM-SA neural network. The method takes segmented time-series data as input and combines the advantages of the three networks to achieve higher recognition accuracy and robustness. The main conclusions are the following:

(1) Simulation results show that for sinusoidal load identification, the CNN-LSTM-SA network has obvious advantages in terms of recognition accuracy and noise immunity. The RMSE and MAE are 0.47 and 0.53 under 0% noise and 8.8 and 8.5 under 20% noise, respectively;

(2) The experimental results show that the CNN-LSTM-SA network achieves high identification accuracies in both sinusoidal and random load identification tasks (RMSE of 0.08 and 0.83; R^2 of 0.98 and 0.93, respectively);

(3) The CNN-LSTM-SA-based load identification method provides researchers with a tool with higher accuracy and noise immunity, as well as a reliable method for structural health monitoring and optimal design.

Author Contributions: Conceptualization, Y.Z. and J.W.; methodology, S.S. and J.W.; software, J.W. and Y.Z.; validation, S.S. and Y.Z.; formal analysis, J.W.; investigation, J.W.; resources, S.S. and C.L.; data curation, S.S.; writing—original draft preparation, Y.Z.; writing—review and editing, J.W. and C.L.; visualization, J.W.; supervision, J.W.; project administration, J.W.; funding acquisition, J.W. All authors have read and agreed to the published version of the manuscript.

Funding: This research received no external funding.

Institutional Review Board Statement: Not applicable.

Informed Consent Statement: Not applicable.

Data Availability Statement: The data provided in this study can be provided at the request of the corresponding author. The research data presented in the paper consist of simulation and experimental results. We have provided the necessary simulation settings and experimental setups within this paper to enable scholars to replicate our results. Therefore, it is not necessary for us to specifically upload the data. Additionally, we have not uploaded the data from this paper to any public datasets. However, if there is a need, we are willing to provide the data upon request.

Conflicts of Interest: The authors declare no conflicts of interest.

References

1. Liu, R.X.; Dobriban, E.; Hou, Z.C.; Qian, K. Dynamic Load Identification for Mechanical Systems: A Review. *Arch. Comput. Methods Eng.* **2022**, *29*, 831–863. [CrossRef]
2. Bartlett, F.D.; Flannelly, W.G. Model Verification of Force Determination for Measuring Vibratory Loads. *J. Am. Helicopter Soc.* **1979**, *24*, 10–18. [CrossRef]

3. Li, K.; Liu, J.; Wen, J.; Lu, C. Time domain identification method for random dynamic loads and its application on reconstruction of road excitations. *Int. J. Appl. Mech.* **2020**, *12*, 2050087. [CrossRef]
4. Kazemi, M.; Hematiyan, M.; Ghavami, K. An efficient method for dynamic load identification based on structural response. In Proceedings of the International Conference on Engineering Optimization, Rio de Janeiro, Brazil, 1–5 June 2008; pp. 1–6.
5. Liu, J.; Meng, X.; Zhang, D.; Jiang, C.; Han, X. An efficient method to reduce ill-posedness for structural dynamic load identification. *Mech. Syst. Signal Process.* **2017**, *95*, 273–285. [CrossRef]
6. Roseiro, L.; Alcobia, C.; Ferreira, P.; Baïri, A.; Laraqi, N.; Alilat, N. Identification of the forces in the suspension system of a race car using artificial neural networks. In *Computational Intelligence and Decision Making: Trends and Applications*; Springer: Berlin, Germany, 2012; pp. 469–477.
7. Tang, Q.; Xin, J.; Jiang, Y.; Zhou, J.; Li, S.; Chen, Z. Novel identification technique of moving loads using the random response power spectral density and deep transfer learning. *Measurement* **2022**, *195*, 111120. [CrossRef]
8. Zhang, X.; He, W.; Cui, Q.; Bai, T.; Li, B.; Li, J.; Li, X. WavLoadNet: Dynamic Load Identification for Aeronautical Structures Based on Convolution Neural Network and Wavelet Transform. *Appl. Sci.* **2024**, *14*, 1928. [CrossRef]
9. Zhou, J.; Dong, L.; Guan, W.; Yan, J. Impact load identification of nonlinear structures using deep Recurrent Neural Network. *Mech. Syst. Signal Process.* **2019**, *133*, 106292. [CrossRef]
10. Yang, H.; Jiang, J.; Chen, G.; Mohamed, M.S.; Lu, F. A recurrent neural network-based method for dynamic load identification of beam structures. *Materials* **2021**, *14*, 7846. [CrossRef] [PubMed]
11. Yang, H.; Jiang, J.; Chen, G.; Zhao, J. Dynamic load identification based on deep convolution neural network. *Mech. Syst. Signal Process.* **2023**, *185*, 109757. [CrossRef]
12. Wang, C.; Chen, D.; Chen, J.; Lai, X.; He, T. Deep regression adaptation networks with model-based transfer learning for dynamic load identification in the frequency domain. *Eng. Appl. Artif. Intell.* **2021**, *102*, 104244. [CrossRef]
13. Lu, W.; Li, J.; Wang, J.; Qin, L. A CNN-BiLSTM-AM method for stock price prediction. *Neural Comput. Appl.* **2021**, *33*, 4741–4753. [CrossRef]
14. Huy, T.H.B.; Vo, D.N.; Nguyen, K.P.; Huynh, V.Q.; Huynh, M.Q.; Truong, K.H. Short-term load forecasting in power system using CNN-LSTM neural network. In Proceedings of the 2023 Asia Meeting on Environment and Electrical Engineering (EEE-AM), Hanoi, Vietnam, 15 November 2023; pp. 1–6.
15. Impraimakis, M. Deep recurrent-convolutional neural network learning and physics Kalman filtering comparison in dynamic load identification. *Struct. Health Monit.* **2024**, *1*, 31. [CrossRef]
16. Zhang, J.; Luo, F.; Quan, X.; Wang, Y.; Shi, J.; Shen, C.; Zhang, C. Improving wave height prediction accuracy with deep learning. *Ocean Model.* **2024**, *188*, 102312. [CrossRef]
17. Hu, X.; Yu, S.; Zheng, J.; Fang, Z.; Zhao, Z.; Qu, X. A hybrid CNN-LSTM model for involuntary fall detection using wrist-worn sensors. *Adv. Eng. Inform.* **2025**, *65*, 103178. [CrossRef]
18. Venkateswaran, D.; Cho, Y. Efficient solar power generation forecasting for greenhouses: A hybrid deep learning approach. *Alex. Eng. J.* **2024**, *91*, 222–236. [CrossRef]
19. Liu, J.; Meng, X.; Jiang, C.; Han, X.; Zhang, D. Time-domain Galerkin method for dynamic load identification. *Int. J. Numer. Methods Eng.* **2016**, *105*, 620–640. [CrossRef]
20. Ilesanmi, A.E.; Ilesanmi, T.O. Methods for image denoising using convolutional neural network: A review. *Complex Intell. Syst.* **2021**, *7*, 2179–2198. [CrossRef]
21. Wen, L.; Gao, L.; Li, X.; Zeng, B. Convolutional Neural Network with Automatic Learning Rate Scheduler for Fault Classification. *IEEE Trans. Instrum. Meas.* **2021**, *70*, 3509912. [CrossRef]
22. Sang, S.; Li, L. A Novel Variant of LSTM Stock Prediction Method Incorporating Attention Mechanism. *Mathematics* **2024**, *12*, 945. [CrossRef]
23. Uchida, K.; Tanaka, M.; Okutomi, M. Coupled convolution layer for convolutional neural network. *Neural Netw.* **2018**, *105*, 197–205. [CrossRef] [PubMed]

Disclaimer/Publisher's Note: The statements, opinions and data contained in all publications are solely those of the individual author(s) and contributor(s) and not of MDPI and/or the editor(s). MDPI and/or the editor(s) disclaim responsibility for any injury to people or property resulting from any ideas, methods, instructions or products referred to in the content.

materials

MDPI

Article

Prediction and Experimental Study of Low-Frequency Acoustic and Vibration Responses for an Aircraft Instrument Compartment Based on the Virtual Material Method

Shaowei Song [1,2], Jun Wang [2], Chang Liu [2] and Rongze Huang [3,*]

1 School of Aerospace Engineering, Xi'an Jiaotong University, Xi'an 710049, China; songshw82@163.com
2 National Key Laboratory of Aerospace Liquid Propulsion, Xi'an Aerospace Propulsion Institute, Xi'an 710100, China; junw83@163.com (J.W.); 13720753215@163.com (C.L.)
3 State Key Laboratory for Strength and Vibration of Mechanical Structures, School of Mechanical Engineering, Xi'an Jiaotong University, Xi'an 710049, China
* Correspondence: rzhuang_xjtu@163.com; Tel.: +86-18-8424-12997

Abstract: Bolted connections are extensively utilized in aircraft structures, and accurately simulating these connections is a critical factor affecting the precision of vibration and noise response predictions for aircraft. This study focuses on an instrument compartment of a specific aircraft model, employing the virtual material method to simulate the bolted joints within the structure. Parameters for the virtual material layer were obtained through theoretical calculations combined with parameter identification methods, achieving precise modeling of the instrument compartment. By comparing the calculated modes with the experimental modes of the instrument compartment, it was found that the first four modal shapes from both calculation and experiment were completely consistent, with the error in natural frequencies within three percent. Subsequently, acoustic and vibration computations were performed using both the virtual material model and the tied constraint model, with comparisons made against experimental results. The findings indicate that the root mean square (RMS) acceleration response of the virtual material model was 11.23 g, closely matching the experimental value of 10.35 g. Additionally, the total sound pressure level inside the acoustic cavity was 136.98 dB, closely aligning with the experimental value of 135.76 dB. These results demonstrate that the virtual material method offers higher accuracy in structural acoustic and vibration calculations.

Keywords: aircraft instrument compartment; virtual material method; modal testing; vibration and noise

Academic Editors: Pawel Wysmulski, Katarzyna Falkowicz and Patryk Rozylo

Received: 13 January 2025
Revised: 26 January 2025
Accepted: 28 January 2025
Published: 20 February 2025

Citation: Song, S.; Wang, J.; Liu, C.; Huang, R. Prediction and Experimental Study of Low-Frequency Acoustic and Vibration Responses for an Aircraft Instrument Compartment Based on the Virtual Material Method. *Materials* **2025**, *18*, 932. https://doi.org/10.3390/ma18050932

Copyright: © 2025 by the authors. Licensee MDPI, Basel, Switzerland. This article is an open access article distributed under the terms and conditions of the Creative Commons Attribution (CC BY) license (https://creativecommons.org/licenses/by/4.0/).

1. Introduction

Bolted connections are prevalent in various aircraft structures, but the complex mechanisms at their interfaces can lead to nonlinear dynamic phenomena such as energy dissipation, stiffness softening, and increased local damping at the bolted interfaces [1–3], which can influence the entire structure's modal and transfer function characteristics [4]. Precise models are required to describe the dynamic characteristics of significant and complex equipment like aircraft devices.

During high-speed flight, the interaction between high-speed fluid and the wall surface forms a turbulent boundary layer on the aircraft's surface, resulting in powerful aerodynamic fluctuation pressures and inducing an aerodynamic noise field with frequencies reaching up to 8000 Hz. This high-frequency noise strongly couples with the aircraft structure, producing vibrations with root-mean-square (RMS) acceleration responses as high as 50 g [5], subjecting the aircraft structure and internal instruments to complex and harsh environments. To understand the response levels and reveal the response patterns to improve the aircraft's resistance to acoustic and vibrational effects, simulation predictions and ground-based acoustic and vibration tests must be conducted [6–8]. The dynamic response calculation of structures with nonlinear characteristics is a complex problem. Karpenko et al. [9] obtained reliable nonlinear material properties by combining experimental research with numerical simulations. Due to the wide frequency spectrum of the noise environment, it is challenging to predict the full-band acoustic and vibration responses using a single method [10]. Typically, the frequency band is divided into low, medium, and high segments, with the finite element method used for the low-frequency band, statistical energy analysis for the high-frequency band, and hybrid methods for the mid-frequency band. Given the numerous components and complex interconnections of aircraft structures, accurately simulating joint connections is crucial for predicting acoustic and vibration behavior.

Currently, scholars have achieved fruitful results in modeling structural joints, with common bolted interface characterization methods including the Iwan equivalent model [11–13], thin-layer element method [14–16], and virtual material method [17–19], among others, the latter being widely applied due to its high modeling accuracy and low computational cost. Although equivalent models of bolted joints can accurately represent the nonlinear dynamic characteristics of simple connection structures, research applying this theory to large, complex assembled structures like aircraft for acoustic and vibration issues is relatively scarce. As the modeling of bolted joints significantly impacts structural dynamics, further investigation into its influence on acoustic and vibration prediction problems is necessary. This study uses an aircraft instrument compartment model as the object of research, establishing models using both the virtual material method and the tied constraint method for predicting low-frequency band acoustic and vibration responses under specified noise excitation conditions, followed by experimental validation. Since the finite element method is suitable for low-frequency band acoustic and vibration response predictions, considering the model's characteristics, the calculation frequency band is selected as 50–400 Hz, investigating the impact of micro-contact characteristics of bolted joints on the low-frequency band acoustic and vibration prediction results, aiming to enhance the precision of low-frequency band acoustic and vibration predictions.

2. Parametric Modeling of Structural Bolted Joint Virtual Material Layer Dynamics

2.1. Basic Principles of Assumptions in the Virtual Material Method

The virtual material method involves adding a layer of virtual material between two component bolted joints, simulating the dynamic characteristics of the bolted joint by altering parameters such as the density, elastic modulus, and Poisson's ratio of the virtual material [20]. The isotropic virtual material theory posits that the interface has varying degrees of roughness, which can be considered as a whole, formed by microasperities distributed according to different features, with the heights of these microasperities following a normal distribution, and being isotropic, with the essence of rough interface contact being the deformation of microasperity contacts.

The parameters of the virtual material layer include elastic modulus E, Poisson's ratio μ, thickness h, and density ρ. These parameters are related to the material properties of the two components and the contact surface data (such as surface roughness, bolt preload, etc.).

2.2. *Determination of Virtual Material Layer Parameters for Structural Bolted Joints*

2.2.1. Determination of Virtual Material Layer Thickness and Density

The thickness h of the virtual material layer is the sum of the thicknesses of two layers of microasperities on the rough surfaces, calculated using the formula:

$$h = h_1 + h_2 \tag{1}$$

where h_1 and h_2 are the thicknesses of the microasperity layers on the contact surfaces of the two parts, which are typically 1 mm.

According to the definition of material density, the density ρ of the virtual material can be obtained by the following equation:

$$\rho = \frac{m_1 + m_2}{V_1 + V_2} = \frac{\rho_1 A_a V_1 + \rho_2 A_a V_2}{A_a(h_1 + h_2)} = \frac{\rho_1 h_1 + \rho_2 h_2}{h_1 + h_2} \tag{2}$$

where A_a is the nominal contact area of the two connectors.

Thickness and density can be relatively simply derived from formulas. However, the calculation of the elastic modulus and Poisson's ratio is more complex and theoretical calculations often contain some error. In engineering applications, parameter identification methods are frequently used to obtain these values [21].

2.2.2. Parameter Identification for Elastic Modulus and Poisson's Ratio of the Virtual Material Layer

Parameter identification involves iteratively adjusting model parameters so that the computational results approach experimental results, ultimately obtaining an optimal set of parameters within an acceptable error range. The specific identification process is illustrated in Figure 1. The basic steps for parameter identification in this paper are as follows:

1. Establish a finite element model of the aircraft instrument compartment structure in Ansys, incorporating a virtual material layer at the bolted joints, and perform simulation calculations to obtain initial computed frequencies and mode shapes.
2. Conduct modal experiments on the aircraft instrument compartment to acquire actual natural frequencies and mode shapes of the structure.
3. Formulate an objective function based on the structural computed frequencies and the experimental frequencies obtained from modal testing, with the elastic parameters of the virtual material layer serving as design variables.
4. Set constraints and apply genetic algorithms to identify the elastic parameters of the virtual material layer.
5. Once the objective function meets the termination criteria, the identified parameters for the virtual material layer model are obtained.

The objective function of parameter recognition is defined as the finite element calculation and the natural frequency difference obtained by the modal experiment is minimized, as shown in the following equation:

$$F_{ul} = min \sum_{j=1}^{3} \left(\frac{f_j^a}{f_j^e} - 1 \right)^2 \tag{3}$$

In the formula: f_j^a represents the calculated natural frequency; f_j^e represents the experimental natural frequency; The objective function is defined to minimize the differences between the first three experimental frequencies and the calculated frequencies of the aircraft compartment. The accuracy of the parameter identification for the equivalent model of the virtual connection layer is validated using the fourth natural frequency, which was not involved in the identification process.

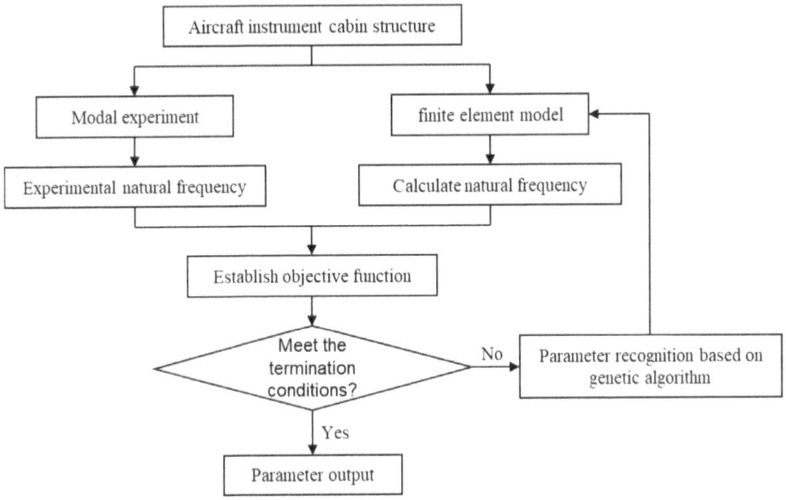

Figure 1. Flowchart of virtual material layer parameter identification.

3. Modal Acquisition of Instrument Compartment and Identification of Elastic Parameters of Virtual Material Layer

The instrument compartment model of the aircraft discussed in this paper features a conical shell structure with a thickness of 5 mm. The lower part is sealed with a circular plate, and the upper part is closed by a second-order curved surface plate. The material used is aluminum alloy, and it includes two bolted interfaces, each connected by six uniformly distributed M16 bolts. The surface roughness of any two component surfaces is considered in the design. According to the parameter identification method presented in this paper, it is necessary to obtain both the experimental modes and calculated modes of the aircraft instrument compartment. By iteratively adjusting the elastic parameters of the virtual material layer, the optimal set of parameters is ultimately obtained.

3.1. Modal Test of Aircraft Instrument Compartment

To conduct free modal testing on the aircraft instrument compartment model, it was suspended as shown in Figure 2. The modal experiment employed the impact hammer method for pulse excitation to obtain the structure's free modes. Considering the structural characteristics, 193 equidistant measurement points were arranged on the second-order curved plate and the conical shell body. One accelerometer is arranged on the structure surface. Since the mass of the accelerometers is much smaller compared to the model mass, their influence on the test can be neglected. Each measurement point was measured four times to ensure data reliability. The Hunter Box (Hanhang (Beijing) Technology Co., Limited, Beijing, China) data acquisition system recorded the experimental data and performed modal analysis. The modal shapes and frequency data obtained will be presented in the following sections for comparative purposes.

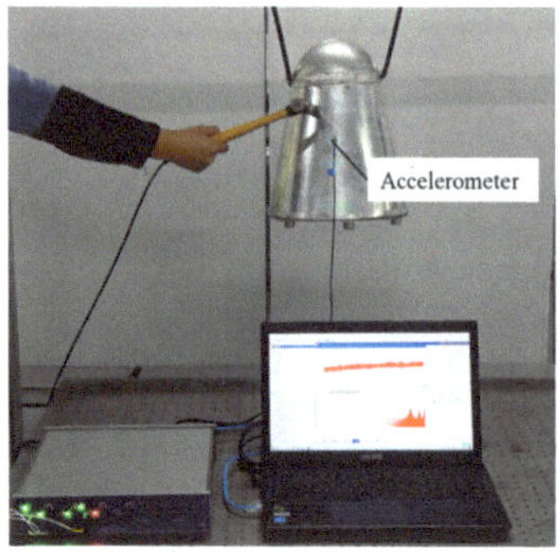

Figure 2. Modal experiment of aircraft instrument cabin.

3.2. Modal Calculation of the Instrument Compartment of the Aircraft

When establishing the finite element model of the instrument compartment, we removed the bolts top hooks, and other parts that have little influence on the structural dynamics, and retained the bolt holes. The finite element model of the instrument compartment obtained based on the virtual material method is shown in Figure 3, with a minimum grid size of 5 mm, including 114,049 elements and 288,505 nodes. The model was analyzed in free mode.

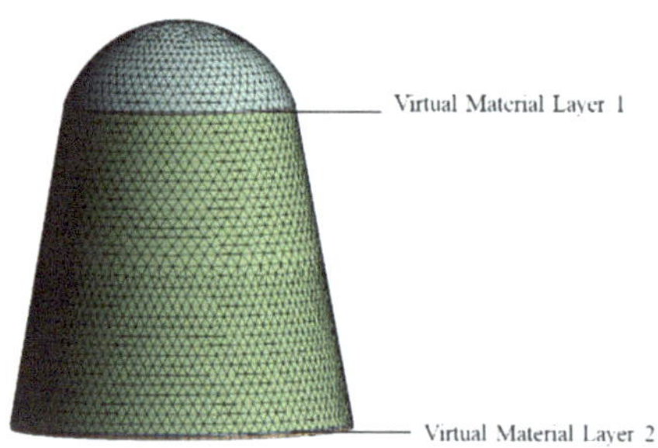

Figure 3. Finite element model of the virtual materials method instrument compartment.

3.3. Identification of Elastic Parameters of Structural Virtual Material Layer Based on Genetic Algorithm

The comparison of the first four modal shapes of the aircraft instrument compartment obtained from experiments and calculations is shown in Figure 4. The calculated modal shapes are fundamentally consistent with the experimental modal shapes. On the premise of consistent modal shapes, according to Equation (2), the objective function for the genetic algorithm is defined as the minimization of the differences between the experimental and simulated values of the structure's first three natural frequencies. With each iteration of the genetic algorithm, the updated elastic modulus and Poisson's ratio after iteration are applied to the connection layer model. The process continues until the termination criteria are met.

The initial values and ranges of variation for the material parameters are provided in Table 1.

Table 1. Initial values and variation range of virtual material layer parameters.

Material Parameters	Initial Value	Lower Limit	Upper Limit
Elastic modulus/GPa	70	0.1	80
Poisson's ratio	0.3	0.15	0.45

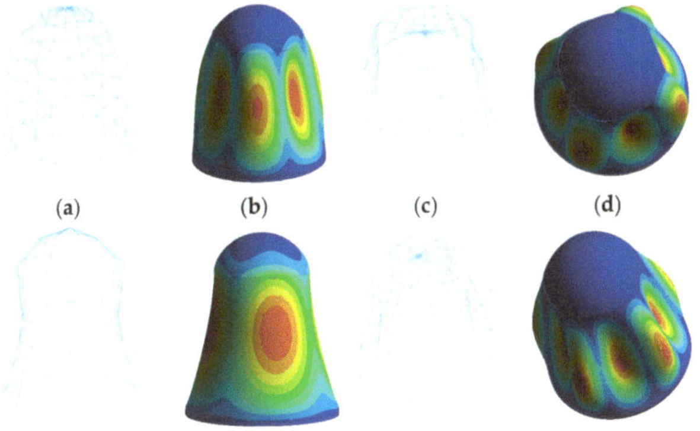

Figure 4. (a) 1st order test mode shape; (b) 1st order simulated mode shape; (c) 2nd order test mode shape; (d) 2nd order simulated mode shape; (e) 3rd order test mode shape; (f) 3rd order simulated mode shape; (g) 4th order test mode shape; (h) 4th order test mode shape.

After calculation, the objective function converges after 22 iterations, and 20 design points are calculated for each iteration. The elastic parameters of the virtual material layer of the structure after parameter identification are shown in Table 2. The calculated modal frequencies obtained by the virtual material model and the binding constraint model are compared with the experimental modal frequencies, as shown in Table 3. The frequency error of the first three modes of the model obtained by using the virtual material method is less than 3%, and the calculation accuracy is high, and the error of the fourth natural frequency that does not participate in the recognition is also less than 3%, which verifies the accuracy of the parameter identification method. Compared with the traditional binding constraint method, the modal frequency error obtained by the virtual material method is lower, which indicates that the virtual material method has higher accuracy than the binding constraint method.

Table 2. Model virtual material layer elastic parameters.

	Elastic Modulus/GPa	Poisson's Ratio
Connection Layer 1	4.836	0.287
Connection Layer 2	3.556	0.265

Table 3. Instrument compartment calculation and test modal frequency comparison.

Order	Test Modal Frequency/Hz	Virtual Materials Method		Binding Constraint Method	
		Calculate Modal Frequency/Hz	Error %	Calculate Modal Frequency/Hz	Error %
1	98.44	101.24	2.84	105.61	7.28
2	201.56	204.66	1.54	217.81	8.06
3	240.63	246.79	2.56	258.51	7.43
4	309.38	315.85	2.09	330.23	6.74

4. Calculation and Test Comparison and Analysis of the Acoustic and Vibrating Response of the Instrument Compartment

To compare the differences in computational results between the virtual material method and the tied constraint method, models of the instrument compartment are created using both approaches. The vibration and noise responses are then analyzed based on these models. The mesh models are imported into acoustic and vibration analysis software, and the frequency band from 50 to 400 Hz, as shown in Figure 5 according to environmental test standard spectra, is selected as the noise excitation spectrum for both computation and experimentation. The calculated values of the structural acoustic and vibration responses are obtained through this process.

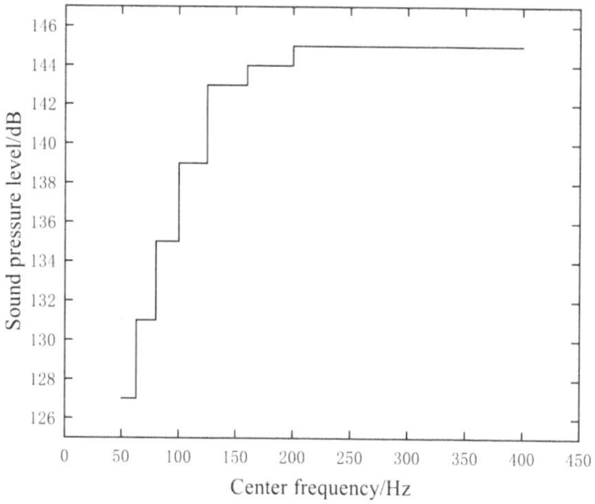

Figure 5. Experimental spectrum of the external acoustic field.

The experiment in this article is based on the acoustic control module of the HAN-HANG testing system and conducted in a simple reverberation room at Xi'an Jiaotong University. The volume of the reverberation room is 70 m^3, and the walls are made of high-density concrete, which can effectively reflect sound waves. The experiment measured the structural vibration response and the internal noise response under a specified noise spectrum excitation, with a noise exposure duration of 30 s. The model was suspended within the reverberation chamber using rubber cords. Three microphones were arranged around the model to employ a three-point averaging method for closed-loop control of the acoustic field. The test site is illustrated in Figure 6.

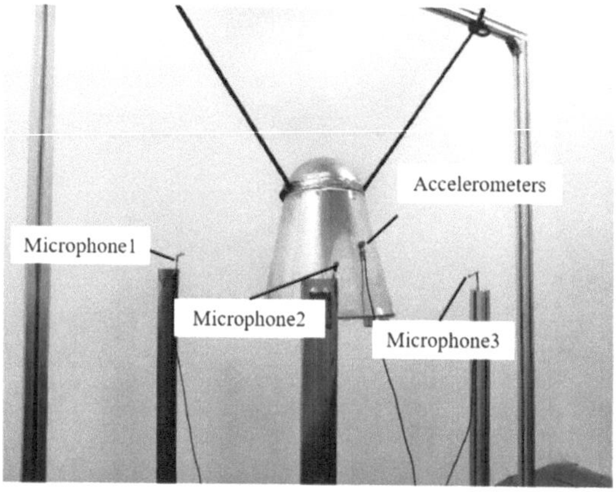

Figure 6. Model noise excitation test site.

4.1. Comparison of the Vibration Response Calculation and Test Results of the Instrument Compartment

Figure 7 shows the comparison of the calculated acceleration power spectral density curves of the two models with the experimental measured acceleration power spectral density curves, and gives the root mean square value of acceleration. As seen in Figure 7, the acceleration power spectral density curve calculated based on the virtual material model is closer to the experimental values, and the frequency position of the vibration peak is very close. Table 4 shows the frequency of the peak in the experimental and simulated response curves, as well as the frequency offset and error percentage of the simulated values relative to the experimental values. The frequency offset of the virtual material model is much lower than that of the binding constraint model, not exceeding 7 Hz, while the frequency offset of the third peak point of the binding constraint model exceeds 20 Hz. This is caused by the natural frequency of the computational model. The binding constraint law rigidly connects the joint surface, increasing the structural stiffness and natural frequency accordingly. In addition, the vibration peak of the virtual material model is also closer to the actual value. The root mean square value of acceleration obtained from the experiment is 10.35 g, and the root mean square values of acceleration calculated by the virtual material model and the binding constraint model are 11.23 g and 11.71 g, respectively. Obviously, the accuracy of the virtual material model is higher.

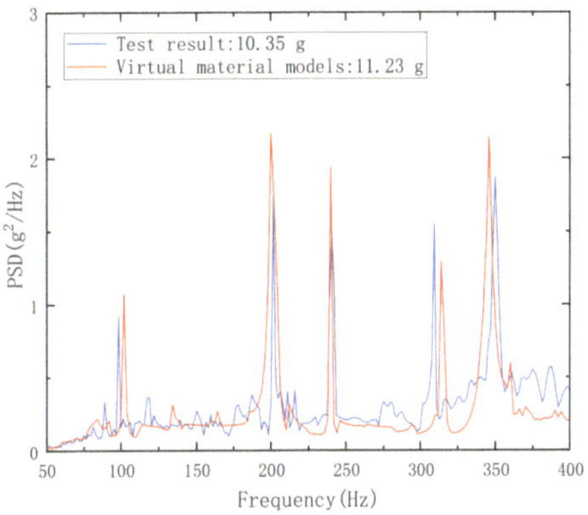

(a)

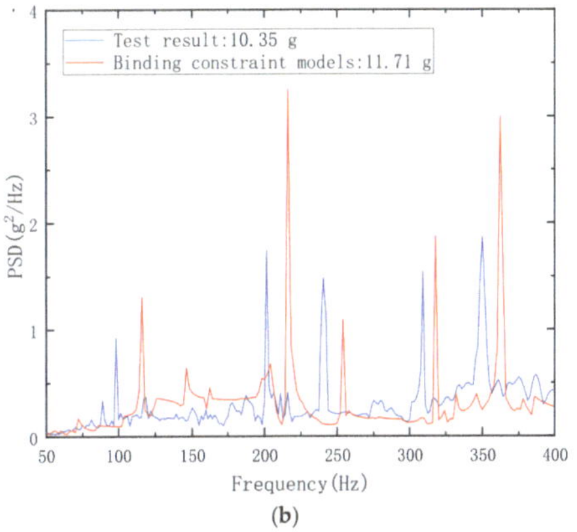

(b)

Figure 7. (**a**) Comparison of Calculated and Experimental Values of Acceleration PSD (Virtual Material Method); (**b**) Comparison of Calculated and Experimental Values of Acceleration PSD (Binding Constraint Method).

Table 4. Comparison of Peak Frequency Calculation and Experiment between Two Models.

Test Peak Frequency/Hz	Virtual Materials Method			Binding Constraint Method		
	Calculate Peak Frequency/Hz	Frequency Offset/Hz	Error %	Calculate Peak Frequency/Hz	Frequency Offset/Hz	Error %
97.52	102.86	5.34	5.48	105.87	8.09	8.30
200.32	198.87	−1.45	−0.72	216.96	17.49	8.73
238.45	237.01	−1.44	−0.60	256.43	21.5	9.02
307.89	314.43	6.54	2.12	328.55	15.8	5.13
349.76	346.67	−3.09	−0.88	361.87	14.87	4.25

4.2. Calculation of Noise Response of Acoustic Cavity Inside the Instrument Compartment and Comparison of Test Results

Table 5 shows the comparison of the calculated and experimental values of the two modeling methods within the 1/3 octave range of the sound pressure level of the internal acoustic cavity. At each center frequency point, the sound pressure level obtained based on the virtual material method is closer to the test value, and the total sound pressure level error is not more than 2 dB. The result obtained by the binding constraint method is larger, and the total sound pressure level error is 2.97 dB, because the binding constraint method makes the structural connection surface rigidly connected, which will amplify the noise response to a certain extent.

Table 5. Comparison of sound pressure level calculation and test in sound cavity.

Center Frequency/Hz	Sound Pressure Level in the Internal Acoustic Cavity/dB		
	Test Values	Virtual Materials Method	Binding Constraint Method
50	108.4	110.4	113.2
63	112.2	113.7	114.9
80	116.7	118.4	119.7
100	119.3	121.3	122.5
125	122.5	124.8	126.9
160	124.2	125.4	129.1
200	127.5	128.2	130.7
250	128.7	129.9	131.1
315	129.9	130.7	131.9
400	129.5	131.1	132.6
Total sound pressure level	135.76	136.98	138.73

5. Conclusions

This paper focuses on the instrument compartment of a specific aircraft model, employing the virtual material method for precise modeling. The virtual material parameters for the bolted joints were obtained through theoretical calculations and parameter identification, further analyzing the impact of joint contact characteristics on the prediction of structural vibration and noise. The main conclusions are as follows:

1. Accurate Simulation with Virtual Material Method: By adding a layer of virtual material at the bolted interfaces, the virtual material method can accurately simulate the contact characteristics of the joints. While theoretically calculating the elastic parameters (elastic modulus, Poisson's ratio) of the virtual material is complex, using parameter identification provides a simpler approach to obtaining these parameters with high precision.
2. High Precision in Modal Analysis: For structures modeled using the virtual material method, the calculated modal shapes are consistent with experimental modal shapes,

and the error between computed frequencies and experimental frequencies is within 3%. Compared to the tied constraint method, this approach offers higher accuracy, indicating that the virtual material method better approximates real-world conditions when simulating bolted connections.

3. Superior Vibration and Acoustic Prediction: The structure modeled using virtual material method has a smaller frequency offset of peak vibration response, not exceeding 7 Hz, while the peak frequency offset of the bound constraint model exceeds 20 Hz; The root mean square value of the vibration response acceleration of the virtual material model is also closer to the experimental results. In terms of noise response, the sound pressure level error at each center frequency point is smaller, and the total sound pressure level error does not exceed 2 dB. The above results indicate that the virtual material method is more accurate in describing the dynamic characteristics of structures.

Author Contributions: Conceptualization, S.S. and J.W.; methodology, S.S. and J.W.; software, J.W. and R.H.; validation, S.S. and R.H.; formal analysis, S.S.; investigation, J.W.; resources, S.S. and C.L.; data curation, S.S.; writing—original draft preparation, R.H.; writing—review and editing, J.W. and C.L.; visualization, S.S.; supervision, S.S.; project administration, S.S.; funding acquisition, S.S. All authors have read and agreed to the published version of the manuscript.

Funding: This research received no external funding.

Institutional Review Board Statement: Not applicable.

Informed Consent Statement: Not applicable.

Data Availability Statement: The data provided in this study can be provided at the request of the corresponding author. The research data presented in the paper consists of simulation and experimental results. We have provided the necessary simulation settings and experimental setups within the paper to enable scholars to replicate our results. Therefore, it is not necessary for us to specifically upload the data. Additionally, we have not uploaded the data from the paper to any public datasets. However, if there is a need, we are willing to provide the data upon request.

Conflicts of Interest: The authors declare no conflict of interest.

References

1. Wang, R.; Crocombe, A.D.; Richardson, G.; Underwood, C.I. Energy dissipation in spacecraft structures incorporating bolted joints operating in macroslip. *J. Aerosp. Eng.* **2008**, *21*, 19–26. [CrossRef]
2. Crocombe, A.D.; Wang, R.; Richardson, G.; Underwood, C.I. Estimating the energy dissipated in a bolted spacecraft at resonance. *Comput. Struct.* **2006**, *84*, 340–350. [CrossRef]
3. De Benedetti, M.; Garofalo, G.; Zumpano, M.; Barboni, R. On the damping effect due to bolted junctions in space structures subjected to pyro-shock. *Acta Astronaut.* **2007**, *60*, 947–956. [CrossRef]
4. Zhao, Y.S.; Yang, C.; Cai, L.G.; Shi, W.M.; Liu, Z.F. Surface contact stress-based nonlinear virtual material method for dynamic analysis of bolted joint of machine tool. *Precis. Eng. J. Int. Soc. Precis. Eng. Nanotechnol.* **2016**, *43*, 230–240. [CrossRef]
5. Zhang, J.; Wang, Y.; Wu, P. Equivalence analysis of acoustic excitation and random vibration exerted on spacecraft units. *J. Phys. Conf. Ser.* **2023**, *2569*, 012007. [CrossRef]
6. Forgrave, J.C.; Man, K.F.; Newell, J.M. Spacecraft acoustic and random vibration test optimization. *Sound Vib.* **1999**, *33*, 28–31.
7. Li, Q.; Xing, L.; Bai, J.; Zou, Y. Method Study on Response Prediction of Structural Vibrations in Spacecraft Acoustic Tests. *Chin. J. Theor. Appl. Mech.* **2019**, *51*, 569–576.
8. Rarata, Z.; Dacko, A.; Barcinski, T.; Polak, S.; Musial, J.; Kubacki, S.; Kowalski, T.; Sikorski, A.; Baran, J. Vibro-acoustic response of spacecraft instrument subjected to diffuse sound field: Numerical simulations and experimental verification. *Appl. Acoust.* **2021**, *184*, 108338. [CrossRef]
9. Karpenko, M.; Stosiak, M.; Deptuła, A.; Urbanowicz, K.; Nugaras, J.; Królczyk, G.; Żak, K. Performance evaluation of extruded polystyrene foam for aerospace engineering applications using frequency analyses. *Int. J. Adv. Manuf. Technol.* **2023**, *126*, 5515–5526. [CrossRef]

10. Sestieri, A.; Carcaterra, A. Vibroacoustic: The challenges of a mission impossible? *Mech. Syst. Signal Proc.* **2013**, *34*, 1–18. [CrossRef]
11. Yu, Y.; Chen, Y.M.; Dou, Y.H.; Yuan, Q.Y.; Tang, J.R.; Cao, Y.P. Parameter Identification and Energy Dissipation Analysis of Premium Connections Based on the Iwan Model. *Processes* **2024**, *12*, 2382. [CrossRef]
12. Brake, M.R.W. A reduced Iwan model that includes pinning for bolted joint mechanics. *Nonlinear Dyn.* **2017**, *87*, 1335–1349. [CrossRef]
13. Mignolet, M.P.; Song, P.C.; Wang, X.Q. A stochastic Iwan-type model for joint behavior variability modeling. *J. Sound Vib.* **2015**, *349*, 289–298. [CrossRef]
14. Chu, Y.P.; Wen, H.; Chen, T. Nonlinear Modeling and Identification of an Aluminum Honeycomb Panel with Multiple Bolts. *Shock Vib.* **2016**, *2016*, 1276753. [CrossRef]
15. Alamdari, M.M.; Li, J.C.; Samali, B.; Ahmadian, H.; Naghavi, A. Nonlinear Joint Model Updating in Assembled Structures. *J. Eng. Mech.* **2014**, *140*, 04014042. [CrossRef]
16. Zhan, M.; Guo, Q.T.; Yue, L.; Zhang, B.Q. Modeling and Stochastic Model Updating of Bolt-Jointed Structure. *Shock Vib.* **2018**, *2018*, 8354249. [CrossRef]
17. Tian, H.L.; Li, B.; Liu, H.Q.; Mao, K.M.; Peng, F.Y.; Huang, X.L. A new method of virtual material hypothesis-based dynamic modeling on fixed joint interface in machine tools. *Int. J. Mach. Tools Manuf.* **2011**, *51*, 239–249. [CrossRef]
18. Sun, Y.Y.; Xiao, H.F.; Xu, J.W. Contact stiffness ratio of tribological interface using the equivalent thin layer and the micro-slip model. *Proc. Inst. Mech. Eng. Part C J. Eng. Mech. Eng. Sci.* **2020**, *234*, 444–456. [CrossRef]
19. Sun, Y.Y.; Chuang, H.C.; Xiao, H.F.; Xu, J.W. Prediction of the normal contact stiffness between elastic rough surfaces in lubricated contact via an equivalent thin layer. *J. Vib. Control* **2020**, *26*, 2060–2069. [CrossRef]
20. Yang, Y.; Cheng, H.; Liang, B.A.; Zhao, D.; Hu, J.S.; Zhang, K.F. A novel virtual material layer model for predicting natural frequencies of composite bolted joints. *Chin. J. Aeronaut.* **2021**, *34*, 101–111. [CrossRef]
21. Liu, P.; Guan, T.; Wang, X. Nonlinear connection layer modeling and parametric identification for single bolt connected beam. *J. Vib. Shock* **2023**, *42*, 190–197.

Disclaimer/Publisher's Note: The statements, opinions and data contained in all publications are solely those of the individual author(s) and contributor(s) and not of MDPI and/or the editor(s). MDPI and/or the editor(s) disclaim responsibility for any injury to people or property resulting from any ideas, methods, instructions or products referred to in the content.

materials

MDPI

Article

Impact Damage Localization in Composite Structures Using Data-Driven Machine Learning Methods

Can Tang [1], Yujie Zhou [1], Guoqian Song [2] and Wenfeng Hao [2,*]

1 College of Civil Science and Engineering, Yangzhou University, Yangzhou 225127, China; tangcanvip@126.com (C.T.)
2 College of Mechanical Engineering, Yangzhou University, Yangzhou 225127, China
* Correspondence: hwf@yzu.edu.cn

Abstract: Due to the uncertainty of material properties of plate-like structures, many traditional methods are unable to locate the impact source on their surface in real time. It is important to study the impact source-localization problem for plate structures. In this paper, a data-driven machine learning method is proposed to detect impact sources in plate-like structures and its effectiveness is tested on three plate-like structures with different material properties. In order to collect data on the localization of the impact source, four piezoelectric transducers and an oscilloscope were utilized to construct an experimental platform for impulse response testing. Meanwhile, the position of the impact source on the surface of the test plate is generated by manually releasing the steel ball. The eigenvalue of arrival time in the time domain signal is extracted to build data sets for machine learning. This paper uses the Back Propagation (BP) neural network to learn the difference in the arrival time of each sensor and predict the location of the impact source. The results demonstrate that the machine learning method proposed in this paper can predict the location of the impact source in the plate-like structure without relying on the material properties, with high test accuracy and robustness. The research work in this paper can provide experimental methods and testing techniques for locating impact damage in composite material structures.

Keywords: damage localization; composite structures; machine learning; back propagation neural network; arrival time

Academic Editor: Yuwen Chen

Received: 4 January 2025
Revised: 14 January 2025
Accepted: 17 January 2025
Published: 19 January 2025

Citation: Tang, C.; Zhou, Y.; Song, G.; Hao, W. Impact Damage Localization in Composite Structures Using Data-Driven Machine Learning Methods. *Materials* **2025**, *18*, 449. https://doi.org/10.3390/ma18020449

Copyright: © 2025 by the authors. Licensee MDPI, Basel, Switzerland. This article is an open access article distributed under the terms and conditions of the Creative Commons Attribution (CC BY) license (https://creativecommons.org/licenses/by/4.0/).

1. Introduction

Plate structures, especially composite ones, can be seen everywhere in various fields, such as aerospace and civil engineering [1,2]. The structural health monitoring (Structure Health Monitoring, SHM) method and non-destructive testing (Non-destructive Testing, NDT) technology for these structures are of great significance [3–5]. In transportation, service, and maintenance, the plate structure will inevitably bear the impact of various impact objects, such as bird impact, hail impact, tool drops, etc. [6]. The impact can usually be divided into high speed and low velocity [7]. An impact velocity of more than 100 m/s is called high-speed impact. The strain rate is exceptionally high, and it is easy to form apparent damage in the plate-like structure. If the impact velocity is below 10 m/s, it is called low-velocity impact, and the strain rate is low at this time. Phenomena such as falling impacts are all low-speed impacts. This type of damage is often invisible and highly concealed and, to some extent, is more harmful than the more apparent high-speed impact damage. When composite structures are subjected to such impacts, it may cause severe damage to the internal structure without obvious damage marks on the surface, resulting in a relatively low probability of visual detection of damage. If this hidden damage cannot be detected and maintained in time, it may cause disastrous consequences if allowed to accumulate [8]. Currently, the identification methods for low-speed impact load localization mainly include identification methods based on time difference in arrival, impact location identification methods based on optimization algorithms, and impact location identification methods based on signal similarity. Gaul et al. [9] used wavelet transform to extract the difference in arrival time at different frequencies to achieve impact localization on aluminum alloy plates. Sai et al. [10] used the wavelet basis function based on Morlet to extract the wave arrival time difference in the impact response signal. They determined the position of the impact point according to the triangular sensor array. Jiang et al. [11] used the Shannon wavelet to extract the narrowband signal difference in the arrival of the impact response signal. They used the MUSIC algorithm to obtain the position of the impact point. Kundu et al. [12] used six sensors to determine the position of the excitation source on the composite plate without knowing the wave velocity in each direction. Zheng et al. [13] provided an effective strategy for extracting second harmonic Lamb waves from dynamic response signals based on wavelet transforms for the detection of closed microcracks, accurately locating closed microcracks in plate-like metals and composite structures in both numerical simulation and experimental environments. The traditional method based on the time difference in arrival has a certain degree of uncertainty, and it is easy to produce significant errors for the impact localization problem. Seydel [14] roughly calculated the location of the impact load by the time-of-arrival method and then used an optimization algorithm to calculate the minimum difference between the predicted response and the actual response to estimate the impact location. Sai [15] established a nonlinear relationship model between the stress wave time difference in arrival and the spatial position of the sensor. They used the quasi-Newton algorithm to solve the nonlinear equations. The wide application of the optimization algorithm has improved the positioning accuracy very well. Shrestha et al. [16] compared the root mean square error and error outliers of the sample impact point response signal and the unknown impact point response signal to locate the impact of the composite plate structure. Kim et al. [17] achieved the impact localization of composite plate structures by comparing the cross-correlation between the sample point response signal and the unknown point response signal. Zhao et al. [18] performed impact localization based on the k-order natural frequency amplitude deviation between the sample point response signal and the unknown point response signal. Limited by the uncertainty of the material properties of some plate-like structures, its velocity model is often difficult to obtain accurately, leading to deviations in positioning results. Various optimization algorithms can improve accuracy to a certain extent but cannot achieve real-time and fast positioning.

Based on the emerging field of computational intelligence in recent years, new methods exist to process the obtained signals to further improve the accuracy and efficiency of health monitoring of plate structures [19]. Cuomo et al. [20] first obtained a baseline consisting of the structural response induced by the impact test and subsequently evaluated the impact location using the highest cross-correlation coefficient, somewhat overcoming the limitations of the current composite impact localization process. Geng et al. [21] used fiber-grating sensors to obtain structural dynamic response signals and trained neural networks by extracting frequency-domain response features to realize impact damage identification of composite laminates. Using numerical simulation, Dipietrangelo et al. [22] used the K-Fold cross-validation procedure to evaluate the performance of polynomial models of different degrees using different combinations of training/test sets and calculated the average radial error. For shallow neural networks, three learning algorithms were compared: Levenberg–Marquardt, Bayesian Regularization, and Scaled Conjugate Gradient, which confirmed the effectiveness of machine learning for impact detection. Liu et al. [23] used the Empirical Mode Decomposition (EMD) method to remove the trend component in the low-speed impact signal, and extracted time-domain features, frequency-domain features and time-frequency domain features from the preprocessed impact signal, building a Hybrid Support Vector Model to determine the location of low-velocity impacts on composite panel structures. Jiang et al. [24] used fiber Bragg gratings to detect low-velocity impact sources on the surface of carbon fiber-reinforced plastic structures, and used Extreme Learning Machines (ELM) for regional localization. Sai et al. [25] established an Extreme Learning Machine (ELM)-based impact localization model with faster training speed and fewer parameters according to the energy of available frequency band signals. Ai et al. [26] collected an Acoustic Emission (AE) data set by performing impact experiments on a full-scale thermoplastic aircraft elevator in a laboratory environment. A data set consisting of AE parametric features and a data set consisting of AE waveforms were assigned to random forest classifiers and deep-learning networks to study their suitability for impact source localization. Chen et al. [27] utilized a Convolutional Recurrent Encoder-Decoder Neural Network (ED-CRNN) and a Deep Convolutional Recurrent Neural Network (DCRNN) for impact load reconstruction and localization. Jierula [28] installed the sensor in the concrete column, applied the Neighborhood Component Analysis (NCA) feature selection method to select important parameters as the input of the neural network, and proposed a machine learning-based method to locate the impact source in the concrete column. The data-driven method can effectively analyze damage characteristics for classification and identification, overcome the complexity and uncertainty of traditional methods, significantly reduce the workload, and effectively improve the process of structural damage identification. However, complex machine learning algorithms make it difficult to provide the required input parameters in practical engineering testing, resulting in poor engineering practicality. BP neural networks are relatively mature in both network theory and performance. Its outstanding advantages are strong nonlinear mapping ability and flexible network structure. The number of intermediate layers and neurons in each layer of the network can be set arbitrarily according to specific situations, and their performance varies with the difference in structure. However, BP neural networks also have some major shortcomings. The learning speed is slow, and even a simple problem usually requires hundreds or even thousands of learning iterations to converge, therefore, it is easy to fall into local minima. There is no corresponding theoretical guidance for the selection of network layers and the number of neurons. The ability to promote online is limited. There have been many improvement measures for the above issues, with the most researched being on how to accelerate the convergence speed of the network and avoid falling into local minima as much as possible [29,30]. In this paper, the BP neural network is used to detect impact sources in plate-like structures and its effectiveness is tested on three plate-like structures with different material properties.

The follow-up content of this paper is divided into four parts. Section 2 introduces the working principle of the BP neural network used in this paper and the preprocessing of time-domain signals. Section 3 presents the experimental setup and data acquisition system. Section 4 gives the training and prediction results of the model. Section 5 concludes the work.

2. Algorithm of BP Neural Network and Signal Preprocessing

2.1. The Mechanism of the BP Neural Network

The BP neural network is relatively mature in terms of its theoretical basis and performance. Its outstanding advantages are its nonlinear solid mapping ability and flexible network structure. A typical BP neural network consists of three parts: input layer, hidden layer, and output layer, as shown in Figure 1. It has a precise learning mechanism, mainly including two processes: forward propagation of signals and backpropagation of errors. When the input layer of the BP neural network receives the signal, it will pass the signal to the hidden layer, and the hidden layer will process the signal according to the weight, bias, and activation function of the connection, and then pass it to the output layer to output the corresponding predicted value. When the error between the predicted and expected values does not meet the preset target accuracy requirements, the network will feed back the error information layer by layer from the output layer to the input layer. When the error is backpropagated, the weights and offsets of every layer connection will be adapted and updated according to the method of gradient descent. Through continuous training and correction, the error between the predicted and actual values will gradually become smaller, and the predicted result will reach expected outcome. Then, the learning process is over.

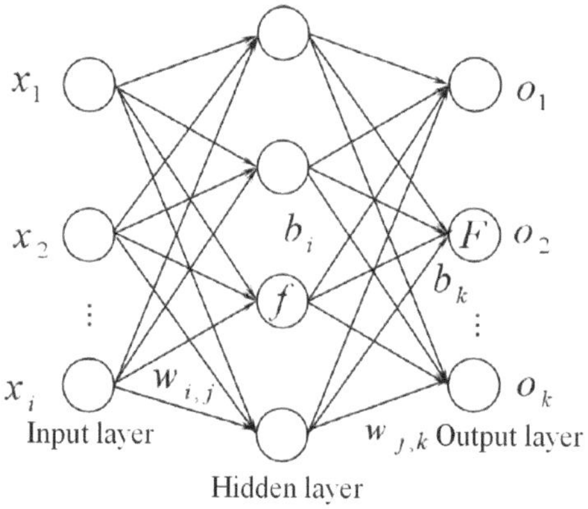

Figure 1. Schematic diagram of BP neural network structure.

The linear function is selected as the activation function of the input layer, and the sigmoid function is used as the activation function of the hidden layer. For the input layer part, the input data are the extracted signal feature value $X = [x_1, x_2, \ldots, x_i, x_n]^T$, and the output $y_i = X_i$ due to the use of a linear function.

For the hidden layer part, the input value of the neuron is equal to the sum of the input value connected to it multiplied by the corresponding weight plus the additional bias of the neuron:

$$net_j = \sum_{i=1}^{n} w_{ij} y_i + b_i \tag{1}$$

n represents the input layer neurons, and j represents the hidden layer neurons.

The output value of the hidden layer is the value calculated by the sigmoid function:

$$y_j = f(net_j) = \frac{1}{1 + exp(-net_j)} \tag{2}$$

For the input layer part, the input value of the neuron is equal to the sum of the input value connected to it multiplied by the corresponding weight and then the additional bias of the neuron:

$$net_k = \sum_{k=1}^{n} w_{jk} y_j + b_k \tag{3}$$

Similarly, the output value of the output layer is the value after the sigmoid function operation:

$$y_k = f(net_k) = \frac{1}{1 + exp(-net_k)} \tag{4}$$

Randomly assign a set of non-zero numbers and small initial values to the weights w_{ij} and w_{jk} between the input layer, the hidden layer, and the output layer. After setting a series of hyperparameters, such as training target, learning rate, the number of iterations, etc., the output y_i, y_j, and y_k of each layer can be calculated from the output layer. Then, calculate the magnitude of the error between the predicted value and the expected value of each neuron in the output layer:

$$\delta_k = \frac{\sum_{k=1}^{n} (y' - y_k)^2}{n} \tag{5}$$

After obtaining the predicted value of the output layer, continue to calculate the error size of each neuron in the hidden layer forward:

$$\delta_j = \sum_j \delta_k w_{kj} \tag{6}$$

Use the resulting error to adjust the weights and biases of each layer:

$$w_{ij}(m+1) = w_{ji}(m) + \Delta w_{ji}(m) \tag{7}$$

In the following:

$$\Delta w_{ji}(m) = m_c \cdot w_{ij}(m-1) + l_r \cdot \delta_j \cdot f'(net_j) \tag{8}$$

In Equation (7), m is the number of iterations, and in Equation (8), m_c is the momentum coefficient; generally 0.9~1. l_r is the hyperparameter learning rate.

After a calculation of forward propagation, compare the error δ_k of the output layer with the maximum error ε of the set training target. If $\delta_k \leq \varepsilon$, stop training. Otherwise, the error is passed forward, the weights and biases between the layers are adjusted, and the forward propagation is performed again. After the conditions for stopping the training are met, the training ends.

The Sigmoid function requires the input value to be in the interval [−1, 1]. Hence, the samples must be normalized to avoid neuron saturation during training and make the model converge faster. This study uses the mapminmax function for normalization, which can map each sample value to between −1 and 1. The model chosen in this study has a three-tier architecture, including one layer of an input layer, a hidden layer, and an output layer, and its diagrammatic sketch is shown in Figure 2. The programming and training of the BP neural network are carried out on the MATLAB platform. Use the trainlm function as the training function and set the following hyperparameters simultaneously: the maximum training times are 1000, the learning rate is 0.0001, and the minimum error goal is 0.000001. There is no clear theory about determining the number of hidden layer nodes in the neural network. When the nodes' number is small, it is easy to cause underfitting and affect the prediction accuracy. When the number of hidden layer nodes is too large, the training time of the model may be prolonged, and overfitting is more likely to occur. Here, this study uses the empirical formula: the hidden nodes number = $\sqrt{(m+n)} + a$ for determining the number of nodes, where m is the amount of input neurons, n is the amount of output neurons, and a is a random number between 1 and 10. Finally, six hidden node numbers are selected according to the formula.

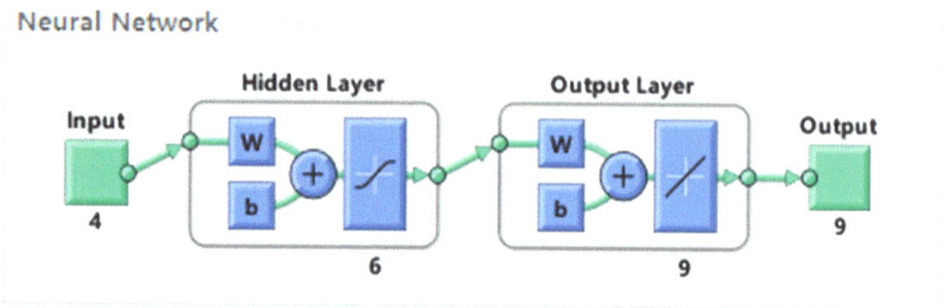

Figure 2. Schematic diagram of the BP neural network model established in this study.

2.2. *Processing of Captured Signals*

The training of the BP neural network needs to choose a typical characteristic value as the input, so we need to process the time domain signal received by the sensor. Clearly, the position of the sound source strongly correlates with the wave propagation time in the plate. Therefore, we choose the time-of-first-arrival parameter in the time-domain signal. According to the set integration time step and total duration, the length of each sensor signal is 1000 samples, as shown in Figure 3. The time-of-first-arrival of the time-domain signals obtained from the four sensors is collected as the input of the model. It is worth noting that to prevent data leakage during neural network training, this paper performs data segmentation before model evaluation. Then, it performs data preparation only on the training data set and then, applies the data preparation techniques to the training and test sets.

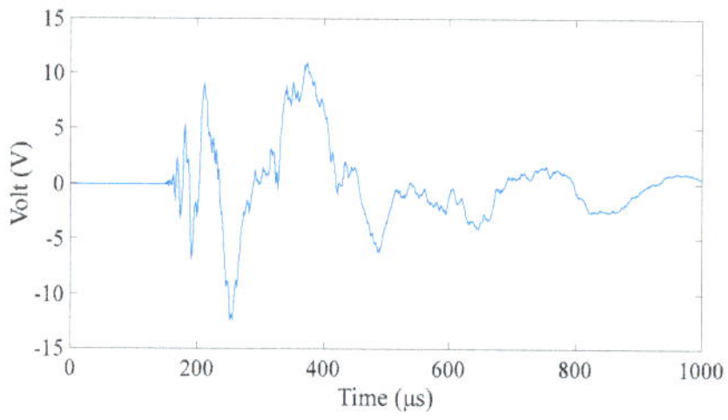

Figure 3. Received signal of a sensor when the aluminum plate is subjected to an impact.

Alt Text: The time-domain signal received by a certain sensor has a signal length of 1000 microseconds. The abscissa of the signal is time in microseconds. The ordinate is the amplitude, and the unit is volts due to the use of an oscilloscope for reception.

3. Experimental Procedure

3.1. *Experimental Setup*

Generate and record impact signals using the experimental setup shown in Figure 4. The experimental objects in this paper are three different plate structures. The aluminum plate is a representative of isotropic materials, while the unidirectional composite plate and orthotropic plate are typical layering representatives of transversely isotropic materials and composite materials. Among them, the size of the isotropic aluminum plate is $400 \times 500 \times 1$ mm^3, which is divided into nine regions with an area of 133.3×166.7 mm^3 on average. The dimensions of the orthotropic laminate and the unidirectional laminate are both $450 \times 450 \times 3$ mm^3, and the thickness of a single layer is 0.125 mm. There are 16 layers in total, equally divided into nine areas. Polymer foam is padded at the bottom of the board to reduce vibration and energy transfer from the environment to the board. A circular piezoelectric transducer (PZT) with a diameter of 10 mm and a thickness of 1 mm is used as the piezoelectric chip, and it is arranged in the middle of the four sides at a distance of

20 mm from the boundary as the receiving source. Before pasting the piezoelectric chip, the surface of the aluminum plate is cleaned with alcohol. Apply the quick-drying adhesive evenly to the designated position and then paste and fix the PZT. A 54820A oscilloscope produced by Agilent was used to record the measurement results. The acquisition signals from all four channels have 16-bit resolution, the sampling rate is set to 1 MHz, the duration of each measurement is 1 ms, and the trigger level is set to 100 mV.

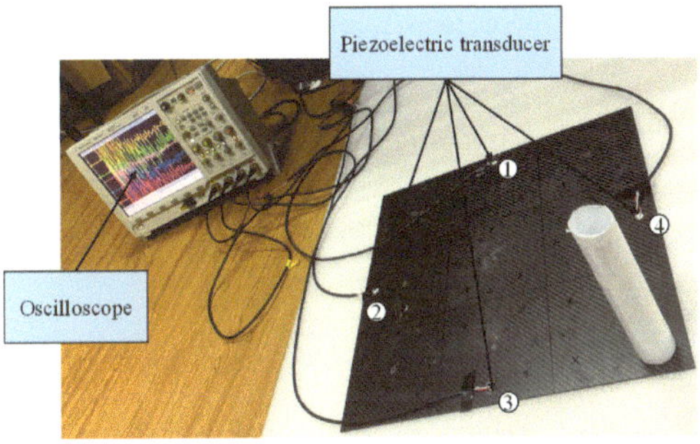

Figure 4. Schematic diagram of the experimental platform.

Alt Text: The steel ball is impacted on the surface of the plate structure by paper guides, and the resulting signal is received by piezoelectric sensors on the four edges of the plate and displayed by an oscilloscope.

Impact signals are generated by impacting pellets of different diameters. Steel balls made of hardened steel excite elastic waves in the plate, which are released from different heights. The diameters of the respective balls were 8 mm, 10 mm, and 15 mm. The repeatability of the impact position and height is ensured by a paper guide, which determines the orientation of the steel ball during free fall. The thin plate entirely absorbs the impact energy of the steel ball, and elastic deformation mainly occurs during the impact process. The acquired sensor signal shows only a single waveform, which is excited by the ball's impact. The lack of other waveforms in the captured time-domain signal indicated that the rebound of the steel ball was not recorded. The model's training needs to input a certain number of samples with specific variability. The resulting waveform after each impact is differently affected by damping, dispersion, and edge reflections, allowing us to obtain a highly variable database.

3.2. Experimental Process

This paper tests the influence of three factors on the accuracy of the BP neural network in predicting the impact area of the falling ball. The experimental procedure is as follows:

(1) At the center points of nine areas, use a small steel ball with a diameter of 10 mm to drop from a height of 200 mm; repeat ten times for each area center, and a total of 90 sets of data are utilized as the training set of the model.

(2) Explore the impact of different drop heights on prediction accuracy. Also, use 10 mm small steel balls to fall at the center points of each area, change the length of the guide rail, and make the small balls fall from the heights of 150 mm and 100 mm, respectively, as shown in Figure 5a. Each regional center is repeated ten times, and a total of 90 sets of data are used as the test set of the neural network.

(3) Explore the influence of different steel ball diameters on the prediction accuracy. Also, use a 200 mm long guide rail at the center point of each area to ensure that the steel balls fall at the same height, change the diameter of the small balls used, and use small steel balls with diameters of 8 mm and 15 mm to drop from the same height, as

shown in Figure 5a. Each regional center is repeated ten times, and a total of 90 sets of data are utilized as the test set of the model.

(4) Explore the impact of different falling positions on prediction accuracy. Steel balls of the same diameter as the test set were dropped from the same height. For the aluminum plate, offset the drop point by 30 mm and 60 mm from the center of the area. For two composite material panels, the landing point positions are randomly selected at 20 mm and 60 mm from the center of the area. As shown in Figure 5. Each drop point is repeated ten times, and 90 sets of data are used as the test set.

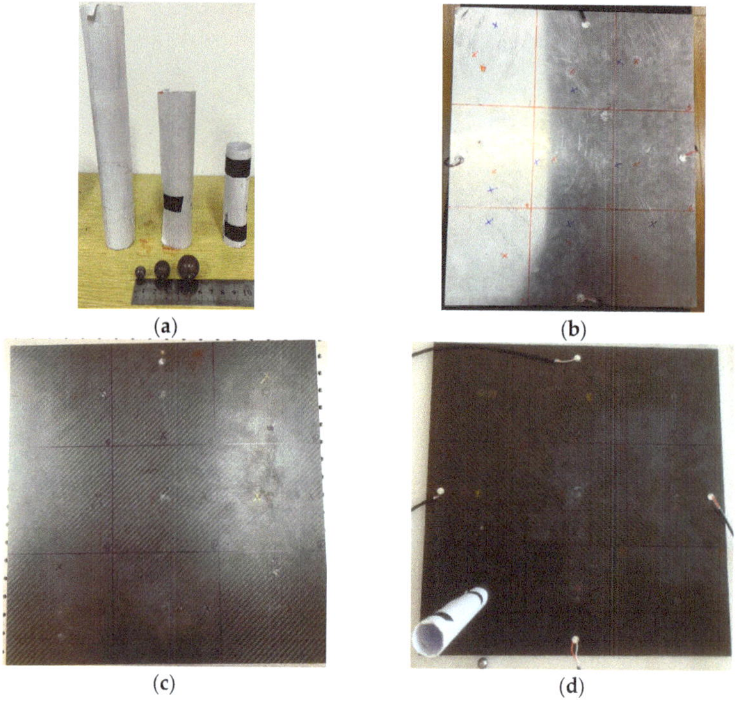

Figure 5. Schematic diagram of different test sets (red marks are 30 mm and 20 mm from the center of the area, and blue is 60 mm): (**a**) balls with different diameters and guide rails with different lengths; (**b**) schematic diagram of the impact position of the aluminum plate; (**c**) schematic diagram of the impact position of the orthotropic laminate; (**d**) schematic diagram of impact position of unidirectional laminate.

Alt Text: The three plate-like structures used in this paper were all divided into nine identical areas, and each area was marked with a marker pen. The red mark is 20 mm from the center of the area, and the blue mark is 60 mm from the center of the area.

4. Prediction Results of Impact Source Localization

4.1. Prediction Results of Impact Source Localization in Aluminum Plate

At nine center points of the area, a small steel ball with a diameter of 10 mm was dropped from a height of 200 mm, each area center was repeated ten times, and a total of 90 samples were used as the training set. It can be seen from Figure 6 that in the isotropic plate, due to the different positions of the sensors, there are apparent differences in the arrival time of the wave generated by the impact source to the four sensors. A neural network model can learn this difference and map the arrival times of the four sensors to the source of the impact. After preprocessing the time of arrival of each time-domain signal as described in Section 2.2, input it into the BP neural network described in Section 2.1 for training. After the training is completed, collect the test set data described in Section 3.2, and input the neural network as the test set according to the same steps. Figure 7 shows the training results of two different data sets when predicting. The correlation coefficient R in the figure indicates the degree of fitting of the data. It can be seen from the figure that the correlation coefficients of all groups in the aluminum plate are above 0.99, the fitting accuracy is high, and the model performs well. At the same time, the fitting degree of the training set and the test set is very high, and there is no over-fitting phenomenon, which also benefits from the relatively simple structure of the BP neural network.

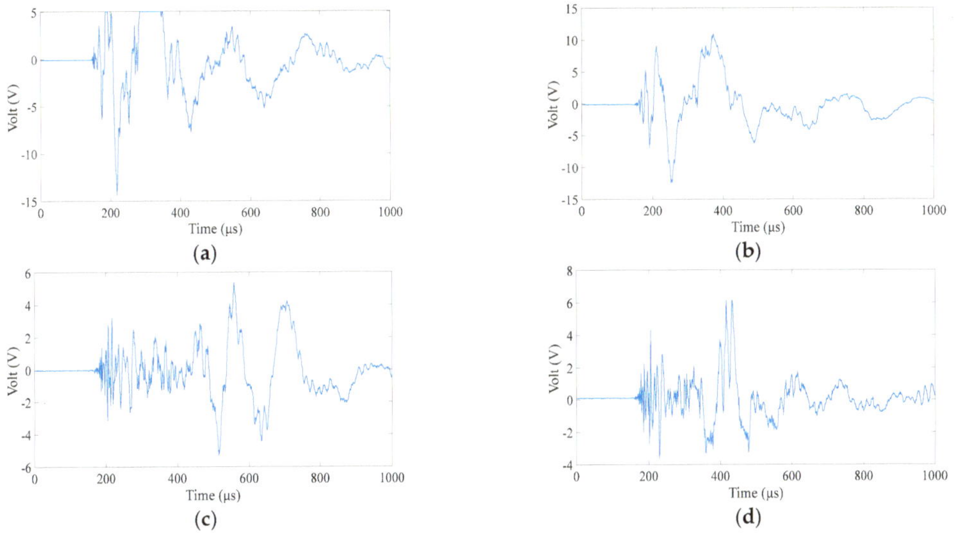

Figure 6. The signals received by the four sensors during a fall in area 1 of the aluminum plate. (**a–d**), respectively, correspond to No. 1–4 piezoelectric sheet (PZT) in Figure 4.

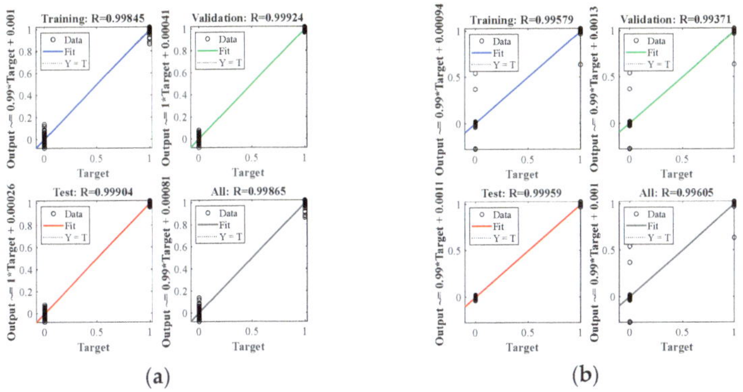

Figure 7. Training results for different test sets in aluminum plates: (**a**) test sets for different steel ball diameters and dropping heights; (**b**) test sets for different impact positions.

Alt Text: There is a significant difference in the arrival time of the waveforms of the signals received by the piezoelectric sensor located at the midpoint of the four sides after an impact in the aluminum plate. The arrival time of the waveform is extracted as the input of the neural network.

Alt Text: The training results of each test are set in the aluminum plate experiment The horizontal and vertical coordinates show the error between the expected value and the actual value, and the fitting coefficient shows the fitting accuracy of each group.

Figure 8 shows the prediction results of the six test sets in the aluminum plate. The blue circle in the figure represents the expected value, that is, the area where the steel ball falls in the experiment, and the red represents the discriminant result of the falling area by the trained neural network based on the eigenvalues input from the test set. It can be seen from the results that the prediction accuracy rate of the first four groups is 100%, and different drop heights and different steel ball diameters do not affect the prediction accuracy. The trained BP neural network can very accurately judge the area where the steel ball falls at this time and has a strong robustness to these two influencing factors. Figure 8e shows that the neural network model trained by the data sets of each regional center can accurately predict the location of the impact source within 30 mm nearby with 100% accuracy, which has specific prospects for practical engineering applications. When we change the training set's position to 60 mm, the position of the impact is close to the area's boundary. Figure 8f shows that the prediction accuracy is 96.7%, and only three misjudgments occur, which is still a more precise result. These three misjudgments were all misjudgments from area 6 to area 5. It can be seen from the blue mark in Figure 5b, the impact point of the steel ball at this time is close to the boundary of these two areas, so a wrong prediction is produced.

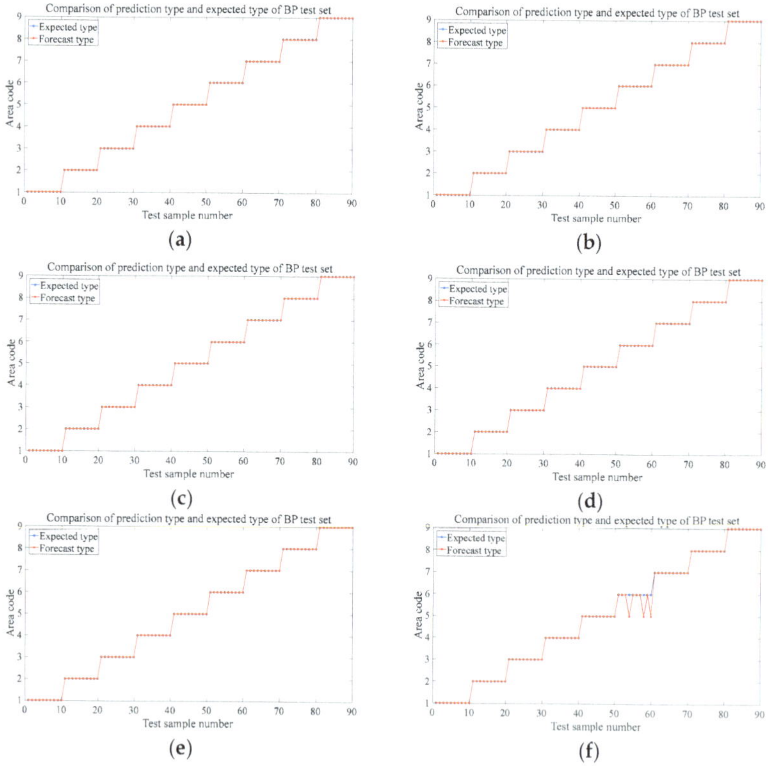

Figure 8. Prediction results for different test sets in aluminum panels. (**a**) The test set at 150 mm in 3.2 (2). (**b**) The test set at 100 mm in 3.2 (2). (**c**) The test set at 8 mm-diameter steel balls in 3.2 (3). (**d**) The test set of 15 mm-diameter steel balls in 3.2 (3). (**e**) 3.2 (4). The test set at 30 mm from the center of the area. (**f**) The test set at 60 mm away from the center of the area in 3.2 (4).

Alt Text: The prediction results of each test set in the aluminum plate experiment. The blue markers represent the expected value, and the red markers represent the output of the neural network. The abscissa represents the sample number of the test set, and the ordinate represents the area code of the output. In the test results of different steel ball diameters and different drop heights, the red mark and the blue mark completely coincide, which means that the prediction accuracy is 100%. There were three wrong predictions in the test set where the impact location was close to the region border.

Because of the situation in Figure 8f, this paper tries to improve the prediction accuracy by expanding the training set. Specifically, as shown in Figure 9, continue to take three points at the same distance of 60 mm from the center of the area and mark them as black. Drop a steel ball with the same diameter from the same height at the black mark and repeat 10 times for each. In this way, the training set of each region is increased by 30, and the entire training set is expanded to 360. After expansion, the neural network is retrained, and the 90 sets of data marked in blue are also used as the test set. The prediction results at this time are shown in Figure 10. When the types of training sets were expanded, the neural network newly learned the characteristics of the arrival time when the impact source was 60 mm from the center of the area, thus improving the accuracy of this part of the test set, reaching 100% like other test sets.

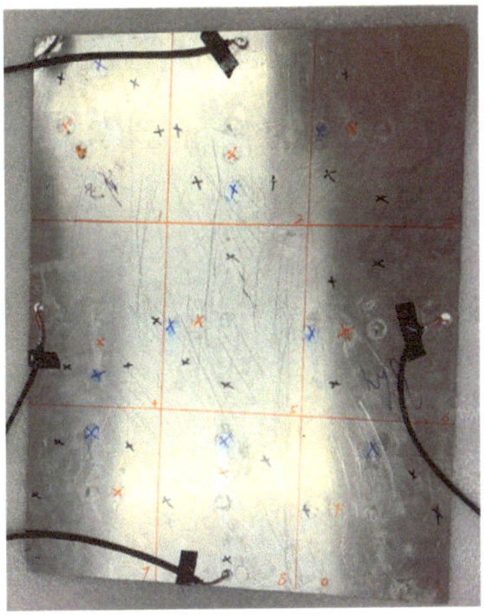

Figure 9. Schematic diagram of the increased impact location on the surface of the aluminum plate. The distance between the black and blue marks is 60 mm from the center of the area.

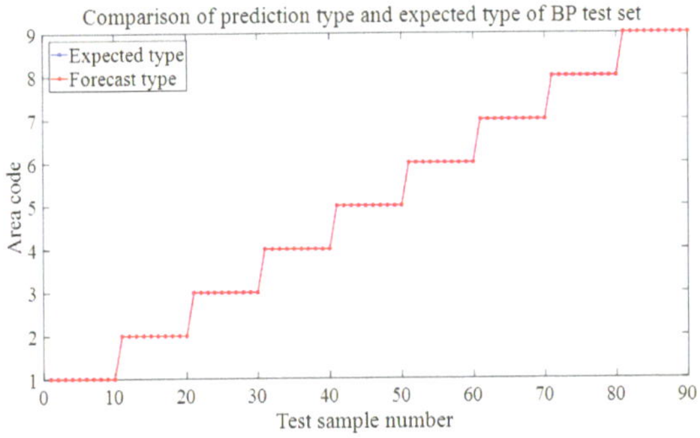

Figure 10. Prediction results after expanding the training set.

Alt Text: On the basis of Figure 6a, continue to use a black marker to mark at the same distance from the center of the area 60 mm. The new sampling points will be used as the expansion of the training set to input the neural network.

Alt Text: The prediction results after adding the samples marked in black in Figure 10 to the training set. The prediction results in Figure 8f are improved, so that the three wrongly predicted samples are correct again.

4.2. Impact Source Localization Results in Composite Panels

The wavefield information in the composite slab was collected using the same experimental platform as in Figure 4. This paper uses two types of carbon fiber-reinforced composite laminates: orthotropic and unidirectional laminates. The geometric parameters are described in Section 3.1. Due to the difference in lay-up direction, the propagation mode of the wave in the two plates is also different, and both are different from the isotropic aluminum plate. This section aims to test whether the neural network can still map arrival times to impact sources under different propagation modes. Figures 11 and 12 show the signals received by the sensors for a ball impact at the center of zone 1 in two panels. Although the steel ball impact is also performed in area 1, the signals captured by the piezoelectric sheets are also different due to the different laying directions of the composite materials. There are apparent differences in the arrival times of the three plates. Traditional methods based on the time difference in arrival often require accurate material properties, so it is difficult to achieve fast and accurate positioning for composite material panels with uncertain material properties. Perform data processing and network testing the same way as in the aluminum plate experiment. Set the same training set and test set according to Section 3.2. The training results of different test sets are shown in Figures 13 and 14. The fitting degree of each data set in the two composite material plates is good. The test is set at 60 mm from the center of each area. Both of them have relatively low correlation coefficients, 0.97474 and 0.97262, respectively, which may affect the model prediction accuracy, but is still an acceptable result. No overfitting was observed.

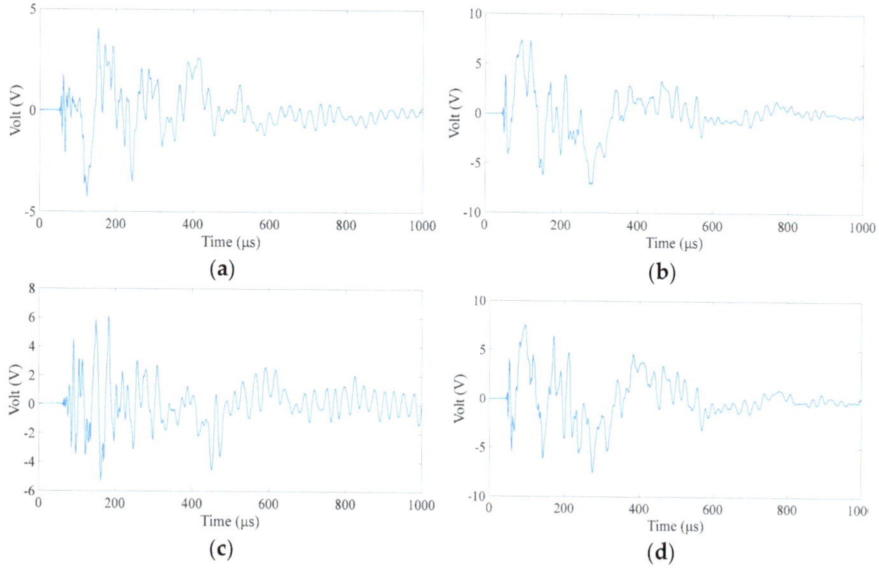

Figure 11. Signals received by four sensors during a fall in area 1 in an orthotropic laminate. (**a**–**d**), respectively, correspond to No. 1–4 PZT in Figure 4.

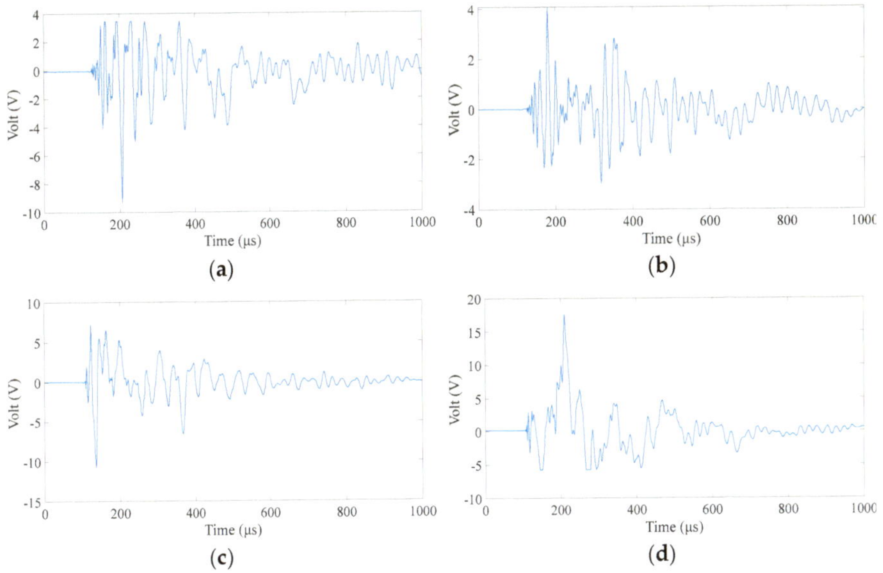

Figure 12. Signals received by four sensors during a fall in zone 1 in a unidirectional laminate. (**a**–**d**), respectively, correspond to No. 1–4 PZT in Figure 4.

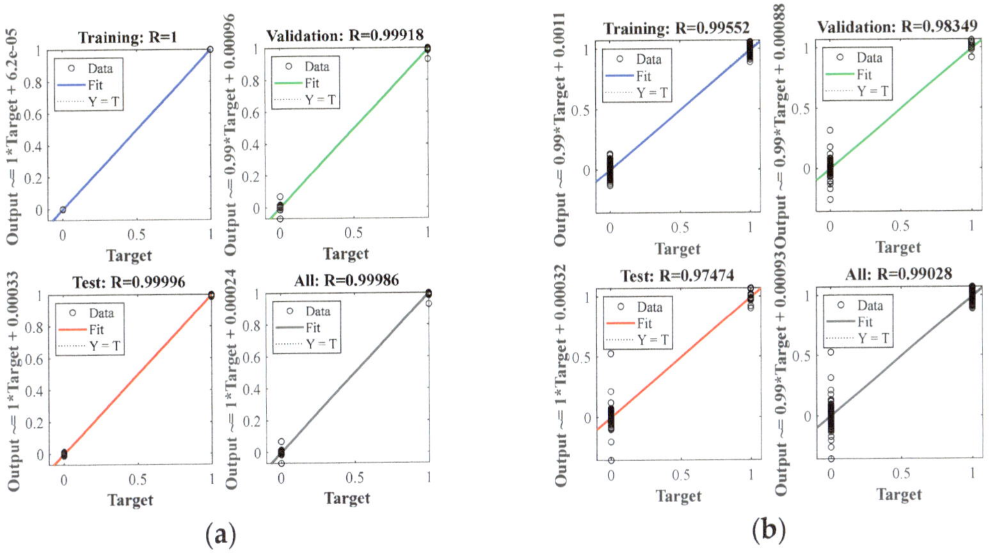

Figure 13. Training results for different test sets in orthotropic laminates: (**a**) test sets for different steel ball diameters and dropping heights; (**b**) test sets for different impact positions.

Alt Text: The signals received by the four sensors during an impact at the center of zone 1 in an orthotropic laminate. Compared with the signal in the aluminum plate in the previous section, there are obvious differences in the arrival time of the waveform, and the relationship between different arrival times is no longer linear, the wave speed is faster, and the arrival time values are smaller.

Alt Text: The signal received by the sensor when the central position of area 1 in the unidirectional laminate receives an impact. Although the impact is also performed in area

1, the signals captured by the piezoelectric sensor are also different due to the different laying directions of the composite material. The wave velocity in unidirectional laminates is between aluminum and orthotropic laminates, and the wave velocity along the ply direction is relatively fast.

Alt Text: Training results for each set of test sets in orthotropic laminates. A similar fitting accuracy to the data set in the aluminum plate was achieved.

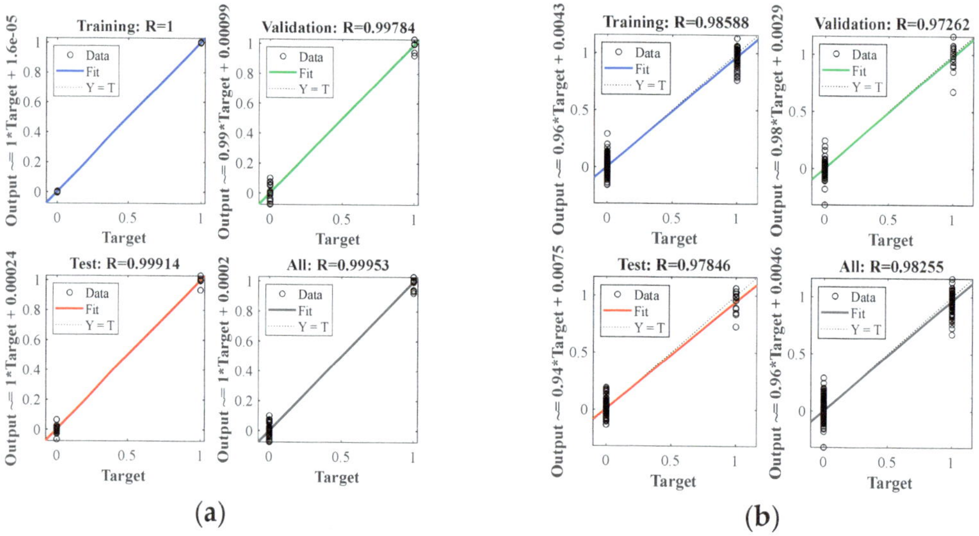

Figure 14. Training results for different test sets in unidirectional laminates: (**a**) test sets for different steel ball diameters and dropping heights; (**b**) test sets for different impact positions.

Alt Text: Training results for each set of test sets in unidirectional laminates. The fitting accuracy of each test set is above 0.97.

Alt Text: The prediction results of each test set in the orthotropic laminate experiment. In the test results of different steel ball diameters and different drop heights, the red mark and the blue mark completely coincide, which means that the prediction accuracy is 100%. There were four wrong predictions in the test set where the impact location was close to the region border.

Alt Text: The prediction results of each test set in the unidirectional laminate experiment. In the test results of different steel ball diameters and different drop heights, the accuracy rate is also 100%. There were four wrong predictions in the test set where the impact location was close to the region border, they both appeared in area 4 and were misjudged to area 7.

Figures 15 and 16 are the prediction results of the trained model for the location of the impact source in two composite laminates. The prediction results of the test set at different drop heights, different steel ball diameters, and 20 mm near the center of the area are the same as the aluminum plate and also have strong robustness, and the prediction accuracy reaches 100%. The prediction accuracy of the steel ball drop position test set at 60 mm from the center of each area is slightly lower than that of the aluminum plate, both of which are 95.6%, and four wrong predictions are different from the expected results. In the orthogonal plate, two samples of the test set in region 1 were mispredicted to region 2; mispredictions also occurred in region 5, and 2 samples were judged to belong to region 6. The four mispredictions in the one-way board all occurred in area 4 and were misjudged in area 7. Combined with the blue marks in Figure 5c,d, it can be seen that because they are very close to the boundaries of each region, the samples with wrong predictions are all mistaken by the network as the impact source of the adjacent region. Like the aluminum plate in Section 4.2, this study also tries to expand the training set to improve prediction accuracy. Figure 17 is the newly added impact position of the falling ball. Three black marks are added in each area of the two laminated plates. All of them are 60 mm away from the center of each area. A small steel ball with a diameter of 10 mm is dropped from 200 mm at each mark, repeated 10 times. This increases the number of samples to train the neural network to 360. Figure 18 shows the predicted results at this time. Similarly to the aluminum plate experiment results, after increasing the diversity of the training set, the model can predict with 100% accuracy the area where the impact source of the small steel ball belongs.

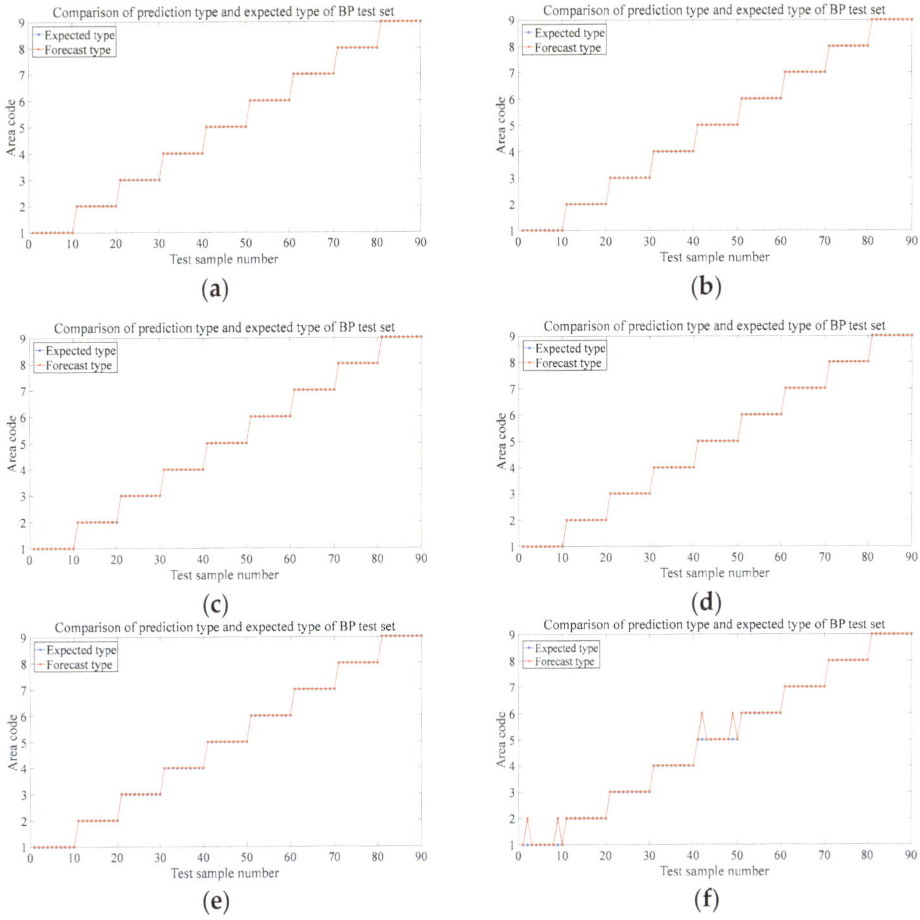

Figure 15. Prediction results for different test sets in orthotropic laminates. (**a**) The test set at 150 mm in 3.2 (2). (**b**) The test set at 100 mm in 3.2 (2). (**c**) The test set at 8 mm-diameter steel balls in 3.2 (3). (**d**) The test set at 15 mm-diameter steel balls in 3.2 (3). (**e**) 3.2 (4) The test is set at 30 mm from the center of the area. (**f**) The test set at 60 mm away from the center of the area in 3.2 (4).

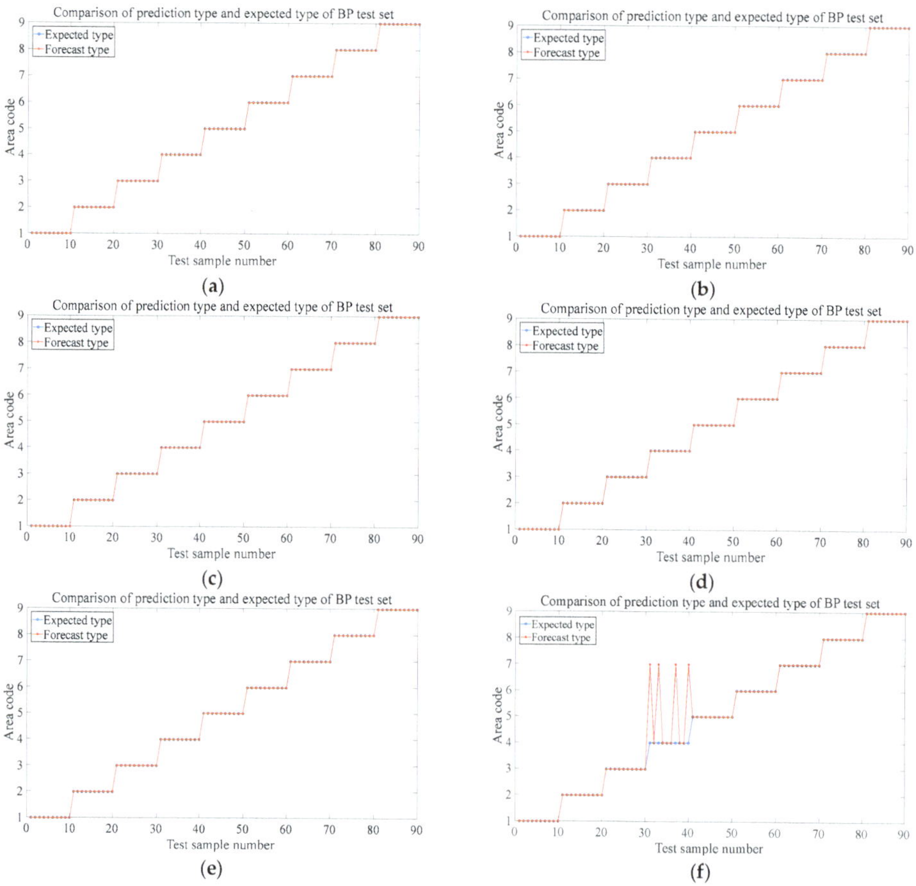

Figure 16. Prediction results for different test sets in unidirectional laminates. (**a**) The test set at 150 mm in 3.2 (2). (**b**) The test set at 100 mm in 3.2 (2). (**c**) The test set at 8 mm-diameter steel balls in 3.2 (3). (**d**) The test set at 15 mm-diameter steel balls in 3.2 (3). (**e**) 3.2 (4) The test set at 20 mm from the center of the area. (**f**) The test set at 60 mm away from the center of the area in 3.2 (4).

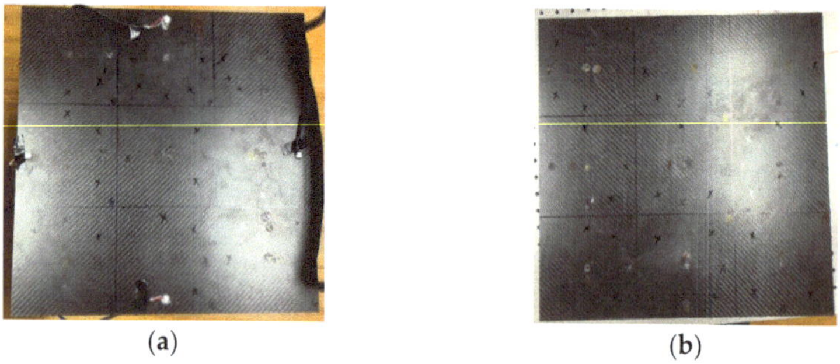

Figure 17. Schematic diagram of the location of the added training set sampling points: (**a**) orthotropic laminates; (**b**) unidirectional laminates.

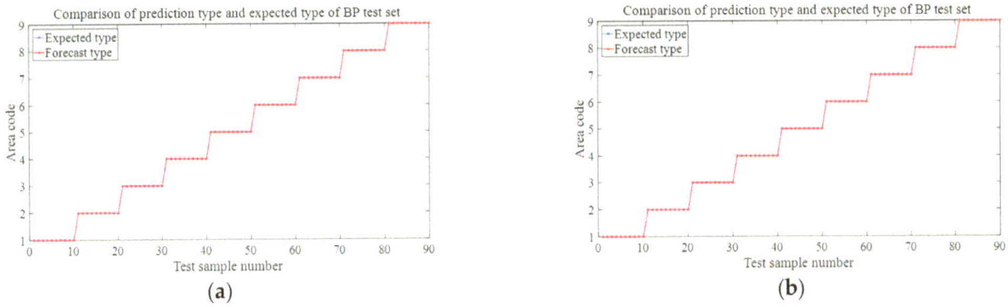

Figure 18. Prediction results of each plate after expanding the training set: (**a**) schematic diagram of increased sampling points of orthotropic laminates; (**b**) schematic diagram of sampling points added to unidirectional laminates.

Alt Text: On the basis of Figure 5c,d, continue to use a black marker to mark at the same distance from the center of the area 60 mm.

Alt Text: The prediction results after adding the samples marked in black in Figure 18 to the training set. The prediction results in Figure 16f are improved, the respective for wrong samples were re-predicted as correct.

5. Conclusions

This paper uses a machine learning approach to locate the source of a falling ball impact on the surface of three panel-like structures with different material properties, including an isotropic aluminum panel and two composite laminates with different ply orientations. In this study, the impact sources are from signals at different positions of the piezoelectric sensor at different times. The most typical three-layer BP neural network is utilized to learn this feature. When the material properties are unknown, four sensors are used to test the low-velocity impact source of the steel ball through regional targeting. A training data set is generated in the center of each area, and the area in which the impact sources in other locations belong is predicted. The robustness of the model is tested at the same time. Experiments were carried out on three plate structures with different material properties to verify the proposed scheme. The results reveal that the trained model can precisely predict the area where the impact source is located, whether in isotropic aluminum panels or composite panels with different ply directions. At the same time, the prediction result has strong robustness. Changing the diameter and drop height of the falling steel ball within a specific range does not affect the prediction accuracy (100%). When the falling position of the steel ball is 20 mm or 30 mm away from the center of the area where the training set data are located, all the training samples in the three boards can accurately determine the area they belong to, with an accuracy rate of 100%. When the drop position continues to be away from each center point to 60 mm, the prediction result in the aluminum plate is better, with an accuracy rate of 96.7%. The accuracy within the two composite panels was slightly lower at 95.6%. In addition, a method of augmenting the training set is adopted to improve the case of misassignment of shocks at the boundaries of each region. After adding samples at the same distance of 60 mm from the center as the training set, the model's performance improved, and the prediction accuracy reached 100%. The above results demonstrate the potential of the method proposed in this study in effectively characterizing the impact source locations on the surface of plate-like structures. This paper did not consider the impact of the environment and the anti-interference ability of the proposed model. Our future research will focus on discussing the impact of environmental factors such as noise and temperature on the results.

Author Contributions: Conceptualization, W.H.; Methodology, C.T., Y.Z. and W.H.; Software G.S.; Validation, Y.Z.; Formal analysis, C.T.; Investigation, Y.Z.; Data curation, C.T. and G.S., Writing—original draft, C.T.; Writing—review & editing, W.H. All authors have read and agreed to the published version of the manuscript.

Funding: The authors are grateful for the financial support provided by the Six Talent Peaks Project in Jiangsu Province (Grant No. 2019-KTHY-059).

Institutional Review Board Statement: Not applicable.

Informed Consent Statement: Not applicable.

Data Availability Statement: The original contributions presented in this study are included in the article. Further inquiries can be directed to the corresponding author.

Conflicts of Interest: The authors declared that they have no conflicts of interest in this work.

References

1. Cheng, H. Composite Material Application and Development in Pressure Vessels. *Adv. Mater. Res.* **2011**, *321*, 222–225. [CrossRef]
2. Huo, X.; Luo, Q.; Li, Q.; Sun, G. Measurement of fracture parameters based upon digital image correlation and virtual crack closure techniques. *Compos. B Eng.* **2021**, *224*, 109157. [CrossRef]
3. Nsengiyumva, W.; Zhong, S.; Lin, J.; Zhang, Q.; Zhong, J.; Huang, Y. Advances, limitations and prospects of nondestructive testing and evaluation of thick composites and sandwich structures: A state-of-the-art review. *Compos. Struct.* **2021**, *256*, 112951. [CrossRef]
4. Dixit, A. Structural Health Monitoring Systems and Methods. *Recent Pat. Mech. Eng.* **2017**, *10*, 118–125. [CrossRef]
5. Cui, L.; Zhao, Y.; Zhao, P.; Sun, J.C.; Jia, Z.Q. Review of Noncontact Ultrasonic Nondestructive Testing for the Solid Materials. *Appl. Mech. Mater.* **2014**, *528*, 346–352. [CrossRef]
6. Dzenis, Y.; Qian, J. Analysis of microdamage evolution histories in composites. *Int. J. Solids Struct.* **2001**, *38*, 1831–1854. [CrossRef]
7. Qiu, Z.; Wang, J.; Fan, H. Low velocity flexural impact behaviors of bamboo fiber reinforced composite beams. *Polym. Test.* **2021**, *94*, 107047. [CrossRef]
8. Coverley, P.T.; Staszewski, W.J. Impact damage location in composite structures using optimized sensor triangulation procedure. *Smart Mater. Struct.* **2003**, *12*, 795. [CrossRef]
9. Gaul, L.; Hurlebaus, S. Identification of the impact location on a plate using wavelets. *Mech. Syst. Signal Process.* **1998**, *12*, 783–795. [CrossRef]
10. Sai, Y.; Jiang, M.; Sui, Q.; Lu, S.; Jia, L. FBG sensor array-based-low speed impact localization system on composite plate. *J. Mod. Opt.* **2016**, *63*, 462–467. [CrossRef]
11. Jiang, M.; Sai, Y.; Geng, X.; Sui, Q.; Liu, X.; Jia, L. Development of an FBG Sensor Array for Multi-Impact Source Localization on CFRP Structures. *Sensors* **2016**, *16*, 1770. [CrossRef] [PubMed]
12. Kundu, T.; Nakatani, H.; Takeda, N. Acoustic source localization in anisotropic plates. *Ultrasonics* **2012**, *52*, 740–746. [CrossRef] [PubMed]
13. Zheng, K.; Shao, S.; Hameed, M.S.; Li, X.; Zhu, W.; Chen, J.; Li, Z.; Shui, G. Damage detection method based on continuous wavelet transform of second harmonic Lamb waves. *Nondestruct. Test. Eval.* **2023**, *38*, 1006–1026. [CrossRef]
14. Seydel, R.; Chang, F.K. Impact identification of stiffened composite panels: II. Implementation studies. *Smart Mater. Struct.* **2001**, *10*, 370. [CrossRef]
15. Sai, Y.; Jiang, M.; Sui, Q.; Lu, S.; Jia, L. Composite plate low energy impact localization system based on FBG sensing network and hybrid algorithm. *Opt. Fiber Technol.* **2015**, *24*, 84–88. [CrossRef]
16. Shrestha, P.; Kim, J.H.; Park, Y.; Kim, C.G. Impact localization on composite structure using FBG sensors and novel impact localization technique based on error outliers. *Compos. Struct.* **2016**, *142*, 263–271. [CrossRef]
17. Kim, J.H.; Kim, Y.Y.; Park, Y.; Kim, C.G. Low-velocity impact localization in a stiffened composite panel using a normalized cross-correlation method. *Smart Mater. Struct.* **2015**, *24*, 045036. [CrossRef]
18. Zhao, G.; Li, S.; Hu, H.; Zhong, Y.; Li, K. Impact localization on composite laminates using fiber Bragg grating sensors and a novel technique based on strain amplitude. *Opt. Fiber Technol.* **2018**, *40*, 172–179. [CrossRef]
19. Salehi, H.; Burgueno, R. Emerging artificial intelligence methods in structural engineering. *Eng. Struct.* **2018**, *171*, 170–189. [CrossRef]
20. Cuomo, C.; De Simone, M.E.; Andreades, C.; Ciampa, F.; Meo, M. Machine learning for impact detection on composite structures. *Mater. Today-Proc.* **2021**, *34*, 93–98. [CrossRef]

21. Geng, X.; Lu, S.; Jiang, M.; Sui, Q.; Lv, S.; Xiao, H.; Jia, Y.; Jia, L. Research on FBG-Based CFRP Structural Damage Identification Using BP Neural Network. *Photonic Sens.* **2018**, *8*, 168–175. [CrossRef]
22. Dipietrangelo, F.; Nicassio, F.; Scarselli, G. Structural Health Monitoring for impact localisation via machine learning. *Mech. Syst. Signal Process.* **2023**, *183*, 109621. [CrossRef]
23. Liu, Q.; Wang, F.; Li, J.; Xiao, W. A hybrid support vector regression with multi-domain features for low-velocity impact localization on composite plate structure. *Mech. Syst. Signal Process.* **2021**, *154*, 107547. [CrossRef]
24. Jiang, M.; Lu, S.; Sui, Q.; Dong, H.; Sai, Y.; Jia, L. Low velocity impact localization on CFRP based on FBG sensors and ELM algorithm. *IEEE Sens. J.* **2015**, *15*, 4451–4456. [CrossRef]
25. Sai, Y.; Zhao, X.; Wang, L.; Hou, D. Impact localization of CFRP structure based on FBG sensor network. *Photonic Sens.* **2020**, *10*, 88–96. [CrossRef]
26. Ai, L.; Soltangharaei, V.; Bayat, M.; Tooren, M.V.; Ziehl, P. Detection of impact on aircraft composite structure using machine learning techniques. *Meas. Sci. Technol.* **2021**, *32*, 084013. [CrossRef]
27. Chen, T.; Guo, L.; Duan, A.; Gao, H.; Feng, T.; He, Y. A feature learning-based method for impact load reconstruction and localization of the plate-rib assembled structure. Struct. *Health Monit. Int. J.* **2021**, *21*, 1590–1607. [CrossRef]
28. Jierula, A.; Wang, S.; Oh, T.M.; Lee, J.W.; Lee, J.H. Detection of source locations in RC columns using machine learning with acoustic emission data. *Eng. Struct.* **2021**, *246*, 112992. [CrossRef]
29. Huang, Y.; Tang, C.; Hao, W.; Zhao, G. Acoustic Source Localization in Metal Plates Using BP Neural Network. *Metals* **2023**, *13*, 755. [CrossRef]
30. Hao, W.; Huang, Y.; Zhao, G. Acoustic sources localization for composite pate using arrival time and BP neural network. *Polym. Test.* **2022**, *115*, 107754. [CrossRef]

Disclaimer/Publisher's Note: The statements, opinions and data contained in all publications are solely those of the individual author(s) and contributor(s) and not of MDPI and/or the editor(s). MDPI and/or the editor(s) disclaim responsibility for any injury to people or property resulting from any ideas, methods, instructions or products referred to in the content.

materials

MDPI

Article

Some Unfamiliar Structural Stability Aspects of Unsymmetric Laminated Composite Plates

Mehdi Bohlooly Fotovat

Department of Strength of Materials, Lodz University of Technology, Stefanowskiego 1/15, 90-537 Lodz, Poland; mehdi.bohlooly-fotovat@p.lodz.pl

Abstract: It is widely recognized that certain structures, when subjected to static compression, may exhibit a bifurcation point, leading to the potential occurrence of a secondary equilibrium path. Also, there is a tendency of deflection increment without a bifurcation point to occur for imperfect structures. In this paper, some relatively unknown phenomena are investigated. First, it is demonstrated that in some conditions, the linear buckling mode shape may differ from the result of geometrically nonlinear analysis. Second, a mode jumping phenomenon is described as a transition from a secondary equilibrium path to an obscure one as a tertiary equilibrium path or a second bifurcation point. In this regard, some non-square plates with unsymmetric layer arrangements (in the presence of extension–bending coupling) are subjected to a uniaxial in-plane compression. By considering the geometrically linear and nonlinear problems, the bucking modes and post-buckling behaviors, e.g., the out-of-plane displacement of the plate versus the load, are obtained by ANSYS 2023 R1 software. Through a parametric analysis, the possibility of these phenomena is investigated in detail.

Keywords: post-buckling; second bifurcation; mode jumping; tertiary equilibrium path

Citation: Bohlooly Fotovat, M. Some Unfamiliar Structural Stability Aspects of Unsymmetric Laminated Composite Plates. *Materials* **2024**, *17*, 3856. https://doi.org/10.3390/ma17153856

Academic Editor: Enrique Casarejos

Received: 2 June 2024
Revised: 31 July 2024
Accepted: 2 August 2024
Published: 4 August 2024

Copyright: © 2024 by the author. Licensee MDPI, Basel, Switzerland. This article is an open access article distributed under the terms and conditions of the Creative Commons Attribution (CC BY) license (https://creativecommons.org/licenses/by/4.0/).

1. Introduction

For square plates without the effect of extension–bending coupling, the mode shapes based on linear buckling (or so-called eigenvalue solution) and nonlinear post-buckling analysis are the same. For instance, Figure 1 demonstrates the equilibrium path of the square, which is a simply supported plate made of glass-fiber-reinforced polymer (GFRP), with the material properties outlined in [1] and the layer arrangement of $[(45/-45)_4]_T$. The plate is subjected to a uniform uniaxial in-plane compression F_{ex}, and the variation in the deflection of the center of the plate w_c with respect to the total thickness of the plate h is plotted in this figure. As seen, both analyses (linear eigenvalue and geometrically nonlinear) lead to one half-wave in both in-plane directions. For tracing the nonlinear analysis, the plate with conditions such as imperfections, lateral load, extension–bending coupling, eccentricity of in-plane loads, and so on may tend to show behavior close to the ideal curve. It should be noted that the ideal curve with a bifurcation point is a rare phenomenon, and usually a tendency to this black solid curve exists by using the above conditions. In Figure 1, there are no extension–bending coupling coefficients (i.e., $B_{11} = B_{12} = B_{22} = 0$). However, an imperfection with different amplitudes ζ is applied. In the case of initial imperfection, there are interesting studies which have focused on different types of imperfections [2–4]. However, the current study is limited to the imperfection of the first linear buckling mode [5]. Another point is that the solution has two curve parts. First, the right curve is the result of a load increment from zero, and the left curve is the result of a load decrement from a large value. However, both curves correspond to the same mode of deflection. These two parts of the solutions are also reported for different structures as truss [6] and general systems [7]. One question that may arise is that is the mode of both linear and nonlinear analysis always the same? In this paper, a special circumstance will be demonstrated where there are possibilities to observe different modes based on different linear and nonlinear analyses.

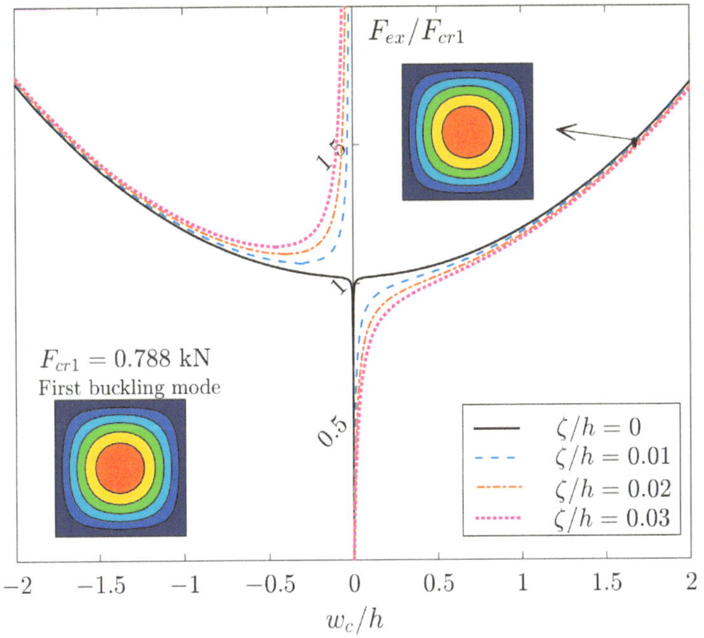

Figure 1. The post-buckling curves of a square GFRP plate $[(45/-45)_4]_T$ with different amplitudes of imperfections.

Concerning the nonlinear analysis of the structures, there is a very rare phenomenon called mode jumping, which is introduced by a limited number of investigators because it usually happens under special circumstances. Among them, Ungureanu et al. [8] presented the possibility of a second bifurcation point for thin-walled steel members with different types of profile sections. Zhang and Murphy [9] worked on the secondary buckling point of the beams by a judicious choice in the beam length. In other words, longer beams have different modes with respect to shorter beams. But at a special length, the possibility of jumping from one mode to another is high. In another case [10], they investigated the effect of a partial elastic foundation on the tertiary equilibrium state of the beams.

Based on different terminology in the literature, mode jumping, second bifurcation point, and tertiary equilibrium path can be significant for those structures for which the conditions of the occurrence of two different modes exist at the same time. However, mode jumping differs from interactive buckling [11,12], where the structure has a mix of more than one mode shape at the same time. In the case of interactive buckling, the beams and columns with I-section profiles have a significant role, where local and global buckling can be seen simultaneously. In particular, the term cellular buckling can be raised in such structures [13–15]. Similar behavior has been discovered in other mechanical systems such as rectangular hollow strut [16] and cylindrical shells [17].

In the case of laminated composites, Pirrera et al. [18] and Coburn et al. [19] investigated the possibility of bi-stability and tri-stability in cylindrical and double curved shells, respectively. However, the effect of the extension–bending coupling matrix for unsymmetric laminated composite structures is one of the interesting topics in the field of the structural stability of composite materials. In this regard, Carrera et al. [20] found that unsymmetric and non-square plates may have a special mode shape in the nonlinear range for which the amplitudes of the half-waves are not same. In other words, one may deal with some larger and smaller half-waves in the post-buckling range. Bohlooly Fotovat and Kubiak [21] presented an analytical solution to understand the reason for the non-bifurcation response in the presence of the extension–bending coupling matrix. Now, there is a good opportunity to combine these two studies [20,21] and focus on the mode jumping of non-square plates.

In this paper, two novel aspects of unsymmetric laminated composite plates under uniaxial compressions are presented. First, the effect of the boundary condition on the deflection response of the plate is presented in Section 2. In the case of simply supported boundary conditions, the mode shape of linear buckling analysis (eigenvalue) and post-buckling (geometrically nonlinear) can be different. The reason for and possibility of this difference are explained in Section 3. Second, the plates with different modes (based on linear and nonlinear analyses) have a possibility of jumping to another mode. This is possible in the presence of imperfection. The details of such a mode jumping are explained in Section 4. In all sections, the results of different analyses are obtained from finite element analysis using ANSYS APDL version 2023 R1 software.

2. Finite Element Analysis: Set-Up and Solver

To obtain the linear buckling results of ANSYS, the plate is modeled with the shell element. It should be noted that a four-node element with a size of 1×1 mm^2 is selected. This size selection is based on a convergence study. In the geometrically nonlinear analysis, the equilibrium paths are obtained by static analysis. The element size is 5×5 mm^2. In this analysis, an initial imperfection with the shape of the first buckling mode is employed. The solver is the load-control Newton–Raphson method. Due to the absence of any snapping conditions, it is not necessary to apply any path-following techniques [22]. In order to avoid rigid body motions, one edge is considered immovable, and the opposite edge is considered movable and then compressed. A coupled boundary condition is applied for two movable edges in both in-plane directions. This is the reason why mode shapes in the current study have straight edges. However, they were curved in a similar study [20]. A verification study of the current results of ANSYS is reported in our previous similar work [21].

3. Unsymmetric Laminated Composite Plates with Simply Supported Edges

According to the classical lamination theory in thin-walled structures, the lack of symmetry with respect to the middle plane results in a behavior of so-called extension–bending coupling [23]. This means that applying in-plane loads can induce deflection in the structure, and conversely, the deflection may influence the in-plane loads. This is due to the nonzero components of the extension–bending coupling matrix. However, this coupling matrix is necessary, but not in a sufficient condition, for plates being deformed before buckling. Another important factor of such a situation is the existence of simply supported boundary conditions. Figure 2 demonstrates a schematic of a two-layer unsymmetric plate [0/90] considering two different kinds of movable boundary conditions (e.g., simply supported and clamped). As seen, the plate is subjected to an external load F_{ex}, and the reaction force of a 0-degree layer F_{in1} is higher than 90-degree layer F_{in2}. According to the free body diagram of a pin in the simply supported edge (located in the middle plane of the plate), this different value of loads leads to a moment M_{in} which is inserted from the plate to the pin. Since the pin has no resistance to any moment, it starts to rotate. The rotation of the pin leads to the deflection of the plate. In contrast, the free body diagram of the attaching layer in the clamped edge shows that it obliterates the moment very easily (by M_{re}), and the plate remains flat during compression (before buckling). Therefore, the next sections of this paper focus on an unsymmetric laminated plate with all simply supported edges.

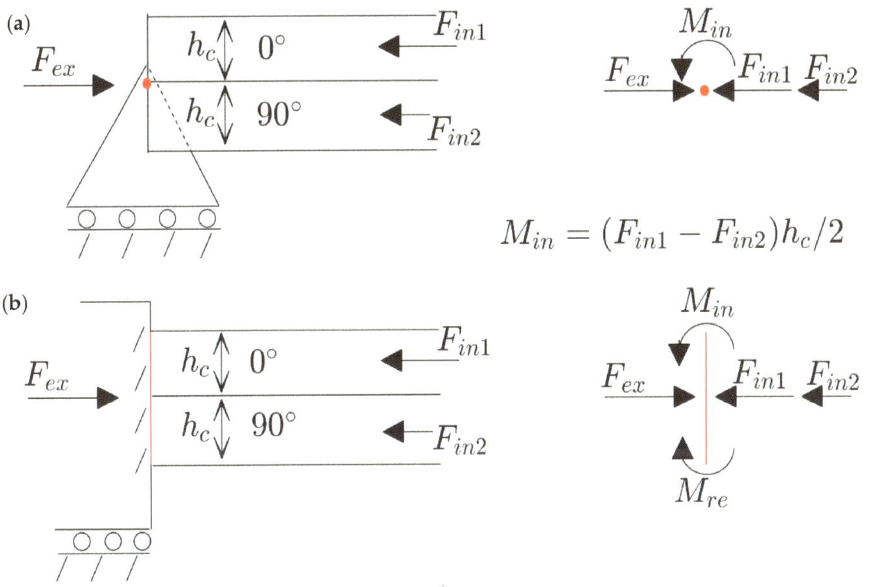

Figure 2. The schematics of reaction forces in a two-layer plate [0/90] with different kinds of boundary conditions and their free body diagrams: (**a**) movable simply support; (**b**) movable clamped.

4. Different Modes in Buckling and Post-Buckling of Non-Square Plates

Usually, the result of buckling mode based on a linear analysis (eigenvalue problem) is the same as the result in the geometrically nonlinear analysis. Therefore, a very small amplitude of this linear mode can be employed as an imperfection to be sure that the plate tends to this mode in a nonlinear analysis. However, one exceptional phenomenon can occur in the unsymmetric and non-square laminated composites. This means that there are possibilities to see different modes of deflection based on linear buckling and nonlinear post-buckling analysis (even with the presence of imperfection of linear buckling mode). The reason for this is illustrated in Figure 3. For instance, it is obvious that a GFRP plate with a lay-up arrangement of $[(0/90)_4]$ and aspect ratio $r = 2$ (i.e., length/width) has a linear buckling mode (or LBM) as $(m, n) = (2, 1)$. In other words, two waves and one half-wave will appear in the loading and perpendicular in-plane directions of the plate, respectively. As seen, due to the existence of edge moments on both of the opposite sides of the plate (see Section 2), the results of the nonlinear analysis indicate that the mode can be either (1,1) or (3,1). Because the directions of these moments do not lead to observing an even number of half-waves, the selection of actual mode (between $m = 1$ or 3) lies in the value of the second buckling load. Therefore, the variations in the buckling loads versus the aspect ratio are plotted in Figure 4a. As seen, blue areas correspond to an even number of half-waves in critical buckling mode (e.g., $m = 2, 4, \ldots$). This means that these areas are for those non-square plates which are prone to having different modes based on linear and nonlinear analyses. For example, in the first blue area, the critical buckling load (in a range r = 1.42~2.42) corresponds to the mode with two half-waves $m = 2$ (in LBM). In this range, the second buckling load plays a vital role. By considering this quantity, the blue area can be divided into two other sub-areas (Figure 4b). First, there is a range as r = 1.42~1.72, where the second buckling mode is $m = 1$ and there is another range as r = 1.72~2.42 for $m = 3$. Therefore, the deflections of the plates based on the geometrically nonlinear analysis of the red sub-area are like $(m, n) = (1, 1)$, and those of the green sub-area are like $(m, n) = (3, 1)$, according to Figure 4b. It should be noted that these modes are for a very initial post-buckling range. Then, a question arises: what is the deflection when it is slightly far away (i.e., not very initially)? By conducting many nonlinear studies in ANSYS software, the evidence proved that for a plate with an aspect ratio in the first sub-area (for this case, r = 1.42~1.72), the possibility of the occurrence of a tertiary equilibrium path is almost zero. This means that the plate keeps the mode $(m, n) = (1, 1)$ and then will be a mixed of $(m, n) = (1, 1)$ and $(3, 1)$ in a very far post-buckling range [24]. However, the second sub-area is unstable, and the occurrence of a tertiary equilibrium path is possible, which is described in the next section.

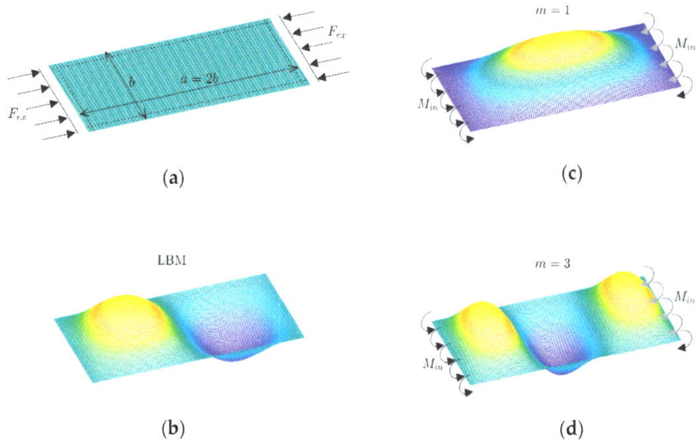

Figure 3. (**a**) A rectangular plate ($r = 2$) under compression, (**b**) linear buckling mode (LBM), and (**c**), (**d**) possible deflections in nonlinear post-buckling analysis (either $m = 1$ or 3).

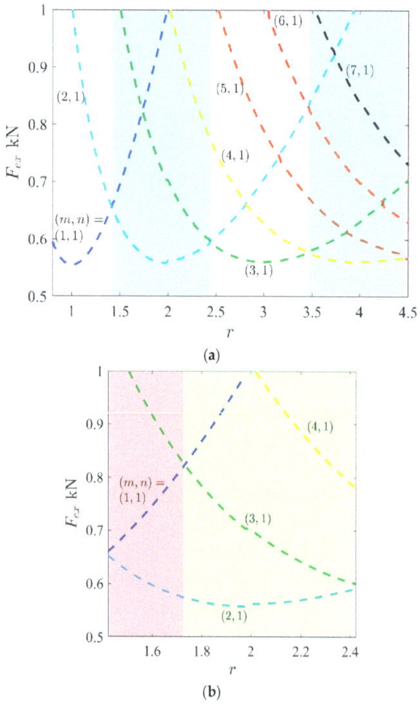

Figure 4. (**a**) The buckling loads of a GFRP plate $[(0/90)_4]$ under compression versus different aspect ratios. (**b**) The first blue area in detail with two sub-areas.

5. Existence of Tertiary Equilibrium Path

Figure 5 demonstrates the equilibrium paths of the GFRP plates ($r = 1.5$) with lay-up arrangements of $[(0/90)_4]_T$ and $[(90/0)_4]_T$. According to Figure 3b, it is obvious that the linear and nonlinear modes of such an aspect ratio are $(m, n) = (2, 1)$ and $(1, 1)$, respectively. In Figure 5, there is not any imperfection. In this case, the solution has no two curve parts (see Figure 1) due to the presence of extension–bending coupling. The right curve is the result of $[(0/90)_4]_T$, and the left curve is the result of $[(90/0)_4]_T$. This order of layer arrangement from the bottom to the top is highly effective on the direction of moment (see Figure 2). However, both curves correspond to the same mode of deflection. In addition, a small value of imperfection of the first buckling mode $(m, n) = (2, 1)$ will not affect the curves.

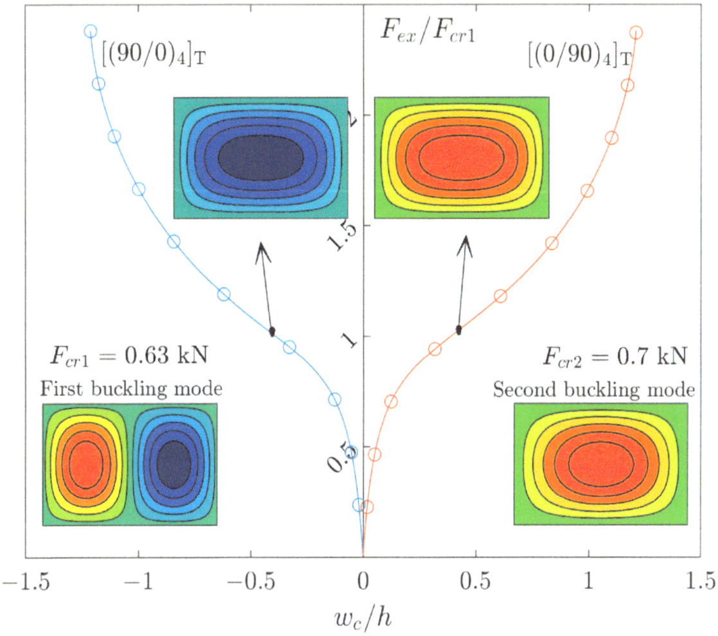

Figure 5. The post-buckling curves of perfect GFRP plates with $r = 1.5$.

Now it is time to select a plate with an aspect ratio higher than 1.72 (see green area in Figure 4b). The post-buckling response of the GFRP plate with $r = 2$ is demonstrated in Figure 6. In this figure, there is no imperfection and, as mentioned previously, the linear buckling mode is $(m, n) = (2, 1)$ and the deflection in post-buckling has a shape like $(m, n) = (3, 1)$. However, this post-buckling curve is unstable, and by adding a small value of imperfection of the first buckling mode, the results will be different.

In this regard, the post-buckling curves of the perfect and imperfect plate are plotted in Figure 7a. In this figure, a plate with $r = 2.2$ and different amplitudes of imperfection of the first buckling mode is analyzed. As seen, the second bifurcation point, tertiary equilibrium path, or jumping mode (based on different terminology in the literature) can occur by applying an amplitude of imperfection larger than 0.006. Also, the moment of separation of different curves is magnified in this figure. The differences between the equilibrium paths of a plate with imperfections of 0.006 and 0.008 can be interpreted as confrontation of two quantities. Both parameters (extension–bending coupling and imperfection) force the plate to have a different mode. When the plate has higher imperfection, it overcomes the extension–bending coupling and the mode shape of plate is the same as the mode shape of imperfection. However, a plate with a lower value of imperfection has a mode shape based on a condition that the extension–bending coupling forces on the plate (see Figure 3d).

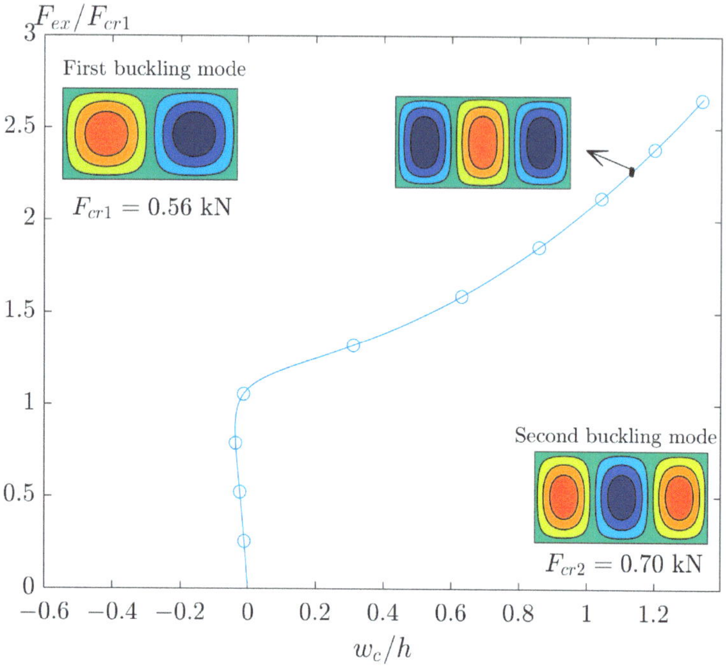

Figure 6. The post-buckling curve of perfect GFRP plate $[(0/90)_4]_T$ with $r = 2$.

The counterpart of Figure 7a for the deflection of another point as d with $(x, y) = (a/4, b/2)$ is plotted in Figure 7b. In this case, the curve corresponding to imperfection $\zeta/h = 0.006$ shows well that the plate tries to adopt mode jumping. However, due to the very small value of imperfection, the moments of extension–bending coupling return the plate to the mode of second buckling load.

Figure 8 presents the different aspect ratios (e.g., $r = 1.9, 2.0,$ 2.1, and 2.2) to illustrate the time of mode jumping. In other words, the time of mode jumping means the location of the point where the red path separates from the blue path. As seen, the location of this point is delayed by increasing the value of the aspect ratio. The physical meaning of this delay comes from Figure 4b. According to the green sub-area of Figure 4b, by increasing the aspect ratio from $r = 1.72$ to 2.42, the second buckling load (with the same mode in blue path) is decreased, and it comes closer to the first buckling load (with the same mode in red path). This means that the second buckling mode is becoming stable, and the time of mode jumping from the second to first buckling mode will be postponed.

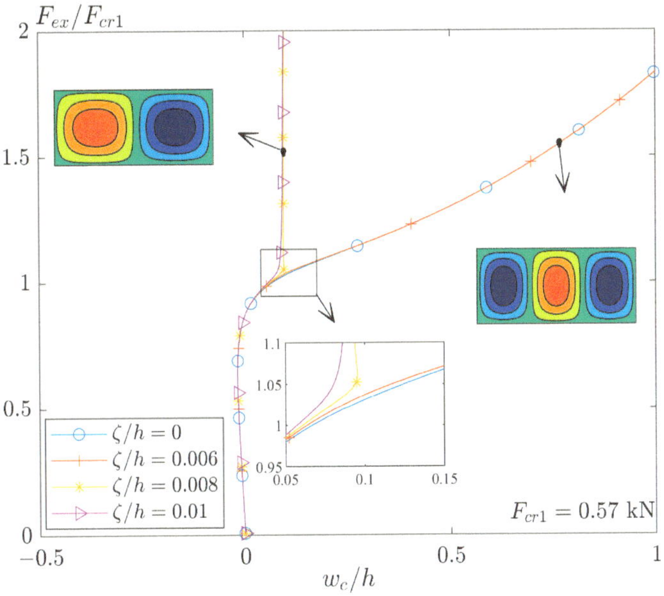

(**a**)

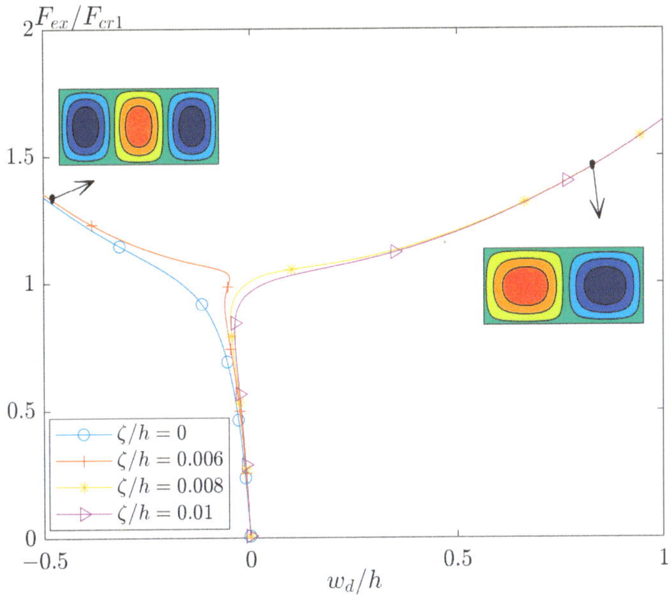

(**b**)

Figure 7. The post-buckling curves of perfect and imperfect GFRP plates $[(0/90)_4]_T$ with $r = 2.2$, (**a**) for displacement at center $(x, y) = (a/2, b/2)$ and (**b**) at $(x, y) = (a/4, b/2)$.

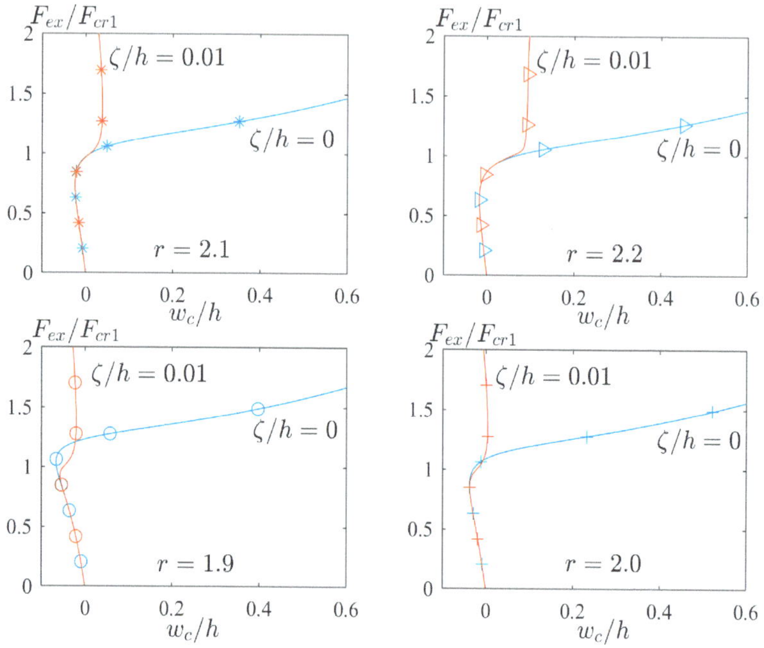

Figure 8. The post-buckling curves of perfect/imperfect GFRP plates with different aspect ratios.

6. Conclusions

In this paper, two linear and geometrically nonlinear analyses of ANSYS software are used and two new topics are presented. (1) A laminated composite plate may have different mode shapes of linear buckling and nonlinear post-buckling analysis. (2) Such a structure may have a behavior called mode jumping in the post-buckling response. In order to observe different modes, the plate should have the following conditions:

(a) Nonzero extension–bending coupling: These couplings, i.e., B_{11} and B_{22}, are present for some unsymmetric laminated composite plates. The plate with cross ply lamination, i.e., $[0/90]_n$ (n = 2, 4, 8, . . .), is one of the practical examples.

(b) Simply supported boundary conditions: The edges should have no resistance against rotation. It is like a simple pin or free edge without any moment reactions. In the current results, a simple pin or so-called simply supported boundary conditions are investigated.

(c) No eccentricity of load and boundary conditions: The resultant of in-plane compressions (through the thickness) and locations of pins should coincide in the middle plane. One of the practical cases is the uniform distribution of compression with which the resultant will coincide in the middle plane.

(d) Aspect ratio: The length of the plate should be larger than the width to have an even number of half-waves in the first linear buckling mode. For GFRP material and a layer arrangement of $[(0/90)_4]$, the aspect ratio should be higher than 1.4.

In the case of the presence of mode jumping, the plate should meet all the above conditions, and the two following cases should be present:

(a) Unstable second buckling mode: The number of half-waves in the second linear buckling mode should be higher than those in the first linear buckling mode. For example, a GFRP plate with $[(0/90)_4]$ and an aspect ratio of 2.1 has first and second buckling mode shapes as (2,1) and (3,1), respectively. So, such a plate has potential to have mode jumping in the post-buckling response.

(b) Imperfection: An amplitude of the imperfection of the first buckling mode should be applied as an initial deflection of the plate. However, it should be a bit larger to overcome the effects of the extension–bending coupling.

If the rectangular plate meets all the above conditions, the extension–bending coupling makes the plate have a mode which is the same as the second buckling mode, and the initial imperfection makes the plate have a mode which is same as the first buckling mode. However, the plate will select a mode shape with a lower value of strain energy, which is the first buckling mode. Therefore, the plate will jump to this mode in the post-buckling response. However, the time of jumping is highly dependent on the value of the aspect ratio. For the last result of the current study, it is concluded that the higher value of the aspect ratio has a severe conflict between imperfection and extension–bending coupling. This is because both modes require strain energy in the same range (both modes are approximately stable). This will result in mode jumping with a longer delay in the response.

Funding: This research received no external funding.

Institutional Review Board Statement: Not applicable.

Informed Consent Statement: Not applicable.

Data Availability Statement: Data are contained within the article.

Conflicts of Interest: The author declares no conflict of interest.

References

1. Kubiak, T.; Bohlooly Fotovat, M. Dynamic response and dynamic buckling of general laminated plates: A semi-inverse method. *Compos. Struct.* **2023**, *324*, 117548. [CrossRef]
2. Cox, B.S.; Groh, R.M.J.; Avitabile, D.; Pirrera, A. Modal nudging in nonlinear elasticity: Tailoring the elastic post-buckling behaviour of engineering structures. *J. Mech. Phys. Solids* **2018**, *116*, 135–149. [CrossRef]
3. Shen, J.; Pirrera, A.; Groh, R.M. Building blocks that govern spontaneous and programmed pattern formation in pre-compressed bilayers. *Proc. R. Soc. A* **2022**, *478*, 20220173. [CrossRef]
4. Shen, J.; Garrad, M.; Zhang, Q.; Leao, O.; Pirrera, A.; Groh, R.M. Active reconfiguration of multistable metamaterials for linear locomotion. *Phys. Rev. B* **2023**, *107*, 214103. [CrossRef]
5. Abramovich, H. *Stability and Vibrations of Thin Walled Composite Structures*; Woodhead Publishing: Thorston, UK, 2017.
6. Wadee, M.A.; Hadjipantelis, N.; Bazzano, J.B.; Gardner, L.; Lozano-Galant, J.A. Stability of steel struts with externally anchored prestressed cables. *J. Constr. Steel Res.* **2020**, *164*, 105790. [CrossRef]
7. Champneys, A.R.; Dodwell, T.J.; Groh, R.M.; Hunt, G.W.; Neville, R.M.; Pirrera, A.; Sakhaei, A.H.; Schenk, M.; Wadee, M.A. Happy catastrophe: Recent progress in analysis and exploitation of elastic instability. *Front. Appl. Math. Stat.* **2019**, *5*, 34. [CrossRef]
8. Ungureanu, V.; Dubina, D.; Crisan, A.; Madeo, A.; Zagari, G.; Zucco, G.; Zinno, R. Koiter asymptotic analysis of thin-walled cold-formed steel members. *Acta Mech. Et Autom.* **2015**, *9*, 245–251. [CrossRef]
9. Zhang, Y.; Murphy, K.D. Secondary buckling and tertiary states of a beam on a non-linear elastic foundation. *Int. J. Non-Linear Mech.* **2005**, *40*, 795–805. [CrossRef]
10. Zhang, Y.; Murphy, K.D. Jumping instabilities in the post-buckling of a beam on a partial nonlinear foundation. *Acta Mech. Solida Sin.* **2013**, *26*, 500–513. [CrossRef]
11. Kołakowski, Z. Static and dynamic interactive buckling of composite columns. *J. Theor. Appl. Mech.* **2009**, *47*, 177–192.
12. Kolakowski, Z.; Jankowski, J. Interactive buckling of steel LC-beams under bending. *Materials* **2019**, *12*, 1440. [CrossRef] [PubMed]
13. Becque, J.; Rasmussen, K.J. Experimental investigation of the interaction of local and overall buckling of stainless steel I-columns. *J. Struct. Eng.* **2009**, *135*, 1340–1348. [CrossRef]
14. Wadee, M.A.; Gardner, L. Cellular buckling from mode interaction in I-beams under uniform bending. *Proc. R. Soc. A Math. Phys. Eng. Sci.* **2012**, *468*, 245–268. [CrossRef]
15. Wadee, M.A.; Bai, L. Cellular buckling in I-section struts. *Thin-Walled Struct.* **2014**, *81*, 89–100. [CrossRef]
16. Shen, J.; Wadee, M.A. Length effects on interactive buckling in thin-walled rectangular hollow section struts. *Thin-Walled Struct.* **2018**, *128*, 152–170. [CrossRef]
17. Groh, R.M.J.; Pirrera, A. On the role of localizations in buckling of axially compressed cylinders. *Proc. R. Soc. A* **2019**, *475*, 20190006. [CrossRef] [PubMed]
18. Pirrera, A.; Avitabile, D.; Weaver, P.M. On the thermally induced bistability of composite cylindrical shells for morphing structures. *Int. J. Solids Struct.* **2012**, *49*, 685–700. [CrossRef]

19. Coburn, B.H.; Pirrera, A.; Weaver, P.M.; Vidoli, S. Tristability of an orthotropic doubly curved shell. *Compos. Struct.* **2013**, *96*, 446–454. [CrossRef]
20. Carrera, E.; Azzara, R.; Daneshkhah, E.; Pagani, A.; Wu, B. Buckling and post-buckling of anisotropic flat panels subjected to axial and shear in-plane loadings accounting for classical and refined structural and nonlinear theories. *Int. J. Non-Linear Mech.* **2021**, *133*, 103716. [CrossRef]
21. Bohlooly Fotovat, M.; Kubiak, T. Non-bifurcation behavior of laminated composite plates under in-plane compression. *Bull. Pol. Acad. Sci. Tech. Sci.* **2024**, *72*, e148874. [CrossRef]
22. Groh, R.M.; Avitabile, D.; Pirrera, A. Generalised path-following for well-behaved nonlinear structures. *Comput. Methods Appl. Mech. Eng.* **2018**, *331*, 394–426. [CrossRef]
23. Jones, R.M. *Mechanics of Composite Materials*; CRC Press: Boca Raton, FL, USA, 2018.
24. Bohlooly Fotovat, M.; Kubiak, T.; Perlikowski, P. Mixed mode nonlinear response of rectangular plates under static and dynamic compression. *Thin-Walled Struct.* **2023**, *184*, 110542. [CrossRef]

Disclaimer/Publisher's Note: The statements, opinions and data contained in all publications are solely those of the individual author(s) and contributor(s) and not of MDPI and/or the editor(s). MDPI and/or the editor(s) disclaim responsibility for any injury to people or property resulting from any ideas, methods, instructions or products referred to in the content.

materials

MDPI

Article

Ultrasonic Welding of Acrylonitrile–Butadiene–Styrene Thermoplastics without Energy Directors

Qian Zhi [1,2], Yongbing Li [1,*], Xinrong Tan [2], Yuhang Hu [2] and Yunwu Ma [1]

1 Shanghai Key Laboratory of Digital Manufacture for Thin-Walled Structure, Shanghai Jiao Tong University, Shanghai 200240, China; zhiqianhnust@163.com (Q.Z.); myw3337@sjtu.edu.cn (Y.M.)
2 Hunan Engineering Research Center of Forming Technology and Damage Resistance Evaluation for High Efficiency Light Alloy Components, Hunan University of Science and Technology, Xiangtan 411201, China; tanxinrong0@163.com (X.T.); 18867379862@163.com (Y.H.)
* Correspondence: yongbinglee@sjtu.edu.cn

Abstract: Ultrasonic welding (USW) of thermoplastics plays a significant role in the automobile industry. In this study, the effect of the welding time on the joint strength of ultrasonically welded acrylonitrile–butadiene–styrene (ABS) and the weld formation mechanism were investigated. The results showed that the peak load firstly increased to a maximum value of 3.4 kN and then dropped with further extension of the welding time, whereas the weld area increased continuously until reaching a plateau. The optimal welding variables for the USW of ABS were a welding time of 1.3 s with a welding pressure of 0.13 MPa. Interfacial failure and workpiece breakage were the main failure modes of the joints. The application of real-time horn displacement into a finite element model could improve the simulation accuracy of weld formation. The simulated results were close to the experimental results, and the welding process of the USW of ABS made with a 1.7 s welding time can be divided into five phases based on the amplitude and horn displacement change: weld initiation (Phase I), horn retraction (Phase II), melt-and-flow equilibrium (Phase III), horn indentation and squeeze out (Phase IV) and weld solidification (Phase V). Obvious pores emerged during Phase IV, owing to the thermal decomposition of the ABS. This study yielded a fundamental understanding of the USW of ABS and provides a theoretical basis and technological support for further application and promotion of other ultrasonically welded thermoplastic composites.

Keywords: ultrasonic welding; ABS; weld formation; simulation; horn amplitude

Citation: Zhi, Q.; Li, Y.; Tan, X.; Hu, Y.; Ma, Y. Ultrasonic Welding of Acrylonitrile–Butadiene–Styrene Thermoplastics without Energy Directors. *Materials* **2024**, *17*, 3638. https://doi.org/10.3390/ma17153638

Academic Editor: Murali Mohan Cheepu

Received: 30 May 2024
Revised: 4 July 2024
Accepted: 15 July 2024
Published: 23 July 2024

Copyright: © 2024 by the authors. Licensee MDPI, Basel, Switzerland. This article is an open access article distributed under the terms and conditions of the Creative Commons Attribution (CC BY) license (https://creativecommons.org/licenses/by/4.0/).

1. Introduction

Lightweight materials have become essential in the automotive industry with the governmental regulations on energy savings and CO_2 emission reduction. The application of lightweight materials, such as thermoplastic composites is considered an effective strategy in structural and manufacturing applications [1]. Acrylonitrile–butadiene–styrene (ABS) is regarded well for use in automobiles due to its advantageous combination of being lightweight, having a high specific strength and good processability [2]. In this context, an effective technique for joining ABS is imperatively important. Mechanical fastening, adhesive bonding, and welding are alternative methods to realize the permanent bonding of thermoplastic materials. The former two techniques have drawbacks of additional weight, stress concentration, complex pre/post-processing, etc., which hinder their further application in automobile manufacturing [3]. Ultrasonic welding (USW), a type of welding, has been demonstrated to be suitable for the production of polymeric joints with solid mechanical strength, but energy directors with various geometries, including triangular, semi-circular, or rectangular, are usually presented on the surfaces of adherends to concentrate the welding energy [4]. However, the existence of an energy director (ED) would easily bring defects, such as incomplete fusion or cracks [5]. Therefore, the ultrasonic welding of thermoplastic materials without EDs has become a research hotspot.

To date, the researches on the USW of thermoplastics without ED mainly focus on the optimization of the welding variables and understanding the heating generation mechanism. Gao [6] et al. investigated the effects of welding parameters on joint strength, microstructure and weld appearance, and concluded that an ultrasonically welded carbon fiber-reinforced polyamide 66 (CF/PA 66) joint without ED can obtain a tensile strength of 5.2 kN with acceptable cosmetic quality and fine compact weld microstructure. Several researchers were devoted to enhancing the mechanical properties of the joint by heat treatment. Prior to the USW of a CF/PA composite, preheating [7] and annealing treatments [8] were applied and the tensile strength increased by approximately 40% compared to that of a normal joint. This phenomenon was correlated with the viscoelastic behavior and crystallinity of the thermoplastic, in which the proper heat pretreatment temperatures (95–125 °C for CF/PA 66 and 180 °C for CF/PA 6) concentrated the welding heat at the contact surface and decreased the dissipation within the workpieces. Accordingly, a proper weld without an obvious porous area was formed. Others also applied mechanical grinding [9], a blank holder [10,11] and a double pulse [12] to improve the welding conditions for the enhancement of tensile strength. These pretreatments aimed to ameliorate the contact behavior at the workpiece/workpiece interface prior to welding, and a robust tensile strength with small data variance was simultaneously obtained.

Finite element analysis (FEM) is a helpful tool in analyzing the weld formation in the USW of polymers, which can provide a comprehensive insight into the dynamic deformation behavior of the workpieces, and the strain/stress distribution at the overlapped region is also possible to predict. Currently, most FEM-related works are concentrated on revealing the heat generation mechanism during the USW process. Tutunjian [13] declared the frictional heat generated in the early stages of the ultrasonic spot welding of 5-HS-woven carbon fabric-reinforced thermoplastic through explicit mechanical 3D FEM analysis. The composite laminates were defined using a 3D continuum shell (only displacement degrees of freedom and no rotation) with eight nodes and reduced integration with the hourglass effect. Friction heating reduces the stiffness of the interface layers inside the weld center, and the applied cyclic strain focuses on the softer interfacial layers and induces a much higher viscoelastic heat. Consequently, the heating process is accelerated and restricted to the weld center with the combination of friction and viscoelastic heat. Li [14] proposed an integrated process–performance model to predict the failure load of the CFRP joint and weld formation information. It was verified that the heat generation mechanisms in the USW of thermoplastics came from friction and viscoelastic dissipation.

In practical USW, the ultrasonic oscillations change quickly, and transient vibrations are difficult to simulate. Currently, several researchers have applied a static pressure and constant horn displacement [14] on the tip of the horn or considered horn displacement changes step by step [15]. However, the experimental and simulated results have shown large differences. The horn vibration during USW was complicated not only because of its high frequency but also due to the resonance effect [16,17]. To date, there is scant research simulating the weld formation in the USW of thermoplastics by applying real-time vibrations in an FEM model. Therefore, there is an urgent need to understand weld initiation and growth by FEM analysis with real-time oscillations.

In this context, the actual vibrations during the USW of ABS were collected by a high-frequency sensor (400 kHz). The real-time oscillations were integrated into the FEM model to investigate the heat generation and weld formation. The effect of the welding time on joint performance and weld area evolution was assessed. The experimental results were also compared with the simulated results.

2. Materials and Methods

2.1. Laminate

ABS laminate with dimensions of 100 mm in length, 30 mm in width and 2 mm in thickness was purchased from Changzhou Xinhejiu Composite Materials Technology Co.,

Ltd. (Changzhou Xinhejiu Composite Materials Technology, Changzhou, China). The properties of ABS provided by the supplier (25 °C) were listed in Table 1.

Table 1. Properties of ABS laminate.

Materials	Density/(kg · m^{-3})	Poisson's Ratio	Elastic Modulus/(MPa)	Thermal Conductivity/(W · $m^{-1}K^{-1}$)
ABS	1100	0.394	2000	0.2256

2.2. Ultrasonic Welding

All the ABS laminates were joined on a KZH-2026 welder (Weihai Kaizheng Ultrasonic Technologies Co., Ltd., Weihai, China) with a nominal frequency of 20 kHz, and a nominal amplitude of 25 μm. The welder had three working modes of time-, energy- and displacement-control, and time-control was selected in this experiment. Prior to USW, the delay time, welding time and holding time were preset. The delay time (from the initial welding process to ultrasonic oscillations began) and holding time (ultrasonic oscillations stopped to the welding horn retracted back) were set as 2 s and 3 s on the basis of preliminary experiments, respectively [18]. Three high-frequency sensors—displacement, pressure and power sensors—at 400 kHz were installed on top of the pneumatic motion axis in the welder to collect real-time USW process information. Then, a data acquisition system (>500 kHz) connected the end of the sensor with a computer as shown in Figure 1a. A single lap joint with an overlap distance of 25 mm was applied.

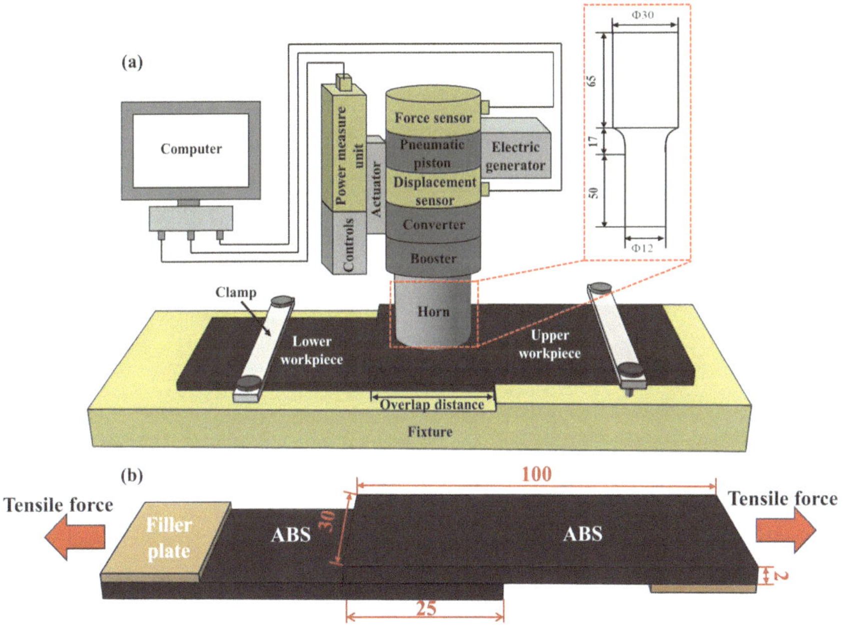

Figure 1. Schematic illustration of ultrasonically welded ABS without energy directors: (**a**) welding configuration; (**b**) single-lapped joint (dimensions in mm).

2.3. Characterization

Tensile testing of the joint was conducted on an MTS E45.105 tester (MTS, Prairie, MN, USA) with a stroke rate of 2 mm/min based on ASTM standards [19]. Two filler plates were attached to the ends of the workpieces to accommodate the offset as shown in Figure 1b. Five sets of joints were welded with identical welding variables and the average joint strength (peak load) was obtained. The microstructure of the joint was examined with scanning electrical (SEM, SU3500 Hitachi, Tokyo, Japan). Prior to the examination, the sample was sputter-coated with gold to increase conductivity.

2.4. Rheological Experiment

The rheological experiment was carried out to measure the viscoelastic properties of ABS by using a shear-strain-controlled rotational rheometer (ARES-G2, Waters, Milford, MA, USA). The ABS specimen, in a circular shape of coin size, was subjected to a frequency sweep in the temperature range from room temperature to 180 °C with an ascending temperature rate of 2 °C/min to study its temperature/frequency-related performance. The storage modulus and loss modulus were obtained as a function of frequency, with a shear strain of 1% by applying the angular frequency in the range between 10^{-1} and 10^{2} rad/s.

3. Modelling

3.1. Material Property

ABS polymer is a typical viscoelastic material, and the Maxwell or Voigt–Kelvin model is usually utilized to describe the elastic and viscous properties. The Maxwell model, consisting of a spring (representing the elastic part) and a dashpot (denoting the viscous part), is selected in this simulation. Ten Maxwell units in parallel are used as also reported in other literature [19].

The constitutive relation of viscoelastic polymer is needed to understand the stress and deformation of ABS. Combing the Maxwell model, the constitutive behavior of ABS can be defined using [20]:

$$\sigma(t) = \varepsilon_0 e(t) + \int_0^t e(t-\lambda)\frac{d\varepsilon(\lambda)}{d\lambda}d\lambda \tag{1}$$

where $\sigma(t)$ is stress, ε_0 is the initial value of strain, t and λ are the current and past time, respectively. $e(t)$ is the relaxation modulus. The experimental frequency-related data are input into the model to define the viscoelastic property of the material. To connect the experimental test of storage and loss modulus with the constitutive model of ABS, the Maxwell series is converted to the frequency domain from the time domain using the Fourier transform and can be expressed as follows [21]:

$$E'(\omega) = G_0\left[1 - \sum_i^N g_i\right] + G_0 \sum_{i=1}^N \frac{e_i \lambda_i^2 \omega^2}{1 + \lambda_i^2 \omega^2} \tag{2}$$

$$E\prime\prime(\omega) = G_0 \sum_{i=1}^N \frac{e_i \lambda_i^2 \omega^2}{1 + \lambda_i^2 \omega^2} \tag{3}$$

$$G_0 = \frac{e_0}{2(1+\mu_0)} \tag{4}$$

where ω is the angular frequency, G_0 is the transient shear modulus, μ is the transient Poisson ratio, and N is the number of Maxwell series. The relaxation spectra of ABS in the range from 10^{-1} to 10^{2} rad/s are shown in Figure 2.

In defining the viscoelastic property of the material, which varies with frequency in Abaqus, the storage/loss moduli-related real and imaginary parts of $\omega\Re(g^*)$ and $\omega\Im(g^*)$, and the volume modulus-related $\omega\Re(k^*)$ and $\omega\Im(k^*)$ are needed. Since the ABS composite is a viscoelastic polymer, the imaginary parts are neglected [22].

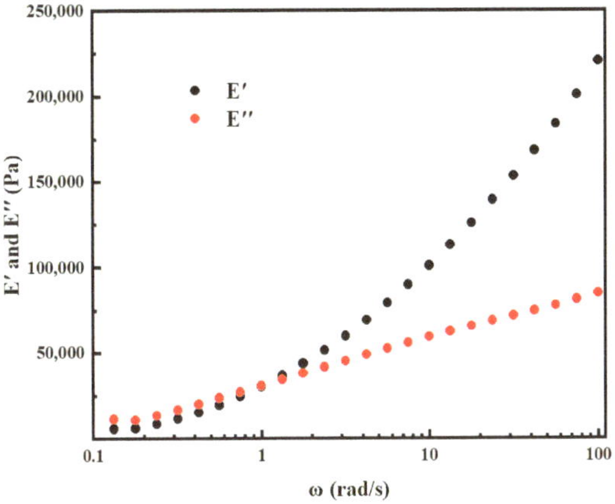

Figure 2. Discrete relaxation spectra of ABS in the frequency range of 0.1–100 Hz.

$$\omega\Re(g^*) = \frac{E\prime\prime}{G_\infty} \tag{5}$$

$$\omega\Im(g^*) = 1 - \frac{E'}{G_\infty} \tag{6}$$

$$G_\infty = G_0[1 - \sum_i^N e_i] \tag{7}$$

where G_∞ is the long-term shear, with 789.09 MPa for ABS material.

The temperature dependence of the materials can be considered with the WLF model [21]:

$$-log\alpha_T = \frac{C_1(T - T_0)}{C_2 + (T - T_0)} \tag{8}$$

where α_T is the horizontal shift factor, T is the temperature, T_0 is the reference temperature and chosen as 180 °C to generate the master curve, and C_1 and C_2 are the fitting parameters, with C_1 = 5.8 and C_2 = 120.8 K in this study.

After defining the viscoelastic property of the material, the heat-transfer-related property of heat capacity, which changes with temperature, is considered. The specific heat (c) is expressed as follows [23]:

$$c = \frac{1}{m} \times \frac{dQ/dt}{dT/dt} \tag{9}$$

where m is the mass of the tested material, dQ/dt is the heat flux and dT/dt is the heating rate during the DSC test. The specific heat curve of ABS varies with temperature as shown in Figure 3 (blue line). The thermal gravity curve is also included in Figure 3, where melting and decomposition of ABS occur at 220 °C and 265 °C (intersections of the green dotted line and black solid line), respectively. Therefore, careful selection of the welding time is crucial in the USW of ABS.

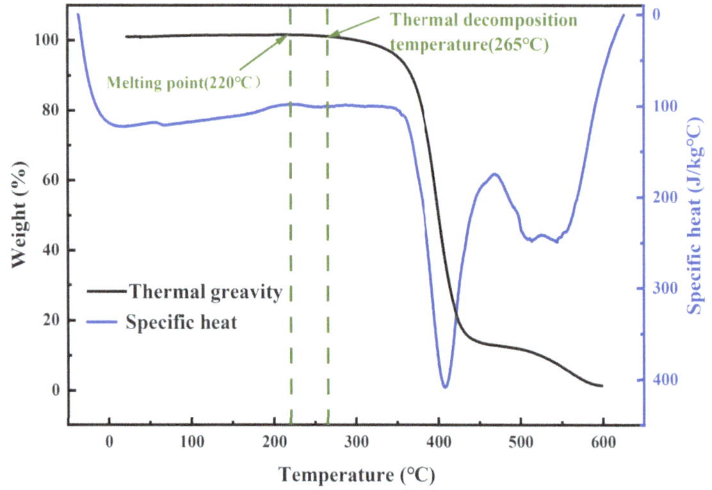

Figure 3. Heat-specific and weight loss curves of ABS.

3.2. Finite Element Modelling

The commercial Abaqus software (version 6.22) is implemented for the finite element modelling of the ultrasonically welded ABS process, and this software enables one to perform geometric modelling, material property definitions, meshing, and visualization. Then, the temperature evolution and stress distribution can be simulated.

In this study, a three-dimensional finite element model, consisting of a 7075 aluminum horn, aluminum anvil and two pieces of 2 mm-thick ABS laminates, is built in Abaqus as shown in Figure 4. The C3D8T hexagonal solid element is utilized to divide the mesh. The total number of meshed grids of the model is 5724, consisting of 9142 nodes. Considering the calculation accuracy and time taken for the analysis, the mesh size for the overlapped region between upper and lower workpieces is 1 while that for the rest of the regions is set as 5.

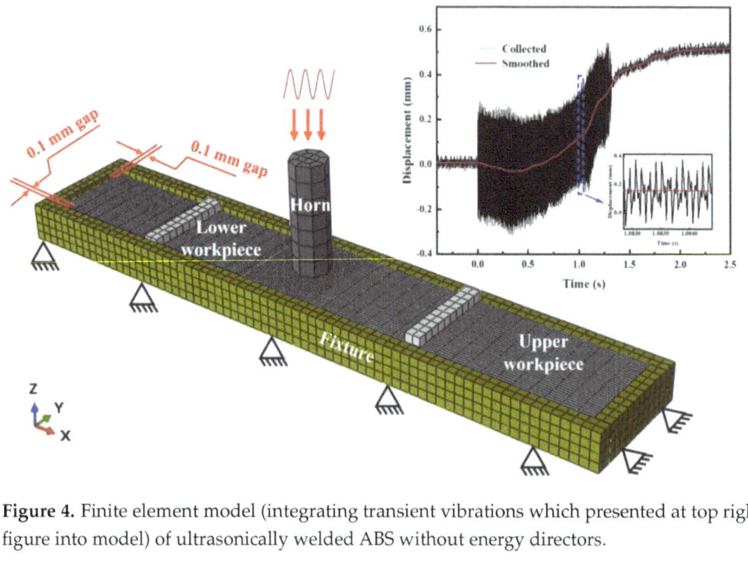

Figure 4. Finite element model (integrating transient vibrations which presented at top right of the figure into model) of ultrasonically welded ABS without energy directors.

Thermal–mechanical coupling analysis is applied to simulate the temperature field in the USW of ABS. The fixture is fixed in the X-/Y-/Z-direction. The workpieces are fixed in the X- and Z-direction, with a small gap of 0.1 mm between the workpiece and fixture set as a boundary condition, and are illustrated in Figure 4. To improve simulation accuracy, real-time vibration (unveiling the amplitude change) during the USW process, which is recorded using a high-frequency sensor and data collector software installed in the welder, is introduced into the model and applied on the tip of the horn.

The initial temperature of the model is set as 20 °C, and the governing equation of heat conduction is expressed as [24]:

$$\rho c \frac{\partial T}{\partial t} = \frac{\partial}{\partial x}\left(\lambda \frac{\partial T}{\partial x}\right) + \frac{\partial}{\partial y}\left(\lambda \frac{\partial T}{\partial y}\right) + \frac{\partial}{\partial z}\left(\lambda \frac{\partial T}{\partial z}\right) + Q \quad (10)$$

where ρ is the material density, λ is the thermal conductivity of ABS, and Q is the welding energy. As for heat convection, the convection boundary conditions between the horn and upper workpiece, lower workpiece and the fixture are set as 90 $W/(m^2\ K)$ since the horn and fixture are both made of 7075 aluminum alloy. The contacting heat conduction coefficient between upper and lower workpieces (R) is defined using [25]:

$$R = \frac{KA}{L} \quad (11)$$

where K is the thermal conductivity of ABS, with a value of 0.226 W/m/K, L is the sheet thickness of 2 mm, and A is the sectional area where the ABS sample is perpendicular to the conduction direction. The contact area in this model is the overlapped region, with a sectional area of 900 mm^2. The calculated contacting thermal conductivity coefficient R is 0.1 $W \cdot m^{-2} \cdot K^{-1}$. Heat conduction between the components and ambient atmosphere is not considered owing to the short welding time (less than 2 s).

Surface-to-surface contact is utilized to define component contact in the model. Hard contact is applied when it is normal to the workpiece where a separation is allowed during the USW process. A penalty function is used to define the friction between two parts. Friction coefficients at the horn to upper workpiece and lower sheet to fixture interfaces are set as 0.1 [7,26,27]. At the workpieces contacting surface, the friction coefficient is 0.3 when the temperature at the contacting interface is below the melting point of ABS (220 °C) while it is defined as 0.1 with a further increase in temperature.

4. Results and Discussion

4.1. The Joint Strength of Ultrasonically Welded ABS

Preliminary experiments show the welding pressure of 0.13 MPa is more suitable for the USW of ABS. Herein, the effect of the welding time on peak load and weld area of the joint welded with a welding pressure of 0.13 MPa is evaluated as shown in Figure 5. With the increase in the welding time, the peak load of the joint increases to the maximum value of 3.4 kN and then decreases, while the weld area (measured by the IPP 6.0 software) expands gradually to a plateau. A peak load of 0.37 kN is obtained for the joint with a 0.1 s welding time, and the joint strength and weld area increase dramatically in the first 0.9 s then increase moderately to their peak values. Combining the peak load and weld area, the optimal welding time is selected as 1.3 s, where it can reach the maximum peak load and weld area. Interestingly, the data variance in joint strength becomes larger when the welding time is extended to above 1.3 s. This characteristic is likely attributed to the thermal decomposition of ABS, where the position and distribution of the resultant porous region affect the joint strength severely [6,28].

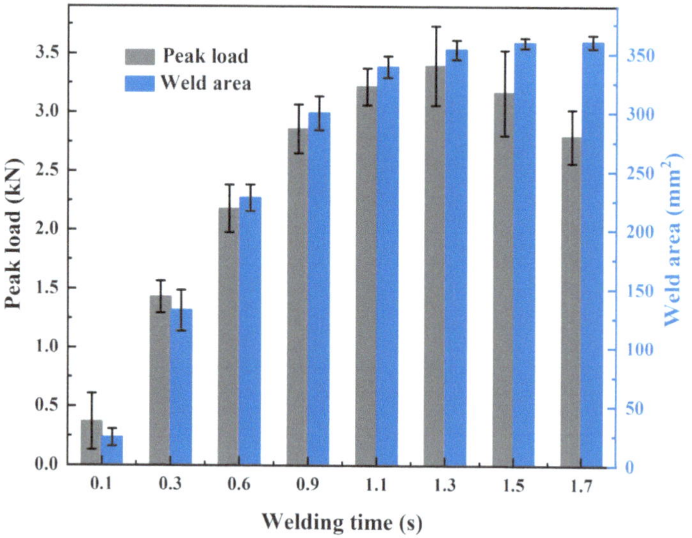

Figure 5. Effect of the welding time on peak load and weld area of ultrasonically welded ABS.

4.2. Microstructure of the Joint

The fractured surfaces of the joints after tensile testing are observed and two main fracture modes of interfacial failure and workpiece breakage are presented as shown in Figure 6. The joint usually fractures at the nugget for underweld joints (welding time shorter than 1.1 s) where the insufficient weld area cannot bear enough tensile force. Prolonging the welding time to above the optimal value, the joint is likely to break at the workpiece (mostly at the upper workpiece), which is closely related to the material property [28] and will be elaborated on later.

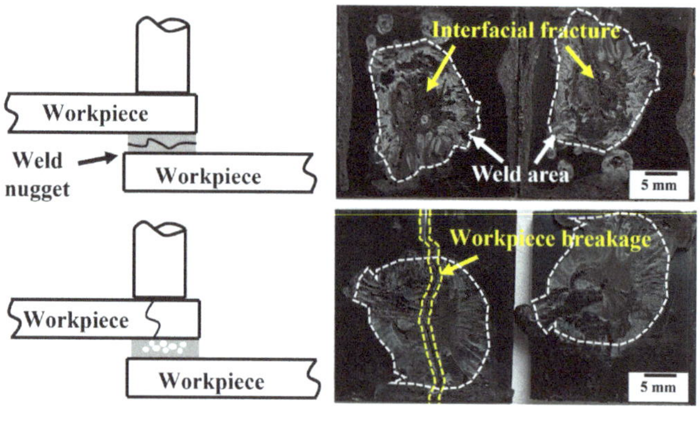

Figure 6. Fracture mode of ultrasonically welded ABS joint.

Careful examination of the fractured morphology of the joint shown in Figure 7 shows that the weld area expands with the welding time and there is a loose microstructure with pores on the surface when it exceeds the optimal time. The effective bonding region is localized in an approximately circular area (directly under the welding horn) and the bonding area can be categorized into two types: compact normal weld area and loose porous area. The microstructure in the normal weld area is dense, while numerous deconsolidation voids appear in the central area of the weld region and are labelled as porous regions in Figure 7. Cross-sectional morphologies of representative joints welded in 1.1 s, 1.3 s, 1.5 s and 1.7 s are observed to analyze the weld nugget. Referring to Figure 8, there is no obvious boundary between the weld zone and matrix when the welding time is less than 1.1 s. The distinction of the weld region is mostly based on the direction of the fracture texture, where the stress conditions for upper and lower workpieces during the tensile test are different. Increasing the welding time to 1.3 s, some pores randomly distribute in the weld zone. The pore increases in density and quantity. The thickness of the porous layer increases with increased welding time, which is harmful to the joint [7,29]. For joints made in 1.7 s, the scale of the porous region enlarges remarkably, and it exhibits a mouth-like shape, where the thickness of the lateral side is small and the central region is relatively large. This behavior is intimately associated with heat generation and melt flow during USW [12,18,29], where thermal decomposition of the polymer releases gases. The pressure in the gases is large, and large pores extend to the faying interface and squeeze the molten materials out. The decomposed material flows bilaterally along the weld area and forms a mouth-like shape (joints with a 1.7 s welding time).

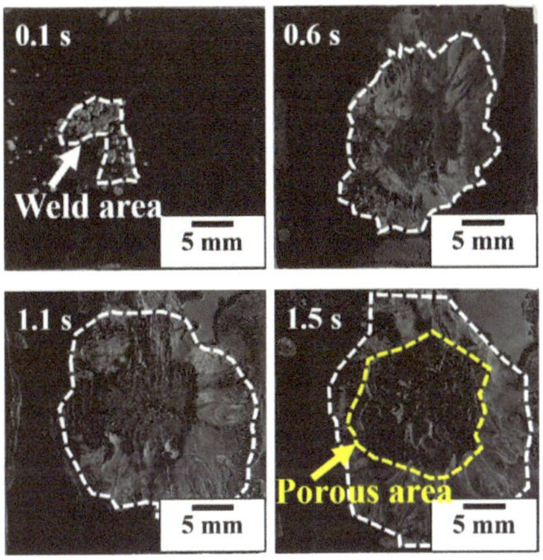

Figure 7. Relation between the welding time and weld area.

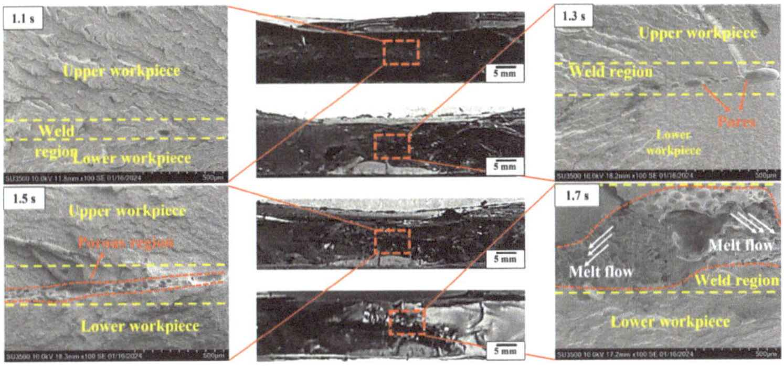

Figure 8. Cross-sectional morphology of representative joints.

A small number of pores in the central region of the weld area slightly influence joint performance, as reported previously [6,28]. Increased pores in the weld zone separate the polymer matrix and weaken its ability to bear loading. Thus, the maximum peak load occurs at a welding time of 1.3 s and drops afterwards.

4.3. Simulation of the USW Process

4.3.1. Heat Generation

To fully understand the weld formation of ultrasonically welded ABS, the joints made with 0.13 MPa and 1.7 s are simulated using Abaqus 6.22 software, and the side views of the simulated weld with various welding times are shown in Figure 9. The melting temperature of the matrix is assigned as T_m while the thermal decomposition point is denoted as T_d. The regions in red or grey color represent the polymer melts or decompose under ultrasonic heat. The highest temperature at the contact interface is in the middle of the overlapped region. Surprisingly, the polymer melts with only a 0.1 s welding time, which is much faster than that of polyamide composites [6,18]. This characteristic verifies that the amorphous polymer is easier to weld than the a polymer with a semi-crystal structure [29–31]. With increased welding time, the red melted region expands gradually into the X/Y/Z plane. The grey decomposed region emerges at joints made in 1.1 s and displays an analogous expansion trend with the melted region.

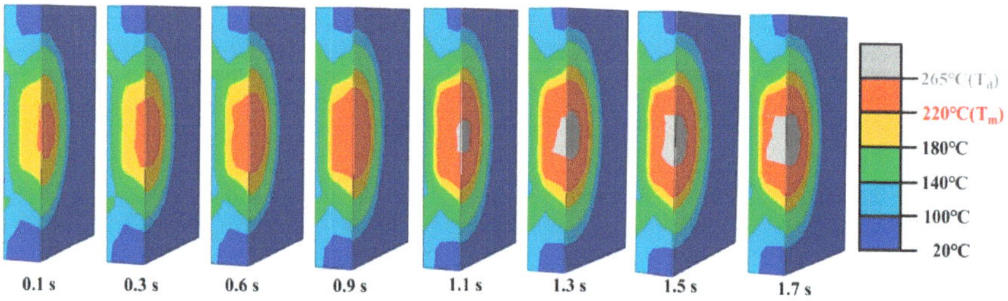

Figure 9. Simulation temperature distribution in the upper workpiece of the joint welded with different times.

The theory that the heat generation in the USW of thermoplastics generally contains frictional heat and viscoelastic dissipation has been well accepted by scholars [16,19,32]. The frictional heat and viscoelastic dissipation in the USW of ABS are derived using "process output" in Abaqus, as presented in Figure 10. The varying tendencies of frictional and viscoelastic dissipations are contrary, which is consistent with the opinion that frictional heat dominates at the early stage of the USW while viscoelastic heating dominates in the following stages [9,19,24].

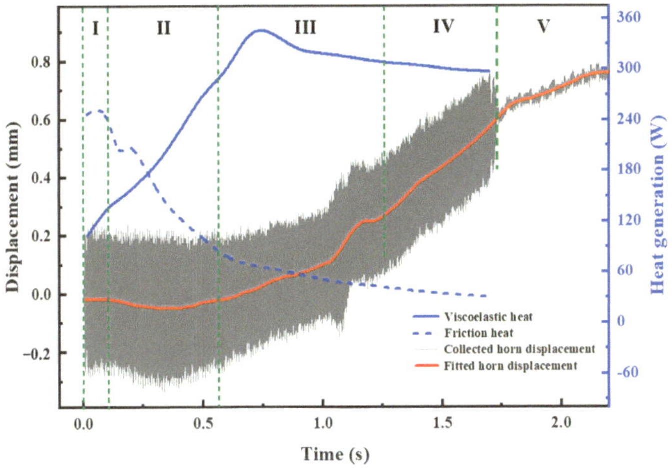

Figure 10. Heat generation and horn displacement curves in ultrasonically welded ABS in 1.7 s.

At the initial stages of USW, the upper and lower workpieces retain their stiffness and toughness, along with the rough surface of the adherends. Before the steady melt forms, the perpendicular vibrations transfer to horizontal deformation at the interface of workpieces. Horizontal deformation adds to slippage and friction at the interface. Friction generates heat on the rough surfaces of workpieces. Once the weld nugget forms, frictional heat greatly decreases. When the temperature of the workpiece rises up to above the glass transition temperature under friction heat, viscoelastic dissipation dominates the ultrasonic welding process, as shown in Figure 10 (blue lines).

4.3.2. The Weld Formation Mechanism

Real-time vibrations for joints with a 1.7 s welding time (the semi-transparent grey region) are also included in Figure 10. As seen from the enlarged view of horn displacement, the vibrations are complicated and change quickly during USW. Hence, simulating weld growth of ultrasonically welded ABS by applying real-time vibration into the model should primarily improve simulation accuracy. Horn displacement in Figure 10 exhibits typical characteristics in the USW of thermoplastics [33,34], and the weld area of the simulated area and measured areas shows similar varying trends in Figures 11–14 as expected. Comprehensively considering the variation characteristics in amplitude, vibration, horn displacement and energy dissipation, the weld formation of ultrasonically welded 2 mm-thick ABS without energy can be divided into five phases.

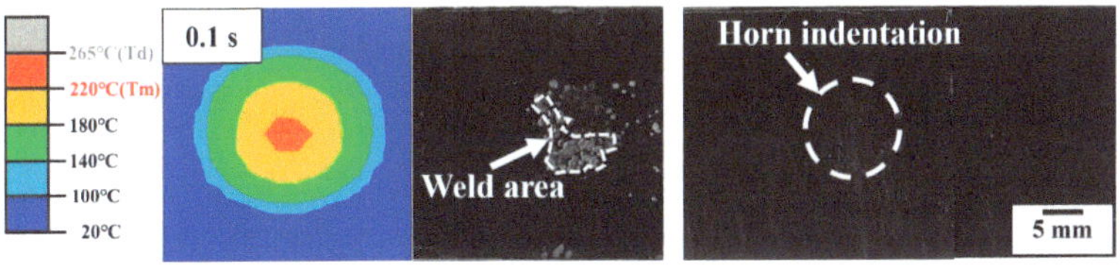

Figure 11. Simulation and experimental weld area; horn indentation of the welded joint in Phase I.

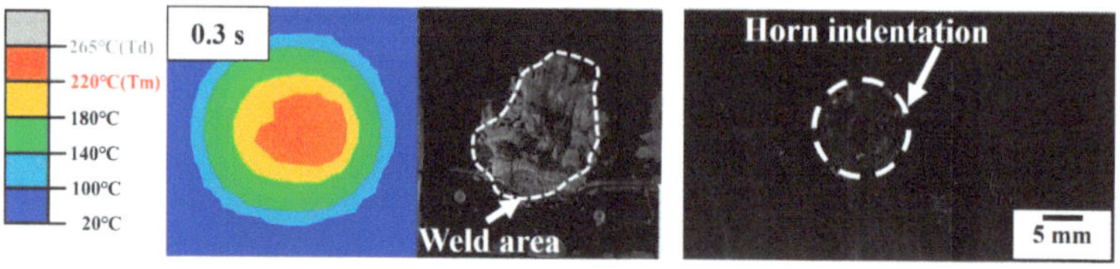

Figure 12. Simulation and experimental weld area; horn indentation of the welded joint in Phase II.

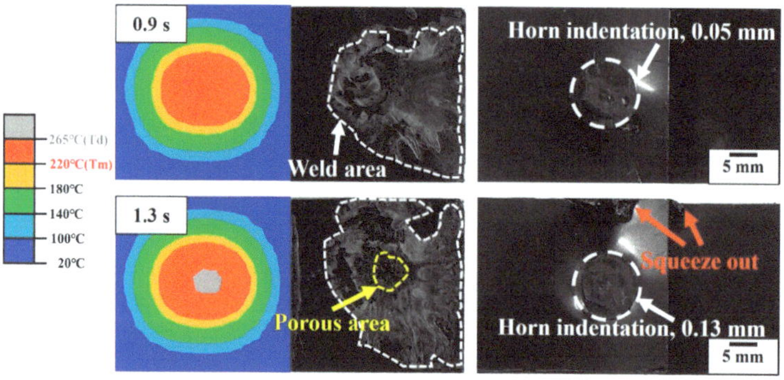

Figure 13. Simulation and experimental weld area; weld appearance of the joint in Phase III.

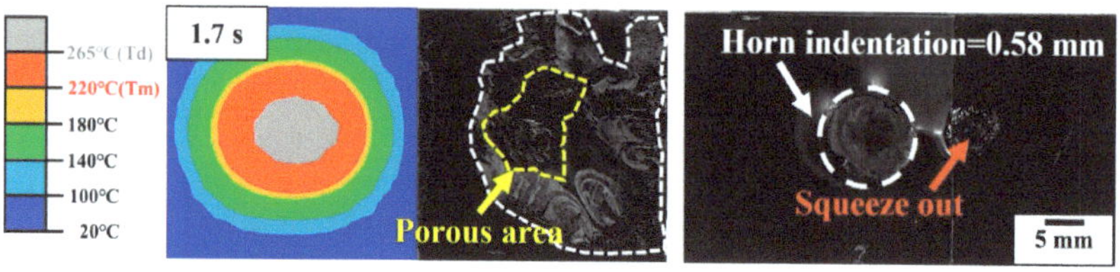

Figure 14. Simulation and experimental weld area; weld appearance of the joint in Phase IV.

Phase I (0~0.1 s): This phase lasts approximately 0.1 s and the fractured surface shows a clear weld initiation with random hot spots as depicted in Figure 11. At this stage, the melting of the matrix is mainly attributed to friction heating, which results from contact point slippage [18]. Since this phase is very short and most of the matrix remains un-melted, no significant horn displacement or indentation is observed. Thus, the amplitude in phase I is larger than the nominal amplitude owing to the resonance in the system [27].

Phase II (0.1~0.6 s): This stage is an unsteady phase. The amplitude changes irregularly and the small asperities at the contact surface melt gradually to form a favorably intimate contact condition for the following phase. Fiction and viscoelastic heat together dominate weld growth at this stage [18]. The heat reduces the stiffness of the interface layers inside the weld apex and the sinusoidal cyclic strain focuses on the softer interfacial area to expand the weld area, as shown in Figure 12. The simulated result is smaller than the measured one owing to the flow and expansion of the melt. A slight horn indentation is presented on the joint. A downward trend in displacement at Phase II is presented in Figure 10, which is mainly due to the thermal expansion of the ABS matrix under the accumulation of ultrasonic heat and causes horn retraction.

Phase III (0.6~1.3 s): The amplitude change is relatively stable in this phase and is characterized by a continuously increasing displacement. The melt rate and flow of the ABS matrix are in equilibrium. Then, weld growth enters into the paramount phase—steady melt flow—which is critical to the joint quality [19,35]. It is worth mentioning that there is a small step-like shape to the smoothed displacement. This phenomenon is correlated with melt flow behavior. It is seen in Figure 10 that viscoelastic heating plays a chief role in weld growth. The simulated weld growth at this stage is also smaller than the measured weld growth as explained in Phase II, while the squeeze out of molten ABS and horn indentation become significant when the welding time exceeds 1.3 s, as illustrated in Figure 13. Before this meltdown, a large amount of ABS melts within the upper plate and is about to be squeezed out. Hence, there is a slight change in the amplitude.

Phase IV (1.3~1.7 s): This phase is characterized by a combination of horn indentation on the upper workpiece and squeezed out of molten ABS. The amplitude at this stage is much more stable and the displacement of the horn increases linearly with the welding time (with a larger increase rate than that in Phase III). At this phase, more viscoelastic dissipation is consumed within the workpiece and leaves numerous voids in the weld region and deep indentation on the joint surface (Figure 14). To thoroughly understand the origin of the pores, Fourier transform infrared spectroscopy (FTIR) tests are conducted in the ABS matrix, weld area and porous region. Referring to Figure 15, the bending vibration of C-H, deformation of C-H for hydrogen atoms and out-of-plane C-H bending in ABS polymer are in the range of 700–1038 cm^{-1}. Stretching vibration peaks of the benzene ring, C≡N and C=C are present at 1450–1600 cm^{-1}, 2237 cm^{-1} and 1630 cm^{-1}, respectively. The aromatic and aliphatic C-H are detected at 3200–2800 cm^{-1} [36,37]. It is clear that the peak positions are similar but the absorption intensity differs significantly. The peak intensity for the porous area is much lower, but similar for the weld area and the matrix, indicating the ABS material decomposes in the porous area [28,38]. Generally, a solid joint should have a dense microstructure, thus the occurrence of this phase is detrimental to the joint strength and should be avoided in actual production.

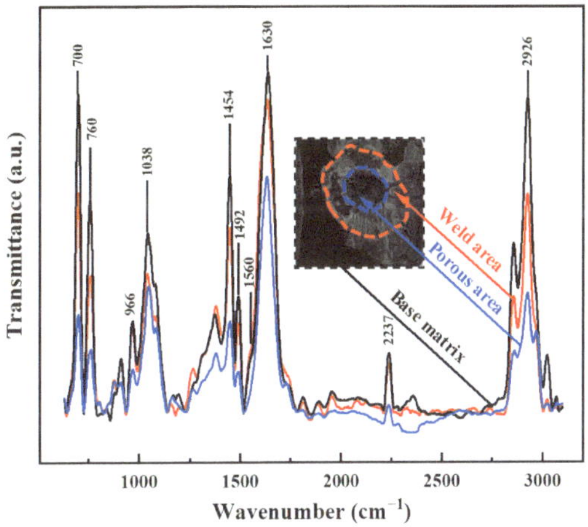

Figure 15. FTIR spectra of the joint within different regions.

It has been verified that the voids result from the thermal decomposition of the ABS matrix. The decomposed ABS releases volatile products, such as HCN, CO, and NO_x. The pressures in these voids are large and will be expanded to the welding interface, accompanied by the squeeze out of molten ABS. With the accumulation of viscoelastic dissipation, the upper workpiece (dissipated the majority of the heat) decomposes consequentially and some of the decomposed ABS flows bilaterally along the weld apex, while the rest of the residues are in the joint. As a result, the joint with a welding time of 1.7 s shows a microstructure with a mouth-like shape and numerous pores, as presented in Figure 8.

Phase V (>1.7 s): Ultrasonic vibration stops at this stage and horn displacement is slightly increased owing to the cold contraction of the weld. The weld solidifies under the welding pressure for 3 s (holding time).

Based on the aforementioned analysis, the characteristics of each phase during the USW of ABS are different. Weld initiations with randomly distributed hotspots are observed in Phase I. The unsteady and steady phases have a sequential pattern with a downward and increasing trend in horn displacement, respectively. Then, thermal decomposition of ABS occurs, with faster and increased horn displacement. When ultrasonic vibration is paused, the process enters Phase V with a slight increase in horn displacement.

5. Conclusions

The USW of ABS without ED was investigated in this study. The effects of the welding time on joint performance, weld area and the weld formation mechanism were analyzed systematically. The following main conclusions were drawn:

(1) The peak load of ultrasonically welded ABS increased with the welding time (less than 1.3 s) and then decreased with a prolonged welding time. The maximum value of 3.4 kN was obtained with an optimal welding time of 1.3 s and 0.13 MPa.
(2) On prolonging the welding time to 1.7 s, the weld areas of joints increased gradually to the maximum value and then reached a plateau. Two typical failure modes of interfacial failure and workpiece breakage appeared during tensile tests.
(3) Integrating real-time horn displacement into the finite element model can improve simulation accuracy in the USW of ABS.
(4) Weld formation of ultrasonically welded ABS without ED (welding time of 1.7 s) consisted of five distinct phases of weld initiation, horn retraction, melt and flow equilibrium, horn indentation and squeeze out, weld solidification based on the variation characteristics, horn displacement and energy dissipation during welding.
(5) An obvious porous area emerged in the joint made with a welding time greater than 1.3 s, which was mainly ascribed to the thermal decomposition of ABS and was detrimental to the joint strength.

Author Contributions: Conceptualization, Q.Z.; methodology, Q.Z. and X.T.; software, Y.H.; validation, Q.Z. and Y.M.; formal analysis, Q.Z. and Y.H.; investigation, Y.H.; resources, Q.Z.; data curation, Y.H.; writing—original draft preparation, Q.Z., X.T., Y.H. and Y.M.; writing—review and editing, Y.L.; visualization, Q.Z. and X.T.; supervision, Y.L.; project administration, Q.Z.; funding acquisition, Q.Z. All authors have read and agreed to the published version of the manuscript.

Funding: This work was financially supported by the Natural Science Foundation of Hunan Province (Grant Nos. 2024JJ4019 and 2023JJ50235) and the China Postdoctoral Science Foundation (Grant No. 2023M732221).

Institutional Review Board Statement: Not applicable.

Informed Consent Statement: Not applicable.

Data Availability Statement: The original contributions presented in the study are included in the article, further inquiries can be directed to the corresponding author.

Conflicts of Interest: The authors declare no conflicts of interest.

References

1. Quan, D.; Ma, Y.; Yue, D.; Liu, J.; Xing, J.; Zhang, M.; Alderliesten, R.; Zhao, G. On the application of strong thermoplastic–thermoset interactions for developing advanced aerospace-composite joints. *Thin Walled Struct.* **2023**, *186*, 110671. [CrossRef]
2. Arif, M.; Kumar, D.; Siddiquee, A.N. Morphological and Mechanical Characterization of Friction Stir Welded Zones in Acrylonitrile Butadiene Styrene (ABS) Polymer. *J. Mater. Eng. Perform.* **2023**, 1–11. [CrossRef]
3. Deng, S.; Djukic, L.; Paton, R.; Ye, L. Thermoplastic–epoxy interactions and their potential applications in joining composite structures–A review. *Compos. Part A Appl. Sci. Manuf.* **2015**, *68*, 121–132. [CrossRef]
4. Wang, Y.; Rao, Z.; Liao, S.; Wang, F. Ultrasonic welding of fiber reinforced thermoplastic composites: Current understanding and challenges. *Compos. Part A Appl. Sci. Manuf.* **2021**, *149*, 106578. [CrossRef]
5. Tao, W.; Su, X.; Wang, H.; Zhang, Z.; Li, H.; Chen, J. Influence mechanism of welding time and energy director to the thermoplastic composite joints by ultrasonic welding. *J. Manuf. Process.* **2019**, *37*, 196–202.
6. Gao, Y.-H.; Zhi, Q.; Lu, L.; Liu, Z.-X.; Wang, P.-C. Ultrasonic welding of carbon fiber reinforced nylon 66 composite without energy director. *J. Manuf. Sci. Eng.* **2018**, *140*, 051009. [CrossRef]

7. Zhi, Q.; Tan, X.; Liu, Z. Effects of preheat treatment on the ultrasonic welding of carbon-fiber-reinforced polyamide 66 composite. *Weld J.* **2017**, *96*, 429S–438S.
8. Wang, K.; Li, Y.; Banu, M.; Li, J.; Guo, W.; Khan, H. Effect of interfacial preheating on welded joints during ultrasonic composite welding. *J. Mater. Process. Technol.* **2017**, *246*, 116–122.
9. Li, Y.; Liu, Z.; Shen, J.; Lee, T.H.; Banu, M.; Hu, S.J. Weld quality prediction in ultrasonic welding of carbon fiber composite based on an ultrasonic wave transmission model. *J. Manuf. Sci. Eng.* **2019**, *141*, 081010. [CrossRef]
10. Tian, Z.; Zhi, Q.; Feng, X.; Zhang, G.; Li, Y.; Liu, Z. Effect of preload on the weld quality of ultrasonic welded carbon-fiber-reinforced nylon 6 composite. *Polymers* **2022**, *14*, 2650. [CrossRef]
11. Li, Y.; Arinez, J.; Liu, Z.; Lee, T.H.; Fan, H.-T.; Xiao, G.; Banu, M.; Hu, S.J. Ultrasonic welding of carbon fiber reinforced composite with variable blank holding force. *J. Manuf. Sci.* **2018**, *140*, 091011. [CrossRef]
12. Zhi, Q.; Li, Y.; Shu, P.; Tan, X.; Tan, C.; Liu, Z. Double-pulse ultrasonicwelding of carbon-fiber-reinforced polyamide 66 composite. *Polymers* **2022**, *14*, 714. [CrossRef] [PubMed]
13. Tutunjian, S.; Eroglu, O.; Dannemann, M.; Modler, N.; Fischer, F. A numerical analysis of an energy directing method through friction heating during the ultrasonic welding of thermoplastic composites. *J. Thermoplast. Compos. Mater.* **2020**, *33*, 1569–1587. [CrossRef]
14. Li, Y.; Lee, T.H.; Banu, M.; Hu, S.J. An integrated process-performance model of ultrasonic composite welding based on finite element and artificial neural network. *J. Manuf. Process.* **2020**, *56*, 1374–1380. [CrossRef]
15. Yang, Y.; Liu, Z.; Wang, Y.; Li, Y. Numerical study of contact behavior and temperature characterization in ultrasonic welding of CF/PA66. *Polymers* **2022**, *14*, 683. [CrossRef] [PubMed]
16. Levy, A.; Le Corre, S.; Fernandez Villegas, I. Modeling of the heating phenomena in ultrasonic welding of thermoplastic composites with flat energy directors. *J. Mater. Process. Technol.* **2014**, *214*, 1361–1371. [CrossRef]
17. Müller, F.W.; Mirz, C.; Weil, S.; Schiebahn, A.; Corves, B.; Reisgen, U. Weld quality characterization by vibration analysis for ultrasonic metal welding processes. *J. Adv. Join. Process.* **2023**, *8*, 100149. [CrossRef]
18. Tan, X.; Zhi, Q.; Ma, J.; Chen, Y.; Li, Y. Simulation of temperature and weld growth mechanism in ultrasonic welding of carbon fiber reinforced polyamide 66 composite: Employing the high frequency real-time horn vibration. *J. Mater. Res. Technol.* **2023**, *27*, 5559–5571. [CrossRef]
19. *ASTM D1002-10*; Standard Test Method for Apparent Shear Strength of Single-Lap-Joint Adhesively Bonded Metal Specimens by Tension Loading (Metal-to-Metal). ASTM International: West Conshohocken, PA, USA, 2010.
20. Christov, I.C.; Christov, C. Stress retardation versus stress relaxation in linear viscoelasticity. *Mech. Res. Commun.* **2016**, *72*, 59–63. [CrossRef]
21. Cha, J.; Kim, M.; Park, D.; Go, J.S. Experimental determination of the viscoelastic parameters of K-BKZ model and the influence of temperature field on the thickness distribution of ABS thermoforming. *Int. J. Adv. Manuf. Technol.* **2019**, *103*, 985–995. [CrossRef]
22. Mohammadi, N.; Nasirshoaibi, M. Simulation of viscoelastic materials by ABAQUS. *Math. Models Eng.* **2015**, *1*, 67–71.
23. Joven, R.; Das, R.; Ahmed, A.; Roozbehjavan, P.; Minaie, B. Thermal properties of carbon fiber-epoxy composites with different fabric weaves. In Proceedings of the SAMPE International Symposium Proceedings, Baltimore, MD, USA, 21–24 May 2012; Volume 14.
24. Chen, K.; Zhang, Y. Numerical analysis of temperature distribution during ultrasonic welding process for dissimilar automotive alloys. *Sci. Technol. Weld. Join.* **2015**, *20*, 522–531. [CrossRef]
25. Liu, Z.; Wang, G.; Yi, J. Study on heat transfer behaviors between Al-Mg-Si alloy and die material at different contact conditions based on inverse heat conduction algorithm. *J. Mater. Res. Technol.* **2020**, *9*, 1918–1928. [CrossRef]
26. Cha, J.; Song, H.Y.; Hyun, K.; Go, J.S. Rheological measurement of the nonlinear viscoelasticity of the ABS polymer and numerical simulation of thermoforming process. *Int. J. Adv. Manuf. Technol.* **2020**, *107*, 2449–2464. [CrossRef]
27. Tsujino, J.; Hongoh, M.; Yoshikuni, M.; Hashii, H.; Ueoka, T. Welding characteristics of 27, 40 and 67 kHz ultrasonic plastic welding systems using fundamental- and higher-resonance frequencies. *Ultrasonics* **2004**, *42*, 131–137. [CrossRef] [PubMed]
28. Zhi, Q.; Tan, X.-R.; Lu, L.; Chen, L.-Y.; Li, J.-C.; Liu, Z.-X. Decomposition of ultrasonically welded carbon fiber/polyamide 66 and its effect on weld quality. *Weld. World* **2017**, *61*, 1017–1028. [CrossRef]
29. Tan, X.; Zhi, Q.; Li, Y.; Chen, Y.; Hu, Y.; Chen, Y.; Shi, X.; Wu, H. Effect of MoS2 on microstructure and mechanical property of ultrasonically welded carbon fiber/polyamide 66. *J. Mater. Res. Technol.* **2024**, *29*, 2857–2867. [CrossRef]
30. Benatar, A.; Cheng, Z. Ultrasonic welding of thermoplastics in the far-field. *Polym. Eng. Sci.* **1989**, *29*, 1699–1704. [CrossRef]
31. Benatar, A.; Eswaran, R.V.; Nayar, S.K. Ultrasonic welding of thermoplastics in the near-field. *Polym. Eng. Sci.* **1989**, *29*, 1689–1698. [CrossRef]
32. Levy, A.; Corre, S.L.; Poitou, A. Ultrasonic welding of thermoplastic composites: A numerical analysis at the mesoscopic scale relating processing parameters, flow of polymer and quality of adhesion. *Int. J. Mater. Form.* **2014**, *7*, 39–51. [CrossRef]
33. Villegas, I.F. Strength development versus process data in ultrasonic welding of thermoplastic composites with flat energy directors and its application to the definition of optimum processing parameters. *Compos. Part A Appl. Sci. Manuf.* **2014**, *65*, 27–37. [CrossRef]
34. Villegas, I.F. In situ monitoring of ultrasonic welding of thermoplastic composites through power and displacement data. *J. Thermoplast. Compos. Mater.* **2015**, *28*, 66–85. [CrossRef]

35. Stoke, V.K. Vibration welding of thermoplastics. Part II: Analysis of the welding process. *Polym. Eng. Sci.* **1988**, *28*, 728–739. [CrossRef]
36. Ma, H.; Wang, J.; Fang, Z. Cross-linking of a novel reactive polymeric intumescent flame retardant to ABS copolymer and its flame retardancy properties. *Polym. Degrad. Stab.* **2012**, *97*, 1596–1605. [CrossRef]
37. Zhou, F.; Tang, W.; Xi, W.; Qian, L.; Wang, J.; Qiu, Y.; Chen, Y. Improving the fracture toughness, flame retardancy and smoke suppression of ABS by core-shell elastic flame retardant particles with P/Si synergistic effect. *Polym. Degrad. Stab.* **2024**, *228*, 110893. [CrossRef]
38. Xu, M.; Liu, B.; Zhao, Y.; Wang, Z.; Dong, Z. Direct joining of thermoplastic ABS to aluminium alloy 6061-T6 using friction lap welding. *Sci. Technol. Weld. Join.* **2020**, *25*, 391–397. [CrossRef]

Disclaimer/Publisher's Note: The statements, opinions and data contained in all publications are solely those of the individual author(s) and contributor(s) and not of MDPI and/or the editor(s). MDPI and/or the editor(s) disclaim responsibility for any injury to people or property resulting from any ideas, methods, instructions or products referred to in the content.

materials

MDPI

Article

Numerical Simulation of Compressive Mechanical Properties of 3D Printed Lattice-Reinforced Cement-Based Composites Based on ABAQUS

Weiguo Wu [1], Jing Qiao [1], Yuanyuan Wei [2], Wenfeng Hao [2] and Can Tang [3,*]

[1] Faculty of Civil Engineering and Mechanics, Jiangsu University, Zhenjiang 212013, China
[2] College of Mechanical Engineering, Yangzhou University, Yangzhou 225127, China
[3] College of Civil Science and Engineering, Yangzhou University, Yangzhou 225127, China
* Correspondence: tangcanvip@126.com

Abstract: Research has established that the incorporation of 3D-printed lattice structures in cement substrates enhances the mechanical properties of cementitious materials. However, given that 3D-printing materials, notably polymers, exhibit varying degrees of mechanical performance under high-temperature conditions, their efficacy is compromised. Notably, at temperatures reaching 150 °C, these materials soften and lose their load-bearing capacity, necessitating further investigation into their compressive mechanical behavior in such environments. This study evaluates the compressibility of cement materials reinforced with lattice structures made from polyamide 6 (PA6) across different structural configurations and ambient temperatures, employing ABAQUS for simulation. Six distinct 3D-printed lattice designs with equivalent volume but varying configurations were tested under ambient temperatures of 20 °C, 50 °C, and 100 °C to assess their impact on compressive properties. The findings indicate that heightened ambient temperatures significantly diminish the reinforcing effect of 3D-printed materials on the properties of cement-based composites.

Keywords: 3D printing; cement-based composites; numerical simulation; mechanical properties

Citation: Wu, W.; Qiao, J.; Wei, Y.; Hao, W.; Tang, C. Numerical Simulation of Compressive Mechanical Properties of 3D Printed Lattice-Reinforced Cement-Based Composites Based on ABAQUS. *Materials* **2024**, *17*, 2370. https://doi.org/10.3390/ma17102370

Academic Editors: Patryk Rozylo, Katarzyna Falkowicz and Pawel Wysmulski

Received: 28 March 2024
Revised: 8 May 2024
Accepted: 9 May 2024
Published: 15 May 2024

Copyright: © 2024 by the authors. Licensee MDPI, Basel, Switzerland. This article is an open access article distributed under the terms and conditions of the Creative Commons Attribution (CC BY) license (https://creativecommons.org/licenses/by/4.0/).

1. Introduction

The rapid expansion of the construction industry has led to the widespread adoption of concrete due to its exceptional compressive strength. Consequently, there has been a growing demand for enhancements in the material's performance attributes. Concrete is recognized for its low tensile strength and susceptibility to brittle fractures, classifying it as a quasi-brittle composite material [1]. This vulnerability allows the formation of cracks under external stresses, facilitating the ingress of corrosive agents and thereby accelerating material degradation and diminishing structural durability. This phenomenon is particularly evident in traditional steel reinforcement, which is more likely to corrode under such conditions, significantly reducing the lifespan of structures. In response to these challenges, polymers fabricated through 3D-printing technology [2,3] have emerged as a novel solution, owing to their superior corrosion resistance compared to conventional rebar. This makes them capable of withstanding the penetration of harmful substances into the cement matrix. The advancement of 3D-printing technology has not only facilitated the creation of polymers with intricate lattice structures [4–7] but also holds the potential to expedite construction timelines and simplify the building process. As a result, 3D-printed polymer-lattice-reinforced cement-based composites have garnered substantial interest from the academic community, highlighting a promising direction for enhancing the durability and performance of concrete structures.

The integration of fiber materials into cement matrices has been demonstrated to effectively mitigate the brittleness inherent in cement-based materials. Notably, investigations into the enhancement of toughness through the inclusion of steel fibers have underscored the beneficial interactions between fibers and the cementitious matrix [8,9]. These interactions bolster the material's energy absorption capacity via robust bonding, thereby augmenting the performance characteristics of cement. Nonetheless, achieving uniform dispersion of fibers within the cement presents significant construction challenges, revealing limitations in this approach. To address these limitations, Nam et al. [10] leveraged 3D-printing technology to design models capable of precisely controlling the orientation and placement of fibers in concrete. Their methodology involved examining the flexural mechanics of fiber-reinforced cement mortars with diverse spatial distributions through three-point bending tests. The advancements in 3D-printing technology have since captivated the research community, prompting extensive experimentation to explore the reinforcing potential of 3D-printed structures as alternatives to fiber materials in cement-based composites. Researchers like Farina [11] have prepared polymer and metal-reinforced 3D-printed structures with varied surface textures, employing them to fortify cement mortar. These studies, which included three-point bending tests, delved into the influence of surface roughness on the mechanical behaviors of cement mortars, revealing how steel bar surface roughness can amplify load-bearing capabilities by altering the adhesion with the cement mortar. Tian et al. [12] conducted experimental assessments on the uniaxial compressive strength of an innovative composite column comprising precast grid-reinforced ultra-high-performance concrete (UHPC) within stay-in-place formwork and post-cast concrete. The findings indicate that polymer grids with carbon fiber reinforcement (CFRP) can foster strain hardening and enhance ductility and toughness through improved lateral confinement compared to stainless steel (SS) grids. Rosewitz et al. [13] designed a 3D-printed biomimetic polymer, utilizing it to reinforce cement-based composites and analyze their mechanical properties and failure mechanisms across different geometric structures. Similarly, Xu et al. [14] and Salazar et al. [15] experimented with varying tactics of 3D printing to reinforce cementitious materials with polymer network structures and investigated the effects on the materials' properties. Particularly, Salazar et al. [15] focused on ultra-high-performance concrete (UHPC) materials reinforced by 3D-printed polymer lattices, assessing their flexural behavior through four-point bending tests. These studies have collectively reinforced the notion that 3D-printed lattice structures can significantly improve the ductility of ultra-high-performance concrete materials. However, the existing research has mainly discussed the mechanical properties of 3D-printed structural reinforcement materials at room temperature, and paid little attention to the effect of high-temperature environment on 3d printed lattice-reinforced cement-based composites. Notably, the compressive mechanical properties of PA6 [16,17] lattice structures fabricated using Multi Jet Fusion (MJF) technology [18,19] degrade at elevated temperatures, potentially leading to softening of the polymer material. Herein, we assess the elastic modulus of PA6 under varying temperatures through experimental samples. The investigation reveals that high temperatures detrimentally affect the material's elastic modulus. Thus, understanding the mechanical behavior of 3D-printed lattice-reinforced cement-based composites in high-temperature conditions is crucial. This study innovatively proposed the influence of a high-temperature environment on the mechanics of 3D-printed lattice-reinforced cement-based materials. Through numerical simulation, a high-temperature uniaxial compression test was carried out on 3D printed lattice-reinforced cement-based composite materials prepared by MJF technology, and its compression mechanical properties and failure mechanism were discussed.

2. Finite Element Model Modeling

2.1. Material Parameters

In this research, the specimens utilized were subjected to uniaxial compression testing. The material properties of the cement matrix are detailed in Table 1, which include a density of 2400 kg/m^3, an elastic modulus of 30 GPa, and a Poisson's ratio of 0.2. Additionally, the dilatancy angle is set at 30°, eccentricity at 0.1, the ratio of biaxial to uniaxial compressive strength at 1.16, and the tension-compression strength ratio (k) at 0.6667. The viscosity parameter is specified as 0.0005, with the failure mode characterized by the concrete damage plasticity (CDP) model [20–23]. The specimens were designed as cubic structures with a side length of 42 mm, featuring an internal lattice prepared using MJF technology. MJF employs powder materials and utilizes a binding agent and detailing agent, sprayed onto the powder; subsequently, energy is applied to fuse the material in the designated area. This 3D-printing technique is noted for producing samples whose mechanical properties are largely invariant to the build direction. Material parameters of the 3D-printed lattice, detailed in Tables 2 and 3, indicate a mass density of 1100 kg/m^3 and a Poisson's ratio of 0.38. To accurately assess the impact of high-temperature environments on the elastic modulus of PA6 material, tensile samples, as illustrated in Figure 1, were designed. The Young's modulus of the lattice material was measured at ambient temperatures of 23 °C (room temperature), 50 °C, and 100 °C. Experimental results demonstrated that at room temperature, the material's Young's modulus was 934 MPa (E1). This value decreased to 564 MPa (E2) at 50 °C and plummeted to 200 MPa (E3) at 100 °C, indicating a significant softening behavior under elevated temperatures.

Table 1. 3D-printing polymer material parameters.

Density (kg/m^3)	2.4×10^{-9}
Young's modulus (GPa)	30
Poisson's ratio	0.2
expansion angle (°)	30
eccentricity ratio	0.1
fb0/fco	1.16
k	0.6667
Viscous parameter	0.0005

Table 2. PA6 material parameters.

Density (kg/m^3)	Indoor Temperature Young's Modulus (MPa)	Poisson's Ratio
1.1×10^{-9}	934	0.38

Table 3. Elastic modulus of PA6 at different ambient temperatures.

Environment Temperature	Indoor Temperature	50 °C	100 °C
elasticity modulus	934 MPa	564 MPa	200 MPa

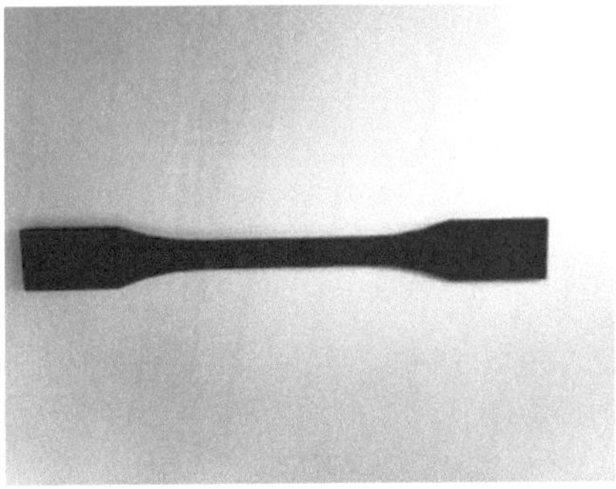

Figure 1. PA6 Elastic modulus measurement sample.

2.2. Structural Design

In this study, the cement-based composite specimen is a cube with each side measuring 42 mm, resulting in a total volume of 74,088 mm^3. Following the guidelines provided in the literature [21], this paper explores six distinct 3D-printed polymer lattice configurations for reinforcement. The selected cellular structures are circular, cubic, Kelvin, octagonal (Oct), rhombicuboctahedron (RO), and a reinforced octagon variant (SO). Each 3D-printed lattice is designed to occupy a volume fraction of 8% within the cement matrix (Table 4). The diameters of the lattice cells, varying according to their structural forms, are determined using the formula provided below:

$$V_{circular} = f(d_1) = 98.125d_1 + 4240.9d_1^2 - 400.94d_1^3 \quad (1)$$

$$V_{cubic} = f(d_2) = 0.0005d_2 + 2356.2d_2^2 - 121.52d_2^3 \quad (2)$$

$$V_{kelvin} = f(d_3) = 0.00005d_3 + 4804.4d_3^2 - 497.67d_3^3 \quad (3)$$

$$V_{Oct} = f(d_4) = 910.78d_4 + 4045.2d_4^2 - 31.93d_4^3 \quad (4)$$

$$V_{RO} = f(d_5) = -2 \times 10^{-5}d_5 + 8115.9d_5^2 - 1151.5d_5^3 \quad (5)$$

$$V_{so} = f(d_6) = 0.0003d_6 + 7072.3d_6^2 - 900.55d_6^3 \quad (6)$$

Table 4. 3D-printed lattice structures.

Structure Serial Number	Lattice Unit Cell Type	Unit Cell Design	Cell Design Parameters (mm)	CAD Southwest View
1	circle	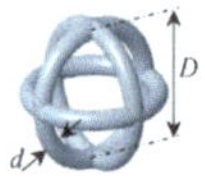	D = 10, d = 1.246	
2	cubic	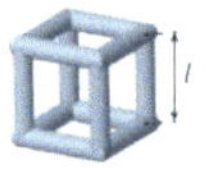	l = 10, d = 1.659	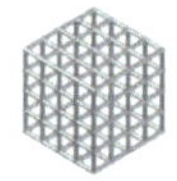
3	kelvin	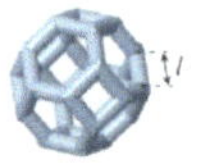	l = 3.54, d = 1.183	
4	octagonal (Oct),		l = 4.14, d = 0.979	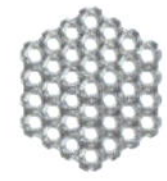
5	rhombicubactahedron (RO)	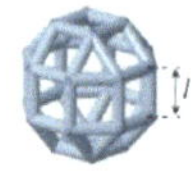	l = 1.414, d = 0.916	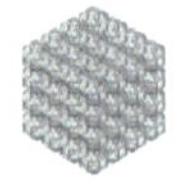
6	strengthened octagon octagonal (SO)		l = 3.54, d = 0.979	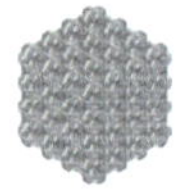

In the numerical simulation, the polymer lattice is integrated within the cement matrix, serving as an internal constraint condition. This configuration ensures that the lattice is encased by the cement matrix, effectively simulating its embedded position.

In conducting numerical simulations with ABAQUS (2022) software, aligning closely with physical laboratory operations improves the accuracy of the simulations [24–26]. This involves designing rigid plates to interact with both the upper and lower surfaces of the cube specimen, thereby restricting its movement. This simulation step is critical for accurately replicating the compressive constraints applied by the laboratory's compression apparatus on the specimen. For the rigidity-enforced plates, it is crucial to apply fixed constraints to the lower surface to simulate the restraining effect of the laboratory compressor's heads. This entails limiting displacement in all directions, effectively preventing any movement of the surface. In ABAQUS, such a scenario is facilitated through the application of boundary conditions that enforce these fixed constraints. Conversely, the plate on the upper surface is subjected to a fixed displacement constraint, which allows for the specification of displacement in certain directions while permitting the application of force or displacement in others. Given the focus on uniaxial compression tests in this simulation, a vertical displacement constraint is applied perpendicular to the sample surface. This approach effectively simulates the compressive force exerted during the actual laboratory test.

2.3. Finite Element Simulation

ABAQUS, a finite element simulation software, is extensively utilized across various engineering disciplines due to its robust capabilities. It facilitates a broad spectrum of analyses, ranging from basic linear to intricate nonlinear problems, enabling the calculation and analysis of the mechanical behavior of complex engineering structures with high precision. This study employs ABAQUS to conduct a numerical simulation of a uniaxial compression experiment on cement-based composites, specifically focusing on specimens without any pre-existing cracks.

Figure 2 illustrates the schematic representation of the finite element model used in the simulation. The process of conducting simulations using ABAQUS (2022) software encompasses several key steps, detailed as follows:

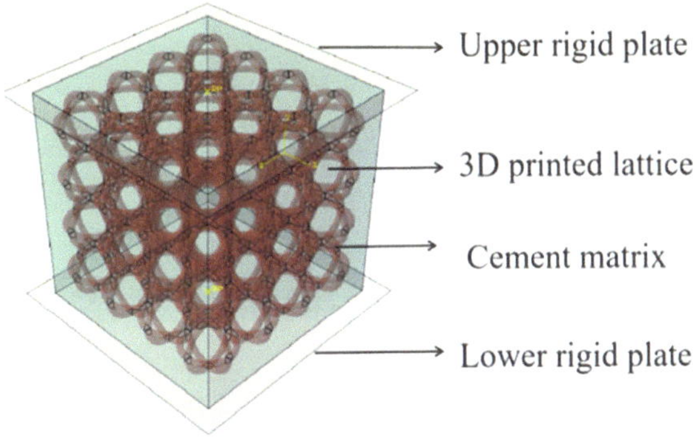

Figure 2. ABAQUS computational modeling.

Firstly, a geometric model is established through 3D modeling of the polymer lattice within the sample, utilizing CAD (2023) software. This model is then exported and imported into ABAQUS, where the component is constructed as a cohesive entity within the assembly module.

The analysis step size is adjusted by configuring the output variables for each step of the analysis.

Regarding interactions, within a specific region and scope, interactions and mutual constraints exist between models. Specifically, contact is established between the upper and lower rigid plates and the corresponding upper and lower surfaces of the sample. Furthermore, the 3D-printed lattice introduces internal constraints within the cement matrix, as depicted in the figure.

Mesh discretization involves dividing the entity into grids. Given the complexity of the 3D-printed lattice structure, alongside its small cellular dimensions, the mesh is configured as tetrahedral. The mesh size is finely adjusted based on the varying structural forms to optimize computational outcomes. In contrast, the cement matrix exhibits a regular cubic structure, which is discretized using a hexahedral mesh.

The loading methodology involves establishing a reference point on the rigid plate located at the model's upper surface. A vertical downward load is applied at this reference point to simulate the pressure dynamics observed in uniaxial compression testing. In this study, a fixed displacement loading approach was employed, with the displacement set at 5% of the sample's side length, to apply the load to the sample.

3. Results and Discussion

This study examined six varieties of cement-based composites reinforced with 3D-printed polymer lattices of different structural configurations through uniaxial compression numerical simulations. By varying the elastic modulus of the polymer materials, the simulation explored how different ambient temperature conditions affect the compressive strength of these 3D-printed polymer lattice-reinforced cement-based composites. The mechanical and deformation properties of the samples were determined by comparing stress–strain curve analyses and strain distribution maps with the results obtained from ABAQUS numerical simulations.

3.1. Stress–Strain Curve Analysis

The stress–strain curve of conventional concrete material is shown in Figure 3, and the numerical simulation results and experimental results of uniaxial compression are shown in Figures 4 and 5, respectively. It can be observed that the stress–strain curve of the 3D printed lattice-reinforced cement-based composite is consistent with that of conventional concrete materials. By comparing the simulation results with the experimental results, it is shown that the compressive mechanical properties of cement-based materials with different mesh structures are similar under uniaxial compression. These findings delineate the mechanical response process of 3D-printed polymer lattice-reinforced cement-based composite specimens under uniaxial compression, characterized by the following stages:

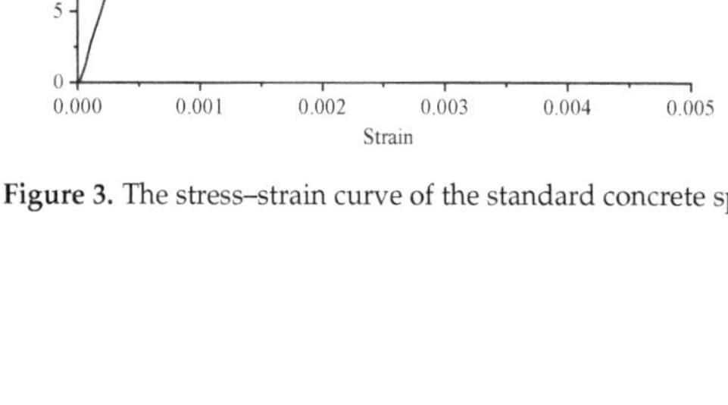

Figure 3. The stress–strain curve of the standard concrete specimen under uniaxial compression test.

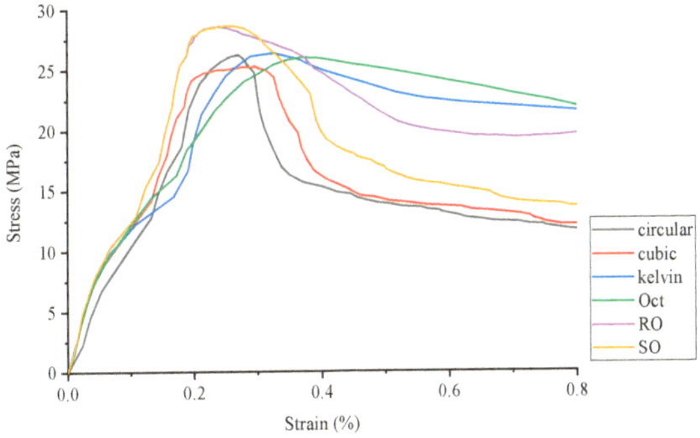

Figure 4. Numerical simulation of stress–strain curves for uniaxial compression of specimens strengthened with different lattice structures.

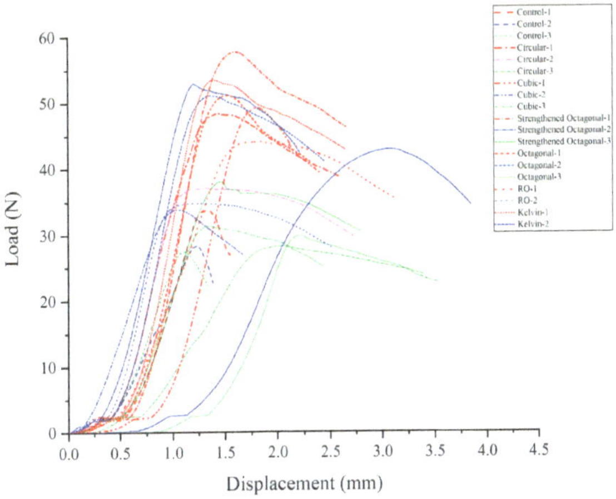

Figure 5. Laboratory-displacement curve for compressive test of cement-based specimens.

(1) Initial Stage (O-A): During this phase, the stress level reaches approximately 70% to 85% of the peak stress. At this point, the deformation observed in the specimen is predominantly elastic, resulting from the interaction between the cement matrix and the polymer lattice. This behavior is interpreted as linear elastic deformation, evident from the near-linear relationship depicted in the stress–strain curve. Concurrently, material displacement changes are minimal, despite the significant alterations in the load.

(2) Second Stage (A-B): At this stage, stress levels range from approximately 85% to 93% of the peak stress. The deformation behavior of the specimen is marked by the gradual emergence and expansion of small cracks within the sample, commensurate with the loading process.

(3) Second Stage (A-B): In this stage, the stress reaches roughly 85% to 93% of the peak load. The stage is characterized by the gradual appearance and steady expansion of small cracks within the specimen.

(4) Third Stage (B-C): Stress levels during this stage approximate 93% to 100% of peak stress. This phase is marked by a deceleration in the material's compressive capacity enhancement. Concurrently, the stress–strain curves exhibit an increasing curvature, transitioning towards a more gradual trend. The predominant deformation observed in the specimen is of an irreversible plastic nature, with a minor component of elastic deformation also present at this stage.

(5) Fourth Stage (C-): Upon exceeding the peak stress, the sample's compressive properties begin to diminish correspondingly. As the external force applied to the material escalates with continued loading, the damage to the specimen progressively worsens.

Additionally, the maximum stress the sample can endure before failure is designated as the peak stress, depicted at point C in Figure 4. The stress–strain curve of concrete material serves as a metric for assessing the effectiveness of cement-based composites reinforced by 3D-printed polymer lattice structures. Notably, 85% of the peak stress is utilized as the demarcation between the plastic deformation and elastic deformation phases, corresponding to point B in Figure 3.

An analysis of Table 2, which presents peak stress data from the numerical simulations of different lattice structures, reveals a consistent trend: a decrease in the elastic modulus of the 3D-printed lattice results in a reduction in the peak stress experienced by all specimens compared to the control specimens. Specifically, in Table 5 when the lattice's elastic modulus decreases to 60.4% of the control modulus, the peak load of the cement-based composite reinforced with a Circular lattice is 0.8% lower than that of the ideal condition. With the use of Cubic, Kelvin, Oct, RO, and SO lattice configurations as reinforcements, the peak loads decrease by 0.8%, 0.14%, 17.82%, 10.57% and 11.73%, respectively.

Table 5. Peak stress of lattice numerical simulation of different structures.

Crystal Structure \ Elasticity Modulus	E_1 = 934 MPa	E_2 = 564 MPa	E_3 = 200 MPa
Circular	26.23	26.04	22.74
Cubic	25.23	25.17	23.16
Kelvin	28.49	28.45	21.79
Oct.	28.57	23.48	20.21
RO	26.02	23.27	19.46
SO	26.26	23.18	19.41

The analysis of the stress–strain curve for 3D-printed Circular lattice reinforced cement-based composite materials, as illustrated in Figure 6 reveals that the contribution of the 3D-printed polymer lattice to the overall cement-based material is minimal, with its volume fraction being only 8%. Consequently, the reduction in its elastic modulus exerts a negligible impact on the peak stress of the cement-based composite samples. Nevertheless, following the failure of the cement matrix, the influence of elastic modulus attenuation on the peak stress of the cement-based composite specimens becomes marginally more significant. It is observed that during the failure stage, denoted as the C-stage, the reduction in the elastic modulus of the 3D-printed lattice has a more pronounced effect on the compressive strength of the sample.

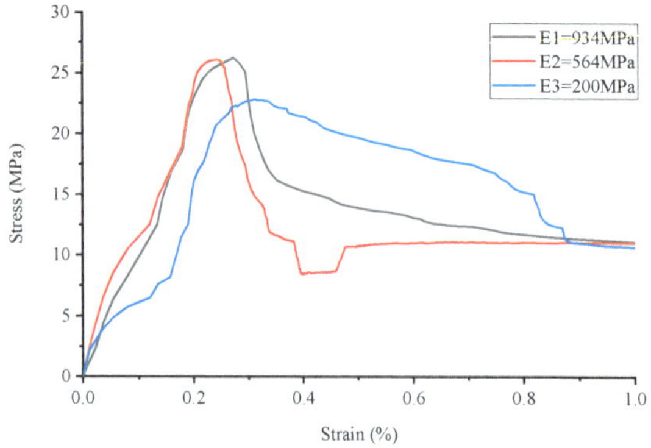

Figure 6. Stress–strain curve of Circular lattice reinforced cement-based composites.

For instance, when the elastic modulus is set at 564 MPa, there is no significant decrease in the peak stress of the sample. However, following the degradation of the cement matrix, a notable trend is observed during the failure stage (referred to as the C-stage); the reduction in the elastic modulus due to environmental temperature fluctuations has a marked impact on the compressive strength of the sample. Conversely, at an elastic modulus of 200 MPa, analysis of the corresponding figure demonstrates a discernible decrease in the peak stress of the material under these experimental conditions.

By comparing the numerical simulation outcomes for five distinct structural configurations, as depicted in Figures 6–11, it becomes evident that variations in ambient temperature lead to a diminution of Young's modulus in the 3D-printed polymer lattice. This reduction directly contributes to a decline in the peak stress of the cement-based material, subsequently impairing its performance.

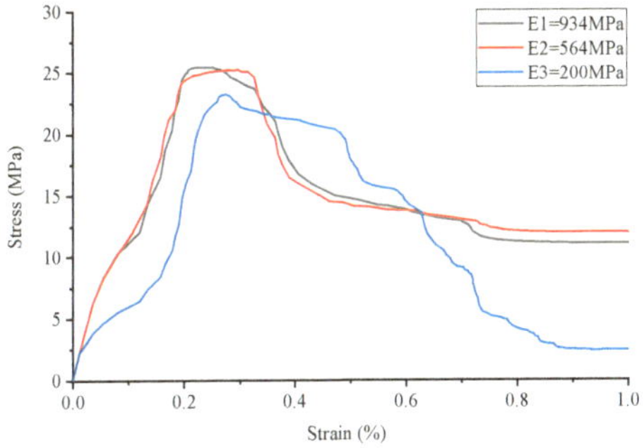

Figure 7. Stress–strain curve of Cubic lattice reinforced cement-based composites.

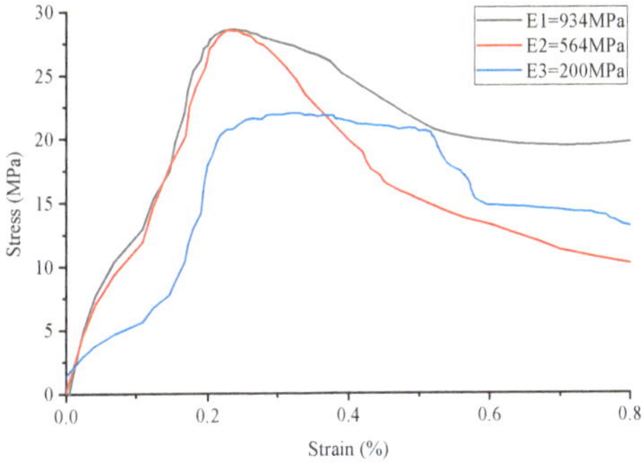

Figure 8. Stress–strain curve of Kelvin lattice reinforced cement-based composites.

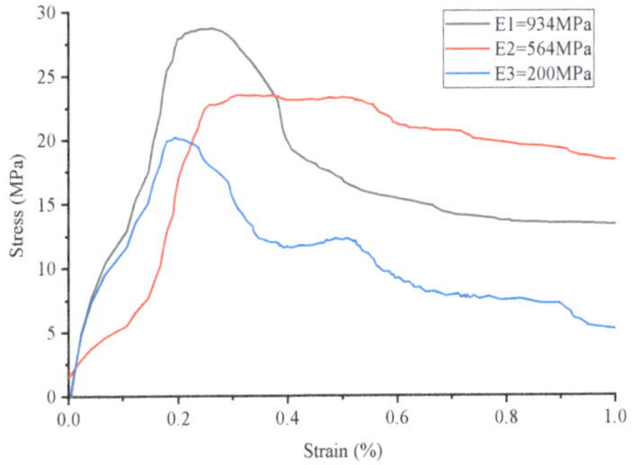

Figure 9. Stress–strain curve of Oct lattice reinforced cement-based composites.

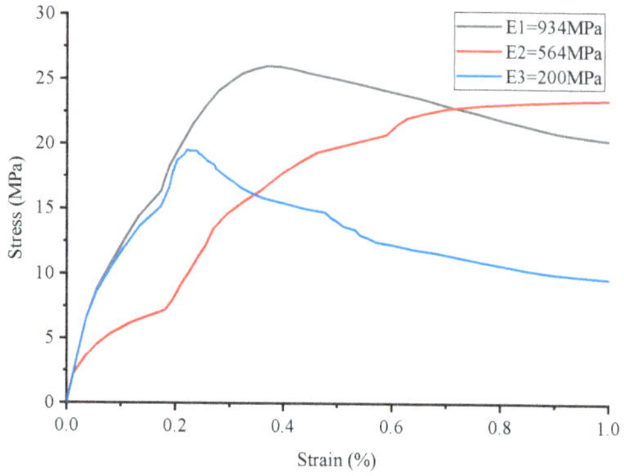

Figure 10. Stress–strain curve of RO lattice reinforced cement-based composites.

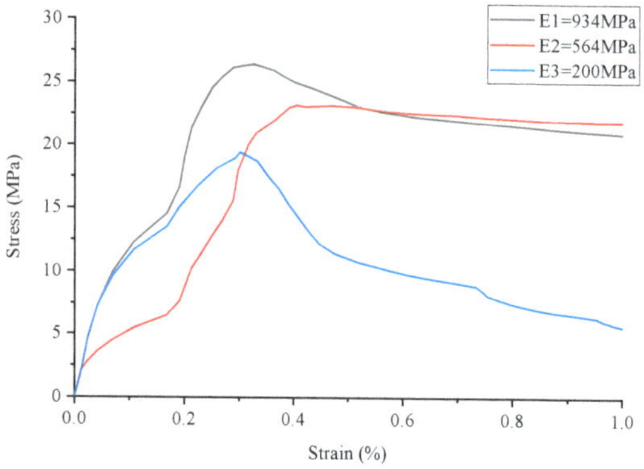

Figure 11. Stress–strain curve of SO lattice reinforced cement-based composites.

3.2. Strain Analysis

In the uniaxial compression experiment, as load is applied, the specimen experiences cracking, leading to its failure. The physical test phase employed the DIC monitoring technique to examine the deformation and the areas of surface cracking under compression. This study simulated the stress–strain relationship of the specimen prior to cracking in a uniaxial compression setting using finite element simulation. Accordingly, this chapter analyzes the strain distribution patterns in the specimen to forecast potential failure points.

Figures 12–17 depict the stress–strain curves for six varieties of cement-based composites reinforced with 3D-printed lattices, derived from numerical simulations of uniaxial compression. Specifically, Figure 12 illustrates the evolution of strain distribution in cement-based composites reinforced with circular lattice specimens under varying ambient temperatures. As the ambient temperature increases, Young's modulus of the polymer lattice diminishes, leading to a reduction in strain capacity and an enlargement of the strain distribution region. Notably, when the ambient temperature reaches 100 °C and Young's modulus of the polymer decreases to 200 MPa, there is a marked increase in the specimen's propensity for failure under pressure. This trend suggests that at elevated temperatures, the compressive properties of the polymer lattice are significantly compromised. Coupled with the cracking of the cement matrix, the lattice becomes increasingly susceptible to ambient temperature effects, heightening the likelihood of specimen failure. This observation underscores the fact that high-temperature environments substantially degrade the stability and load-bearing capabilities of lattice-reinforced cement-based materials, elevating the risk of material failure.

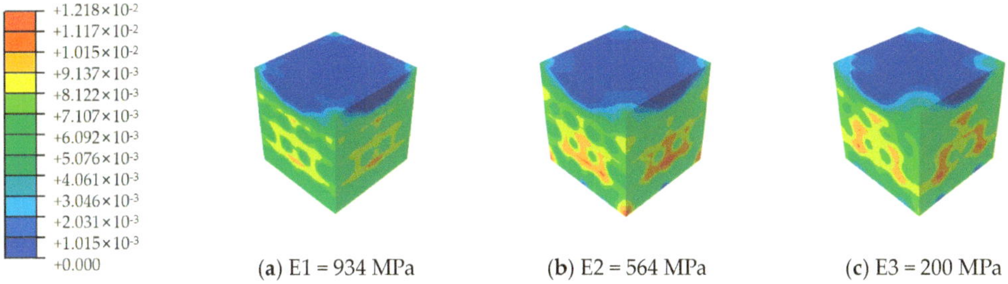

Figure 12. Strain cloud image of cement-based samples reinforced with circular lattice structures at different ambient temperatures.

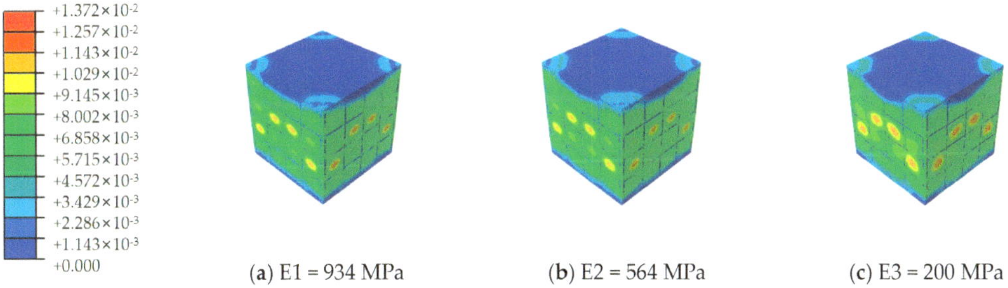

Figure 13. Strain cloud image of cement-based samples reinforced with cubic lattice structures at different ambient temperatures.

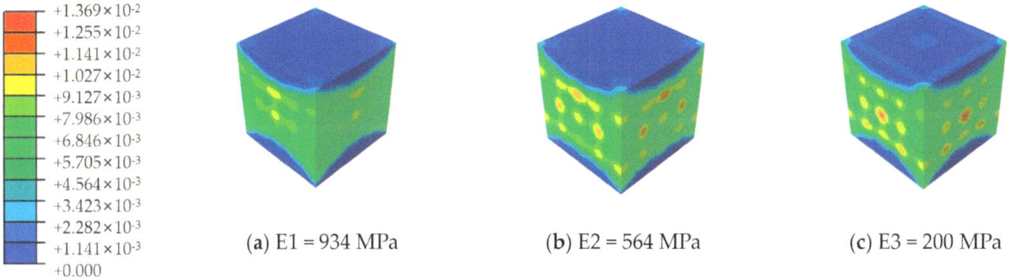

Figure 14. Strain cloud image of cement-based samples reinforced with kelvin lattice structures at different ambient temperatures.

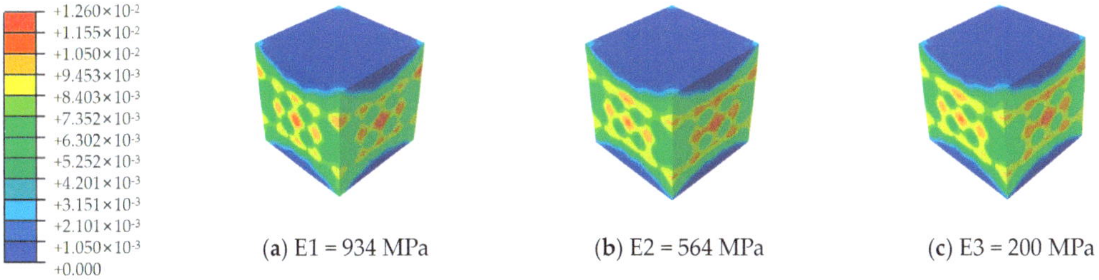

Figure 15. Strain cloud image of cement-based samples reinforced with Oct lattice structures at different ambient temperatures.

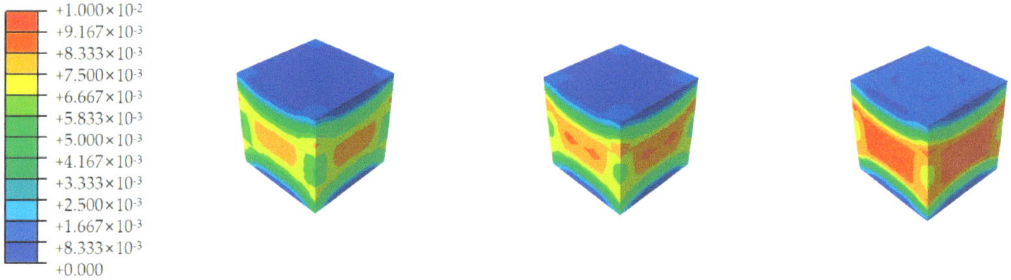

(**a**) E1 = 934 MPa (**b**) E2 = 564 MPa (**c**) E3 = 200 MPa

Figure 16. Strain cloud image of cement-based samples reinforced with RO lattice structures at different ambient temperatures.

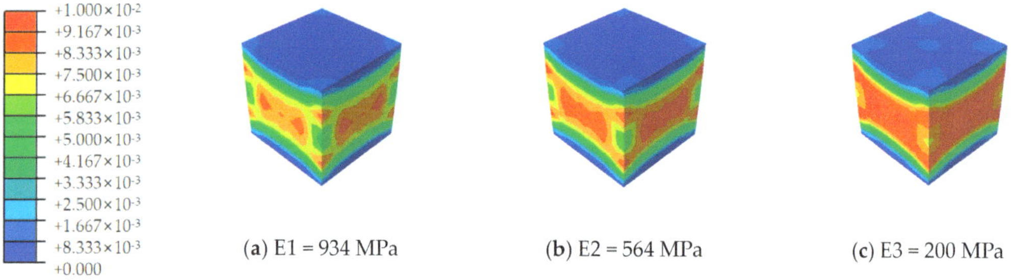

(**a**) E1 = 934 MPa (**b**) E2 = 564 MPa (**c**) E3 = 200 MPa

Figure 17. Strain cloud image of cement-based samples reinforced with SO lattice structures at different ambient temperatures.

For cement matrix composites, the stress–strain curves under vertical compression can generally be segmented into two phases: the ascending and descending phases. In the initial compression phase, the exerted load on the specimen is minimal, and the corresponding deformation primarily results from the elastic deformation of the material's internal structure, rendering the stress–strain curve nearly linear. At this juncture, the specimen predominantly exhibits compression zones. Strain distribution analyses reveal that the cement base test block is subjected to relatively low pressure without significant stress concentration zones. This suggests a fairly uniform stress distribution throughout the material during the initial loading phase, indicating an absence of marked stress concentration.

As the loading process progresses, the stress–strain curve transitions into a descending phase where the material begins to exhibit plastic deformation and the impact of stress concentration on the specimen becomes increasingly evident. In high-temperature conditions, the thermal expansion of the material's various phases and the reduction in Young's modulus of the polymer lattice contribute to diminished compressive performance, leading to a more uneven strain distribution and elevating the risk of material damage. The alterations observed in the strain distribution maps from numerical simulations further elucidate the deterioration in compressive mechanical properties of cement-based composites reinforced with 3D-printed lattices when subjected to high temperatures. These simulations provide insights into the failure dynamics and mechanical property degradation of materials across different ambient temperature conditions.

4. Conclusions

In this study, ABAQUS (2022) software facilitated the numerical simulation of uniaxial compression tests on composite materials featuring 3D-printed polymer lattices embedded within a cement matrix. This approach yielded a series of insightful conclusions, which are instrumental in elucidating the compressive mechanical properties of cement-based composites under varying high-temperature conditions. Presented below is an analytical summary of the key research findings:

(1) Through the creation of a precise finite element model and the simulation of laboratory uniaxial compression tests, this study has successfully validated the accuracy of its numerical analysis model. This achievement not only furnishes a dependable simulation methodology for subsequent research endeavors but also lays a robust groundwork for the enhanced analysis and application of experimental data. Moreover, the process of developing and validating the model serves as a valuable benchmark for the numerical simulation of analogous materials and structures.

(2) The study reveals that variations in ambient temperature markedly influence the elastic modulus of 3D-printed polymer materials, subsequently altering the compressive mechanical properties of cement-based composites. This finding underscores the necessity of accounting for material performance shifts under diverse temperature conditions in practical engineering applications to ensure the reliability and safety of structures.

(3) This study demonstrates that the compressive strength of composite materials tends to decrease as the elastic modulus of polymer materials is reduced. This observation holds significant implications for the optimization of composite material design and the enhancement of their structural characteristics.

(4) The findings indicate that as Young's modulus of the polymer decreases, the strain region widens while the maximum strain diminishes, suggesting an impact on both the ductility and load-bearing capacity of the structure. Furthermore, when the elastic modulus falls to a specific critical threshold, specimen cracking occurs at the onset of the compression test. This highlights the necessity for meticulous attention to the lower limits of material properties during the design process to prevent premature failure.

In conclusion, this study not only introduces a novel approach for examining the mechanical properties of cement-based composites but also lays a crucial theoretical and practical foundation for material design and structural optimization in related domains. Building upon this foundation, future investigations can delve into the properties of various material combinations, structural configurations, and loading scenarios, thereby broadening the application spectrum of cement-based composites in diverse fields.

Author Contributions: W.W.: Conceptualization, Investigation, Data curation, Writing—Original draft preparation. J.Q.: Investigation, Data curation, Writing—Original draft preparation. Y.W.: Software, Data curation, Validation, Methodology. W.H.: Con-ceptualization, Methodology, Writing–Reviewing and Editing. C.T.: Conceptualization, Methodology, Writing—Reviewing and Editing. All authors have read and agreed to the published version of the manuscript.

Funding: The authors are grateful for the financial support provided by the Six Talent Peaks Project in Jiangsu Province (Grant No. 2019-KTHY-059) and Self-made experimental instruments and equipment project of Yangzhou University (Grant No. zzyq2022zy05).

Institutional Review Board Statement: Not applicable.

Informed Consent Statement: Not applicable.

Data Availability Statement: Data are contained within the article.

Conflicts of Interest: The authors declare no conflict of interest.

References

1. Carpinteri, A. *Mechanical Damage and Crack Growth in Concrete: Plastic Collapse to Brittle Fracture*; Springer Science & Business Media: Berlin/Heidelberg, Germany, 2012.
2. Hrițuc, A.; Mihalache, M.A.; Dodun, O.; Nagîț, G.; Beșliu-Băncescu, I.; Rădulescu, B.; Slătineanu, L. Propagation of Sounds through Small Panels Made of Polymer Materials by 3D Printing. *Polymers* **2023**, *16*, 5. [CrossRef] [PubMed]
3. Khosravani, M.R.; Ayatollahi, M.R.; Reinicke, T. Effects of post-processing techniques on the mechanical characterization of additively manufactured parts. *J. Manuf. Process.* **2023**, *107*, 98–114. [CrossRef]
4. Johnson, L.K.; Richburg, C.; Lew, M.; Ledoux, W.R.; Aubin, P.M.; Rombokas, E. 3D Printed lattice microstructures to mimic soft biological materials. *Bioinspir. Biomim.* **2018**, *14*, 016001. [CrossRef] [PubMed]

5. Kang, J.D.; Kim, S. Development of a 3D printing method for the textile hybrid structure. *Int. J. Cloth. Sci. Technol.* **2021**, *34*, 262–272. [CrossRef]
6. Park, S.; Shou, W.; Makatura, L.; Matusik, W.; Fu, K. 3D printing of polymer composites: Materials, processes, and applications. *Matter* **2022**, *5*, 43–76. [CrossRef]
7. Xu, Y.; Zhang, H.; Šavija, B.; Figueiredo, S.C.; Schlangen, E. Deformation and fracture of 3D printed disordered lattice materials: Experiments and modeling. *Mater. Des.* **2019**, *162*, 143–153. [CrossRef]
8. Qin, S.; Cao, S.; Yilmaz, E. Employing U-shaped 3D printed polymer to improve flexural properties of cementitious tailings backfills. *Constr. Build. Mater.* **2022**, *320*, 126296. [CrossRef]
9. Balla, V.K.; Kate, K.H.; Satyavolu, J.; Singh, P.; Tadimeti, J.G.D. Additive manufacturing of natural fiber reinforced polymer composites: Processing and prospects. *Compos. Part B Eng.* **2019**, *174*, 106956. [CrossRef]
10. Nam, Y.J.; Hwang, Y.K.; Park, J.W.; Lim, Y.M. Feasibility study to control fiber distribution for enhancement of composite properties via three-dimensional printing. *Mech. Adv. Mater. Struct.* **2019**, *26*, 465–469. [CrossRef]
11. Farina, I.; Fabbrocino, F.; Carpentieri, G.; Modano, M.; Amendola, A.; Goodall, R.; Feo, L.; Fraternali, F. On the reinforcement of cement mortars through 3D printed polymeric and metallic fibers. *Compos. Part B Eng.* **2016**, *90*, 76–85. [CrossRef]
12. Tian, H.; Zhou, Z.; Zhang, Y.; Wei, Y. Axial behavior of reinforced concrete column with ultra-high performance concrete stay-in-place formwork. *Eng. Struct.* **2020**, *210*, 110403. [CrossRef]
13. Rosewitz, J.A.; Choshali, H.A.; Rahbar, N. Bioinspired design of architected cement-polymer composites. *Cem. Concr. Compos.* **2019**, *96*, 252–265. [CrossRef]
14. Xu, Y.; Šavija, B. Development of strain hardening cementitious composite (SHCC) reinforced with 3D printed polymeric reinforcement: Mechanical properties. *Compos. Part B Eng.* **2019**, *174*, 107011. [CrossRef]
15. Salazar, B.; Aghdasi, P.; Williams, I.D.; Ostertag, C.P.; Taylor, H.K. Polymer lattice-reinforcement for enhancing ductility of concrete. *Mater. Des.* **2020**, *196*, 109184. [CrossRef]
16. Ding, S.; Xiang, Y.; Ni, Y.Q.; Thakur, V.K.; Wang, X.; Han, B.; Ou, J. In-situ synthesizing carbon nanotubes on cement to develop self-sensing cementitious composites for smart high-speed rail infrastructures. *Nano Today* **2022**, *43*, 101438. [CrossRef]
17. Han, B.; Sun, S.; Ding, S.; Zhang, L.; Yu, X.; Ou, J. Review of nanocarbon-engineered multifunctional cementitious composites. *Compos. Part A Appl. Sci. Manuf.* **2015**, *70*, 69–81. [CrossRef]
18. Ráž, K.; Chval, Z.; Thomann, S. Minimizing Deformations during HP MJF 3D Printing. *Materials* **2023**, *16*, 7389. [CrossRef]
19. Robar, J.L.; Kammerzell, B.; Hulick, K.; Kaiser, P.; Young, C.; Verzwyvelt, V.; Cheng, X.; Shepherd, M.; Orbovic, R.; Fedullo, S.; et al. Novel multi jet fusion 3D-printed patient immobilization for radiation therapy. *J. Appl. Clin. Med. Phys.* **2022**, *23*, e13773. [CrossRef]
20. Shalchy, F. An Investigation on Interfacial Adhesion Energy between Polymeric and Cellulose-Based Additives Embedded in CSH Gel. Master's Thesis, Worcester Polytechnic Institute, Worcester, MA, USA, 2016.
21. Kakavand, M.R.; Taciroglu, E.; Hofstetter, G. Evaluation of the Performance of an Enhanced Damage Plasticity Model for Predicting the Cyclic Response of Plain Concrete under Multiaxial Loading Conditions. *Int. J. Struct. Civ. Eng. Res.* **2020**, *9*, 273–277. [CrossRef]
22. Voyiadjis, G.Z.; Taqieddin, Z.N. Elastic plastic and damage model for concrete materials: Part I-theoretical formulation. *Int. J. Struct. Chang. Solids* **2009**, *1*, 31–59.
23. Lee, J.; Fenves, G.L. Plastic-damage model for cyclic loading of concrete structures. *J. Eng. Mech.* **1998**, *124*, 892–900. [CrossRef]
24. Jiang, L.; Zhao, C.; Zhang, J. Calculation and Analysis of Sluice Bottom Seepage with Two Methods. *IOP Conf. Ser. Earth Environ. Sci.* **2021**, *643*, 012103. [CrossRef]
25. Merzuki, M.N.M.; Ma, Q.; Rejab, M.R.M.; Sani, M.; Zhang, B. Experimental and numerical investigation of fibre-metal-laminates (FMLs) under free vibration analysis. *Mater. Today Proc.* **2022**, *48*, 854–860. [CrossRef]
26. Rizwan, M.; Liang, Q.Q.; Hadi, M.N.S. Fiber-based computational modeling of rectangular double-skin concrete-filled steel tubular short columns including local buckling. *Eng. Struct.* **2021**, *248*, 113268. [CrossRef]

Disclaimer/Publisher's Note: The statements, opinions and data contained in all publications are solely those of the individual author(s) and contributor(s) and not of MDPI and/or the editor(s). MDPI and/or the editor(s) disclaim responsibility for any injury to people or property resulting from any ideas, methods, instructions or products referred to in the content.

MDPI AG
Grosspeteranlage 5
4052 Basel
Switzerland
Tel.: +41 61 683 77 34

Materials Editorial Office
E-mail: materials@mdpi.com
www.mdpi.com/journal/materials

Disclaimer/Publisher's Note: The title and front matter of this reprint are at the discretion of the Guest Editors. The publisher is not responsible for their content or any associated concerns. The statements, opinions and data contained in all individual articles are solely those of the individual Editors and contributors and not of MDPI. MDPI disclaims responsibility for any injury to people or property resulting from any ideas, methods, instructions or products referred to in the content.

www.ingramcontent.com/pod-product-compliance
Ingram Content Group UK Ltd.
Pitfield, Milton Keynes, MK11 3LW, UK
UKHW061544051026
15625UKWH00012B/603
* 9 7 8 3 7 2 5 8 3 9 0 6 3 *